NON-AUTONOMOUS KATO CLASSES AND FEYNMAN–KAC PROPAGATORS

NON-AUTONOMOUS KATO CLASSES AND FEYNMAN–KAC PROPAGATORS

Archil Gulisashvili
Ohio University, USA

Jan A van Casteren
University of Antwerp, Belgium

World Scientific

Published by

World Scientific Publishing Co. Pte. Ltd.
5 Toh Tuck Link, Singapore 596224
USA office: 27 Warren Street, Suite 401-402, Hackensack, NJ 07601
UK office: 57 Shelton Street, Covent Garden, London WC2H 9HE

British Library Cataloguing-in-Publication Data
A catalogue record for this book is available from the British Library.

NON-AUTONOMOUS KATO CLASSES AND FEYNMAN-KAC PROPAGATORS

Copyright © 2006 by World Scientific Publishing Co. Pte. Ltd.

All rights reserved. This book, or parts thereof, may not be reproduced in any form or by any means, electronic or mechanical, including photocopying, recording or any information storage and retrieval system now known or to be invented, without written permission from the Publisher.

For photocopying of material in this volume, please pay a copying fee through the Copyright Clearance Center, Inc., 222 Rosewood Drive, Danvers, MA 01923, USA. In this case permission to photocopy is not required from the publisher.

ISBN-13 978-981-256-557-0
ISBN-10 981-256-557-4

Printed in Singapore

To my wife, Olga Molchanova, and my two sons,
Oleś and Misha

Archil Gulisashvili

To my wife, Riet Wesselink, my daughter Linda,
my son Wilfred, and my granddaughter Elly

Jan A. van Casteren

Preface

This book covers selected topics from propagator theory. Propagators and backward propagators are two-parameter families of linear operators satisfying special conditions. Propagators satisfy the "flow" conditions, while backward propagators satisfy the "backward flow" conditions. Examples of propagators abound in mathematical physics, partial differential equations, and probability theory. For instance, solutions to Cauchy problems for non-autonomous evolution equations are generated by propagators. Important special examples are initial value problems for the heat equation perturbed by low order terms with time-dependent coefficients. A rich source of backward propagators is probability theory, where backward propagators arise as families of integral operators associated with transition probability functions. Such backward propagators are called free backward propagators. They admit a probabilistic characterization in terms of non-homogeneous Markov processes associated with transition probability functions.

Although propagators have many similarities with semigroups, propagator theory is not yet as complete as semigroup theory. One of the main obstacles in our understanding of propagators is their non-commutativity. Even the term "propagator" is not standard, and several other names have appeared in the mathematical literature, e.g., evolution families, solution operators, non-autonomous semigroups, etc. The present book is mainly devoted to free propagators, Feynman-Kac propagators, and related topics such as non-homogeneous Markov processes, reciprocal processes, and non-autonomous Kato classes of functions and measures. Since the selection of topics covered in the book was influenced by the research interests of the authors, many important subjects have been omitted. We refer the reader to Notes and Comments sections at the end of every chapter for additional information and the lists of references. In our opinion, the book is accessible to advanced graduate students interested in semigroup theory, partial differential equations, and probability theory. It is also suited for researches in these areas.

Chapter 1 contains basic facts about non-homogeneous Markov processes. The chapter includes stochastic processes with values in separable locally compact spaces, measurable and progressively measurable processes, one-sided continuity and continuity of sample paths, and space-time processes. We also discuss time-reversal, reciprocal processes, and Brownian and Cauchy bridges.

Chapter 2 introduces propagators and backward propagators. It begins with a discussion of the continuity properties of propagators. Topics included in this chapter are right and left generators of propagators, Kolmogorov's forward and backward equations, Howland semigroups associated with propagators, free backward propagators generated by transition probability functions, the strong Markov property of non-homogeneous stochastic processes, and Feller type properties of propagators. Several sections are devoted to standard processes associated with backward Feller-Dynkin propagators.

Chapter 3 studies non-autonomous Kato classes of functions and measures. These classes are generalizations of the Kato class of potential functions introduced by M. Aizenman and B. Simon. After discussing additive functionals of stochastic processes associated with time-dependent measures we give a probabilistic description of the Kato classes using these functionals. Various exponential estimates for non-autonomous additive functionals are obtained in the chapter. We also discuss a known fact that under certain restrictions fundamental solutions to second order conservative parabolic partial differential equations are transition probability densities. The chapter contains basic information on stochastic integration, stochastic differential equations, and diffusion processes.

Chapter 4 studies Feynman-Kac propagators. It starts with a brief review of Schrödinger semigroups with Kato class potentials. Topics covered in the chapter include Feynman-Kac propagators and Howland semigroups associated with them, the integral kernels of Feynman-Kac propagators, and Duhamel's formula. We also discuss the following problem: Determine what properties of free propagators are inherited by their Feynman-Kac perturbations. Among the properties appearing in the inheritance problem are the L^r-boundedness, the $(L^r - L^q)$-smoothing property, the Feller property, the Feller-Dynkin property, and the BUC-property. We end the book by proving that under certain restrictions the Feynman-Kac propagators generate viscosity solutions to non-autonomous Cauchy problems.

Most of the book was written during our visits to the University of Antwerp, Ohio University, and Centre de Recerca Matematica (CRM) in Bellaterra (Barcelona), Spain. We would like to thank the faculty and staff of the Department of Mathematics and Computer Science at the University of Antwerp, the Department of Mathematics at Ohio University, and CRM for their wonderful hospitality. We acknowledge the financial support of FWO Flanders (FWO Research Network WO.011.96N). The first-named author's research at CRM was supported by the grant "Beca de profesores e investigadores extranjeros en régimen de año sabático del Ministerio de Educación, Cultura y Deporte de España, referencia SAB2002-0066", and the second-named author acknowledges the financial support of the Department of Mathematics at Ohio University during one of his visits to Ohio University. We are very grateful to these universities and research institutions for their support.

A. Gulisashvili and J. A. van Casteren

Contents

Preface   vii

1. **Transition Functions and Markov Processes**   1
   - 1.1 Introduction 1
     - 1.1.1 Notation 1
     - 1.1.2 Elements of Probability Theory 2
     - 1.1.3 Locally Compact Spaces 4
     - 1.1.4 Stochastic Processes 5
     - 1.1.5 Filtrations 7
   - 1.2 Markov Property 8
   - 1.3 Transition Functions and Backward Transition Functions . 10
   - 1.4 Markov Processes Associated with Transition Functions . . 13
   - 1.5 Space-Time Processes 17
   - 1.6 Classes of Stochastic Processes 25
   - 1.7 Completions of σ-Algebras 28
   - 1.8 Path Properties of Stochastic Processes: Separability and Progressive Measurability 33
   - 1.9 Path Properties of Stochastic Processes: One-Sided Continuity and Continuity 44
   - 1.10 Reciprocal Transition Functions and Reciprocal Processes . 55
   - 1.11 Path Properties of Reciprocal Processes 79
   - 1.12 Examples of Transition Functions and Markov Processes . . 89
     - 1.12.1 Brownian motion and Brownian bridge 89
     - 1.12.2 Cauchy process and Cauchy bridge 95
     - 1.12.3 Forward Kolmogorov representation of Brownian bridges 97

1.13 Notes and Comments 98

2. Propagators: General Theory   101

    2.1 Propagators and Backward Propagators on Banach Spaces . 101
    2.2 Free Propagators and Free Backward Propagators 104
    2.3 Generators of Propagators and Kolmogorov's Forward and Backward Equations 106
    2.4 Howland Semigroups 121
    2.5 Feller-Dynkin Propagators and the Continuity Properties of Markov Processes 124
    2.6 Stopping Times and the Strong Markov Property 134
    2.7 Strong Markov Property with Respect to Families of Measures 150
    2.8 Feller-Dynkin Propagators and Completions of σ-Algebras . 172
    2.9 Feller-Dynkin Propagators and Standard Processes 174
    2.10 Hitting Times and Standard Processes 178
    2.11 Notes and Comments 190

3. Non-Autonomous Kato Classes of Measures   193

    3.1 Additive and Multiplicative Functionals 193
    3.2 Potentials of Time-Dependent Measures and Non-Autonomous Kato Classes 195
    3.3 Backward Transition Probability Functions and Non-Autonomous Kato Classes of Functions and Measures 200
    3.4 Weighted Non-Autonomous Kato Classes 203
    3.5 Examples of Functions and Measures in Non-Autonomous Kato Classes 207
    3.6 Transition Probability Densities and Fundamental Solutions to Parabolic Equations in Non-Divergence Form 217
    3.7 Transition Probability Densities and Fundamental Solutions to Parabolic Equations in Divergence Form 222
    3.8 Diffusion Processes and Stochastic Differential Equations . 232
    3.9 Additive Functionals Associated with Time-Dependent Measures 255
    3.10 Exponential Estimates for Additive Functionals 269
    3.11 Probabilistic Description of Non-Autonomous Kato Classes 275
    3.12 Notes and Comments 276

4.	Feynman-Kac Propagators	279
	4.1 Schrödinger Semigroups with Kato Class Potentials	279
	4.2 Feynman-Kac Propagators	283
	4.3 The Behavior of Feynman-Kac Propagators in L^p-Spaces . .	285
	4.4 Feller, Feller-Dynkin, and BUC-Property of Feynman-Kac Propagators	293
	4.5 Integral Kernels of Feynman-Kac Propagators	298
	4.6 Feynman-Kac Propagators and Howland Semigroups	304
	4.7 Duhamel's Formula for Feynman-Kac Propagators	307
	4.8 Feynman-Kac Propagators and Viscosity Solutions	311
	4.9 Notes and Comments	323
5.	Some Theorems of Analysis and Probability Theory	325
	5.1 Monotone Class Theorems	325
	5.2 Kolmogorov's Extension Theorem	326
	5.3 Uniform Integrability	327
	5.4 Radon-Nikodym Theorem	328
	5.5 Vitali-Hahn-Saks Theorem	329
	5.6 Doob's Inequalities	329

Bibliography   331

Index   341

Chapter 1

Transition Functions and Markov Processes

1.1 Introduction

This introductory section contains preliminary material from the theory of stochastic processes with values in locally compact second countable topological spaces. We also define and discuss basic notions and facts from probability theory.

1.1.1 Notation

Throughout the book, $\mathbb{N}$ denotes the set of nonnegative integers, $\mathbb{Z}$ is the set of all integers, $\mathbb{R}$ denotes the set of real numbers, and $\mathbb{R}^d$ is d-dimensional Euclidean space. The space $\mathbb{R}^d$ is equipped with the Borel σ-algebra $\mathcal{B}_{\mathbb{R}^d}$ and the Euclidean norm

$$|x| = \left\{ \sum_{j=1}^{d} |x_j|^2 \right\}^{\frac{1}{2}}, \quad x = (x_1, \ldots, x_d) \in \mathbb{R}^d.$$

The symbols $\vee$ and $\wedge$ stand for the following operations:

$$a \vee b = \max(a, b), \quad a \wedge b = \min(a, b), \quad (a, b) \in \mathbb{R}^2.$$

We will also use the floor function $\lfloor \cdot \rfloor$ and the ceiling function $\lceil \cdot \rceil$. If $a \in \mathbb{R}$, then $\lfloor a \rfloor$ is the greatest integer k such that $k \leq a < k+1$, while $\lceil a \rceil$ is the smallest integer ℓ such that $\ell - 1 < a \leq \ell$. It is easy to see that $\lfloor a \rfloor \leq a < \lfloor a \rfloor + 1$, $\lceil a \rceil - 1 < a \leq \lceil a \rceil$, $\lceil a \rceil - \lfloor a \rfloor = 1$ for $a \notin \mathbb{Z}$, and $\lceil a \rceil - \lfloor a \rfloor = 0$ for $a \in \mathbb{Z}$.

Let f and g be real-valued functions defined on the same set E. Then the functions $f \vee g$, $f \wedge g$, $\lfloor f \rfloor$, and $\lceil g \rceil$ can be defined exactly as in the case of numbers. For instance,

$$f \wedge g(x) = f(x) \wedge g(x) = \min(f(x), g(x)), \quad \lceil g \rceil(x) = \lceil g(x) \rceil, \quad x \in E.$$

1.1.2 Elements of Probability Theory

Let Ω be a set. A family of subsets of Ω is called a σ-algebra if the following conditions hold:

(1) The empty set $\emptyset$ belongs to $\mathcal{F}$.
(2) If $A \in \mathcal{F}$, then $\Omega \backslash A \in \mathcal{F}$.
(3) If $A_i \in \mathcal{F}$, $i \geq 1$, then $\bigcup_i A_i \in \mathcal{F}$.

The pair $(\Omega, \mathcal{F})$ is called a measurable space. A probability measure on $\mathcal{F}$ is a set function $\mathbb{P} : \Omega \mapsto [0, 1]$ such that

(1) $\mathbb{P}(\Omega) = 1$.
(2) If $A_i \in \mathcal{F}$, $i \geq 1$, are disjoint sets, then $\mathbb{P}\left(\bigcup_i A_i\right) = \sum_i \mathbb{P}(A_i)$.

A probability space is a triple $(\Omega, \mathcal{F}, \mathbb{P})$, where Ω is a set, $\mathcal{F}$ is a σ-algebra of subsets of Ω, and $\mathbb{P} : \mathcal{F} \to [0, 1]$ is a probability measure on $\mathcal{F}$. The set Ω is often called the sample space, and the elements of the σ-algebra $\mathcal{F}$ are called events.

Let $(E, \mathcal{E})$ be a measurable space. It is said that an E-valued function $X : \Omega \to E$ is an E-valued random variable if X is $\mathcal{F}/\mathcal{E}$-measurable. The space E is called the state space. Given E, an E-valued random variable is called a state variable. If $E = \mathbb{R}$ and $\mathcal{E}$ coincides with the Borel σ-algebra $\mathcal{B}_\mathbb{R}$ of $\mathbb{R}$, then X is called a random variable.

Let X be an E-valued random variable. Then the σ-algebra $\sigma(X)$ generated by X is defined as follows: $\sigma(X) = \{X^{-1}(A) : A \in \mathcal{E}\}$. The σ-algebra $\sigma(X_\lambda : \lambda \in \Lambda)$ generated by a family X_λ, $\lambda \in \Lambda$, of E-valued random variables is the smallest σ-algebra containing all the events of the form $X_\lambda^{-1}(A)$ with $\lambda \in \Lambda$ and $A \in \mathcal{E}$.

Let $(\Omega, \mathcal{F}, \mathbb{P})$ be a probability space, and let $(E, \mathcal{E})$ be a state space. Suppose that $X : \Omega \to E$ is an E-valued random variable. The distribution $\mu : \mathcal{E} \to [0, 1]$ of X is defined by the following formula:

$$\mu(B) = \mathbb{P}[X \in B], \quad B \in \mathcal{E}.$$

It is clear that μ is a Borel measure on $\mathcal{E}$ satisfying the condition $\mu(E) = 1$. Conversely, if $(E, \mathcal{E}, \mu)$ is a measure space of total mass 1, then there exists a probability space $(\Omega, \mathcal{F}, \mathbb{P})$ and an E-valued random variable X such that the measure μ coincides with the distribution of X. Indeed, we can take $\Omega = E$ and $\mathcal{F} = \mathcal{E}$. The E-valued random variable X is defined by $X(\omega) = \omega$, $\omega \in \Omega$, and the probability measure $\mathbb{P}$ by $\mathbb{P}[X \in B] = \mu(B)$, $B \in \mathcal{F}$.

Suppose that $(\Omega, \mathcal{F}, \mathbb{P})$ is a probability space. The mathematical expectation with respect to the probability measure $\mathbb{P}$ is denoted by $\mathbb{E}$. If $X : \Omega \to E$ is an E-valued random variable with distribution μ and $f : E \to \mathbb{R}$ is a Borel measurable function on E that is integrable with respect to the measure μ, then the following formula holds:

$$\mathbb{E}[f(X)] = \int f(x) d\mu(x).$$

It is said that two random variables F_1 and F_2 are equal $\mathbb{P}$-almost surely if $\mathbb{P}[F_1 \neq F_1] = 0$. A random variable F is called integrable if $\mathbb{E}[|F|] < \infty$. We denote by $L^1(\Omega, \mathcal{F}, \mathbb{P})$ the space of all equivalence classes of integrable random variables with respect to the following equivalence relation: Two random variables are equivalent if they are equal $\mathbb{P}$-almost surely.

Let $(\Omega, \mathcal{F}, \mathbb{P})$ be a probability space, and let $\mathcal{F}_1$ and $\mathcal{F}_2$ be sub-σ-algebras of $\mathcal{F}$. Then $\mathcal{F}_1$ and $\mathcal{F}_2$ are called independent if the equality

$$\mathbb{P}(A_1 \cap A_2) = \mathbb{P}(A_1)\mathbb{P}(A_2)$$

holds for all $A_1 \in \mathcal{F}_1$ and $A_2 \in \mathcal{F}_2$. If $(E, \mathcal{E})$ is a state space, then it is said that E-valued random variables F_1 and F_2 are independent if the σ-algebras $\sigma(F_1)$ and $\sigma(F_2)$ are independent. An equivalent condition is as follows:

$$\mathbb{E}[f(F_1)g(F_2)] = \mathbb{E}[f(F_1)]\mathbb{E}[g(F_2)]$$

for all bounded real valued $\mathcal{E}$-measurable functions f and g on E.

If $A \in \mathcal{F}$ is such that $\mathbb{P}[A] \neq 0$, then the conditional expectation of $F \in L^1(\Omega, \mathcal{F}, \mathbb{P})$ given the event A is defined by

$$\mathbb{E}[F \mid A] = \frac{\mathbb{E}[F\chi_A]}{\mathbb{P}[A]}.$$

We will often use the symbol $\mathbb{E}[F, A]$ instead of $\mathbb{E}[F\chi_A]$. Let $\mathcal{F}_0$ be a sub-σ-algebra of $\mathcal{F}$. The conditional expectation of an integrable random variable F given the σ-algebra $\mathcal{F}_0$ is denoted by

$$Z = \mathbb{E}[F \mid \mathcal{F}_0].$$

It is defined as follows: Z is a $\mathcal{F}_0/\mathcal{B}_\mathbb{R}$-measurable random variable such that
$$\mathbb{E}[Z, A] = \mathbb{E}[F, A] \quad \text{for all } A \in \mathcal{F}_0.$$
Note that the conditional expectation is defined $\mathbb{P}$-almost surely. Some properties of conditional expectations of integrable random variables are listed below:

(1) If $F \in L^1(\Omega, \mathcal{F}, \mathbb{P})$ and $F \geq 0$, then $\mathbb{E}\left[F \mid \mathcal{F}_0\right] \geq 0$ for any sub-σ-algebra $\mathcal{F}_0$ of $\mathcal{F}$.

(2) If $F_1 \in L^1(\Omega, \mathcal{F}, \mathbb{P})$ and $F_2 \in L^1(\Omega, \mathcal{F}, \mathbb{P})$ are integrable random variables and α_1 and α_2 are real numbers, then
$$\mathbb{E}\left[\alpha_1 F_1 + \alpha_2 F_2 \mid \mathcal{F}_0\right] = \alpha_a \mathbb{E}\left[F_1 \mid \mathcal{F}_0\right] + \alpha_2 \mathbb{E}\left[F_2 \mid \mathcal{F}_0\right].$$

(3) If $F_1 \in L^1(\Omega, \mathcal{F}, \mathbb{P})$ and if F_2 is an $\mathcal{F}_0$-measurable random variable such that $F_1 F_2 \in L^1(\Omega, \mathcal{F}, \mathbb{P})$, then
$$\mathbb{E}\left[F_1 F_2 \mid \mathcal{F}_0\right] = F_2 \mathbb{E}\left[F_1 \mid \mathcal{F}_0\right].$$

(4) If $F \in L^1(\Omega, \mathcal{F}, \mathbb{P})$ and if F is independent of the σ-algebra $\mathcal{F}_0$, then
$$\mathbb{E}\left[F \mid \mathcal{F}_0\right] = \mathbb{E}[F].$$

(5) Suppose that $\mathcal{F}_0$ and $\mathcal{F}_1$ are two sub-σ-algebras of $\mathcal{F}$ such that $\mathcal{F}_0 \subset \mathcal{F}_1$. Then for every $F \in L^1(\Omega, \mathcal{F}, \mathbb{P})$,
$$\mathbb{E}\left[\mathbb{E}\left[F \mid \mathcal{F}_1\right] \mid \mathcal{F}_0\right] = \mathbb{E}\left[F \mid \mathcal{F}_0\right].$$

Property (5) is called the tower property of conditional expectations. For more information on conditional expectations see [Yeh (1995)].

1.1.3 Locally Compact Spaces

The state spaces considered in this book are locally compact Hausdorff topological spaces satisfying the second axiom of countability. If E is such a space, then the symbol $\mathcal{E}$ will stand for the Borel σ-algebra of E.

Next we formulate Urysohn's Lemma (see, e.g., [Folland (1999)]).

Lemma 1.1 *Let E be a locally compact Hausdorff topological space, and let a compact set K and an open set U be such that $K \subset U \subset E$. Then there exists a function $\varphi \in C_0$ satisfying the following conditions: $0 \leq \varphi \leq 1$, $\varphi = 1$ on K, and $\varphi = 0$ outside a compact set K' for which $K \subset K' \subset U$.*

In Lemma 1.1, the set K' can be chosen as follows. Let U_0 be a relatively compact open subset of E for which $K \subset U_0 \subset \overline{U}_0 \subset U$, where the symbol $\overline{U}_0$ stands for the closure of the set U_0. Then we can take $K' = \overline{U}_0$. In addition, if the space E is second countable, then the following function can be chosen as the function φ in Urysohn's Lemma:

$$\varphi(x) = \min\left(\beta^{-1} \inf\{\rho(x,y) : y \in E \setminus U_0\}, 1\right)$$

where $\beta = \inf\{\rho(x_1, x_2) : (x_1, x_2) \in K \times (E \setminus U_0)\}$.

Let E be a state space. The symbol C_0 will stand for the vector space of all real valued continuous functions on E such that for every $\varepsilon > 0$, the set $\{x \in E : |f(x)| \geq \varepsilon\}$ is a compact subset of E. Equipped with the supremum norm, C_0 is a Banach space. Since E is second countable, there exists a countable collection $\mathcal{C}$ of open relatively compact subsets of E such that every open subset O of E is a countable union of open sets from $\mathcal{C}$. Let $\mathcal{C} = \{U_j : j \geq 1\}$ be an enumeration of $\mathcal{C}$ and choose $a_j \in U_j$, $j \geq 1$. Let $\varphi_j \in C_0$, $j \geq 1$ be a sequence of continuous functions on E such that $0 \leq \varphi_j \leq 1$ and $\varphi_j(a_j) = 1$ for all $j \geq 1$. Then the algebra generated by the functions φ_j, $j \geq 1$, separates points. By the Stone-Weierstrass theorem, this algebra is uniformly dense in the Banach space $(C_0, \|\cdot\|_\infty)$. It follows that a sequence x_n, $n \in \mathbb{N}$, of elements of E converges to $x \in E$ if and only if

$$\lim_{n \to \infty} \varphi_j(x_n) = \varphi_j(x)$$

for all $j \geq 1$. It is not hard to see that the function $\rho : E \times E \to [0,1]$ defined by

$$\rho(x,y) = \sum_{j=1}^{\infty} 2^{-j} |\varphi_j(x) - \varphi_j(y)|, \quad (x,y) \in E \times E,$$

is a metric on the space E. This metric generates the topology of E. Hence, the space E is metrizable. It is also clear that E is a σ-compact space.

1.1.4 Stochastic Processes

Let $(\Omega, \mathcal{F}, \mathbb{P})$ be a probability space, and let $(E, \mathcal{E})$ be a state space. Suppose that $X_s : \Omega \mapsto E$, $s \in I$, is a family of random variables where I is an index set. Then the family X_s, $s \in I$, is called a stochastic process. Throughout the book, we will use a bounded interval $[a,b]$ or the half-line $[0, \infty)$ as the index set I, unless specified otherwise. The stochastic process X_s, $s \in I$, can be identified with the mapping $X : (s, \omega) \mapsto X_s(\omega)$, $(s, \omega) \in I \times \Omega$. In addition, the process X_s can be considered as a random variable $\mathbf{X} : \Omega \to E^I$ defined by $\mathbf{X}(\omega) = (X_s(\omega))_{s \in I}$. For every $\omega \in \Omega$, the function $s \mapsto X_s(\omega)$, $s \in I$, is called a sample path.

The path space E^I is equipped with the product σ-algebra $\otimes_I \mathcal{E}$. This σ-algebra is generated by finite-dimensional cylinders of the form $\prod_{s \in I} E_s$, where $E_s \in \mathcal{E}$, $s \in I$, and $E_s \neq E$ for only a finite number of elements $s \in I$. Let X_s, $s \in I$, be a stochastic process, and let $J = (s_1, \ldots, s_n)$ be a finite subset of I. The finite-dimensional distribution of the process X_s corresponding to the set J is the distribution of the random variable $X_J : \Omega \to E^J$ given by $X_J(\omega) = (X_{s_1}(\omega), \ldots, X_{s_n}(\omega))$. It is defined as follows:

$$\mu_J(B) = \mathbb{P}[X_J \in B], \quad B \in \mathcal{B}_{E^J}.$$

Here $\mathcal{B}_{E^J}$ denotes the Borel σ-algebra of the space E^J. The distribution of the process X_s, $s \in I$, that is, the distribution of the random variable $\mathbf{X}$ is determined by the family of all finite-dimensional distributions. Each member of the family

$$\left\{ (E^J, \mathcal{B}_{E^J}, \mu_J) : J \subset I, \ J \text{ finite} \right\}$$

is a probability space, and the family $\{\mu_J : J \text{ finite}\}$ is a projective (or consistent) system of probability measures. This means that for finite subsets J and K of I with $J \subset K$, the equality

$$\mu_K \left[\left(p_J^K\right)^{-1}(B) \right] = \mu_J(B)$$

holds for all $B \in \mathcal{B}_{E^J}$. Here the projection mapping $p_J^K : E^K \to E^J$ is given by

$$p_J^K \left((\omega_s)_{s \in K}\right) = (\omega_s)_{s \in J}, \quad (\omega_s)_{s \in K} \in E^K$$

(see Section 5.2).

Let X_s^1 and X_s^2, $s \in I$, be stochastic processes on the probability spaces $(\Omega_1, \mathcal{F}_1, \mathbb{P}_1)$ and $(\Omega_2, \mathcal{F}_2, \mathbb{P}_2)$, respectively, and suppose that both processes have the same state space $(E, \mathcal{E})$. The processes X_t^1 and X_t^2 are called stochastically equivalent if their finite-dimensional distributions coincide, i.e., if the equality

$$\mathbb{P}_1 \left[X_J^1 \in B \right] = \mathbb{P}_2 \left[X_J^2 \in B \right]$$

holds for all finite subsets J of I and all $B \in \mathcal{B}_{E^J}$. Note that in the definition of the stochastic equivalence, it is not necessary to assume that the processes X_s^1 and X_s^2 are defined on the same probability space. If the processes X_s^1 and X_s^2 are defined on the same probability space $(\Omega, \mathcal{F}, \mathbb{P})$, then it is said that the process X_s^2 is a modification of the process X_s^1 if for every $s \in I$, $\mathbb{P}\left[X_s^1 \neq X_s^2\right] = 0$. The processes X_s^1 and X_s^2 are called indistinguishable if $\mathbb{P}\left[X_s^1 \neq X_s^2\right] = 0$ for all $s \in I$. It is not hard to see that if X_s^1 and X_s^2 are continuous stochastic processes and X_s^2 is a modification of X_s^1, then X_s^1 and X_s^2 are indistinguishable (see [Revuz and Yor (1991)], p. 18).

1.1.5 *Filtrations*

Let $(\Omega, \mathcal{F})$ be a measurable space. A filtration $\mathcal{G}_t$, $t \in [0, T]$, is a family of sub-σ-algebras of $\mathcal{F}$ such that if $0 \leq t_1 \leq t_2 \leq T$, then $\mathcal{G}_{t_1} \subset \mathcal{G}_{t_2}$. A two-parameter filtration $\mathcal{G}_t^\tau$, $0 \leq \tau \leq t \leq T$, is a family of sub-$\sigma$-algebras of $\mathcal{F}$ that is increasing in t and decreasing in τ. A stochastic process X_s, $s \in [0, T]$, with state space $(E, \mathcal{E})$ is called adapted to the filtration $\mathcal{G}_t$, $t \in [0, T]$, provided that for every $s \in [0, T]$, X_s is $\mathcal{F}_s/\mathcal{E}$-measurable. The following filtration is associated with the process X_s:

$$\mathcal{F}_t = \sigma\left(X_s : 0 \leq s \leq t\right), \ 0 \leq t \leq T.$$

Recall that $\sigma\left(X_s : 0 \leq s \leq t\right)$ is the smallest σ-algebra containing all sets of the form $\bigcap_{s \in J} X_s^{-1}(B_s)$, where J is any finite subset of $[0, t]$, and $B_s \in \mathcal{E}$, $s \in J$. Instead of Borel subsets of E, one can take, e.g., open subsets, or any other π-system which generates the σ-algebra $\mathcal{E}$ (see Subsection 5.1 for the definition of a π-system). The σ-algebra $\sigma\left(X_s : 0 \leq s \leq t\right)$ is the smallest σ-algebra such that all the state variables X_s with $0 \leq s \leq t$ are measurable. The process X_s, $s \in [0, T]$, generates the following two-parameter filtration:

$$\mathcal{F}_t^\tau = \sigma\left(X_s : \tau \leq s \leq t\right), \ 0 \leq \tau \leq t \leq T.$$

The filtration $\mathcal{F}_t^\tau$ is sometimes called the internal history of the process X_s, $s \in [0, T]$.

Let $(\Omega, \mathcal{F}, \mathbb{P})$ be a probability space, and let $\mathcal{G}_t$ be a filtration. It will be often implicitly assumed that for every $t \in [0, T]$, the σ-algebra $\mathcal{G}_t$ is complete with respect to the measure $\mathbb{P}$. This means that if $A \in \mathcal{F}$, $B \subset A$, and $\mathbb{P}(A) = 0$, then $B \in \mathcal{G}_t$. In other words, all subsets of $\mathbb{P}$-negligible sets belong to every σ-algebra $\mathcal{G}_t$ with $t \in [0, T]$. If the filtration $\mathcal{G}_t$ is not

complete, then we can always augment it by the family
$$\mathcal{N} = \{B \subset \Omega : B \subset A,\ A \in \mathcal{F},\ \mathbb{P}(A) = 0\}.$$
More precisely, the augmentaion means that one passes from the σ-algebra $\mathcal{G}_t$ to the σ-algebra $\bar{\mathcal{G}}_t = \sigma\left(\mathcal{G}_t, \mathcal{N}\right)$. We refer the reader to Section 1.7 for more information on completions of σ-algebras.

1.2 Markov Property

In this section we introduce Markov processes and formulate several equivalent conditions for the validity of the Markov property for a general stochastic process. Let $(\Omega, \mathcal{F}, \mathbb{P})$ be a probability space, and let X_t with $t \in [0, T]$ be a stochastic process on Ω with state space E. Recall that the state space E is a locally compact Hausdorff topological space satisfying the second axiom of countability.

Definition 1.1 A stochastic process X_t is called a Markov process if
$$\mathbb{E}\left[f(X_t) \mid \mathcal{F}_s\right] = \mathbb{E}\left[f(X_t) \mid \sigma(X_s)\right] \tag{1.1}$$
$\mathbb{P}$-almost surely for all $0 \leq s \leq t \leq T$ and all bounded Borel functions f on E.

The next lemma is well-known. It provides several equivalent descriptions of the Markov property. For the definition of two-parameter filtrations see Subsection 1.1.5.

Lemma 1.2 Let X_s be a stochastic process on $(\Omega, \mathcal{F}, \mathbb{P})$ with state space E. Then the following are equivalent:

(1) Condition (1.1) holds.
(2) For all $t \in [0, T]$, all finite sets $\{r_1, \ldots, r_n\}$ with $0 \leq r_1 < r_2 < \cdots < r_n < t$, and all bounded Borel functions f on E, the equality
$$\mathbb{E}\left[f(X_t) \mid \sigma\left(X_{r_1}, \ldots, X_{r_n}\right)\right] = \mathbb{E}\left[f(X_t) \mid \sigma(X_{r_n})\right]$$
holds $\mathbb{P}$-a.s.
(3) For all $s \in [0, T]$ and all bounded real-valued $\mathcal{F}_T^s$-measurable random variables F, the equality
$$\mathbb{E}\left[F \mid \mathcal{F}_s\right] = \mathbb{E}\left[F \mid \sigma(X_s)\right]$$
holds $\mathbb{P}$-a.s.

(4) For all $s \in [0,T]$, and all bounded real-valued random variables G and F such that G is $\mathcal{F}_s$-measurable and F is $\mathcal{F}_T^s$-measurable, the equality
$$\mathbb{E}[GF] = \mathbb{E}\left[G\mathbb{E}\left[F \mid \sigma(X_s)\right]\right]$$
holds.

(5) For all $s \in [0,T]$, $A \in \mathcal{F}_s$, and $B \in \mathcal{F}_T^s$, the equality
$$\mathbb{P}\left[A \cap B \mid \sigma(X_s)\right] = \mathbb{P}\left[A \mid \sigma(X_s)\right]\mathbb{P}\left[B \mid \sigma(X_s)\right]$$
holds $\mathbb{P}$-a.s.

We refer the reader to [Blumenthal and Getoor (1968)] for the proof of Lemma 1.2. A similar lemma concerning the reciprocal Markov property will be established in Section 1.10 (see Lemma 1.20).

Condition (5) in Lemma 1.2 states that for a Markov process the future and the past are conditionally independent, given the present. The σ-algebra $\mathcal{F}_s$ is often interpreted as information from the past before time s, while the σ-algebra $\mathcal{F}_T^s$ contains the future information. The time s is considered as the present time. In a sense, a Markov process forgets its past history.

Remark 1.1 If X_t, $0 \leq t \leq T$, is a Markov process with respect to the probability measure $\mathbb{P}$, then the time reversed process $\widehat{X}^t = X_{T-t}$, $t \in [0,T]$, is also a Markov process with respect to the same measure $\mathbb{P}$. Indeed, it is not hard to see that condition (5) in Lemma 1.2 is invariant with respect to time-reversal. Therefore, Lemma 1.2 implies that the time-reversed process $\widehat{X}_t$ possesses the Markov property.

The next assertion follows from Lemma 1.2:

Lemma 1.3 Let X_s be a stochastic process on $(\Omega, \mathcal{F}, \mathbb{P})$ with state space E. Then the following are equivalent:

(1) For all $0 \leq s \leq t \leq T$, and all bounded Borel functions f on E, the equality
$$\mathbb{E}\left[f(X_s) \mid \mathcal{F}_T^t\right] = \mathbb{E}\left[f(X_s) \mid \sigma(X_t)\right]$$
holds $\mathbb{P}$-a.s.

(2) For all t with $0 \leq t \leq T$, all finite sets $\{u_1,\ldots,u_n\}$ with $t < u_1 < u_2 < \cdots < u_n \leq T$, and all bounded Borel functions f on E, the equality
$$\mathbb{E}\left[f(X_t) \mid \sigma\left(X_{u_1},\ldots,X_{u_n}\right)\right] = \mathbb{E}\left[f(X_t) \mid \sigma(X_{u_1})\right]$$

holds $\mathbb{P}$-a.s.

(3) For all s with $s \in [0,T]$ and all bounded real-valued $\mathcal{F}_s$-measurable random variables F, the equality

$$\mathbb{E}\left[F \mid \mathcal{F}_T^s\right] = \mathbb{E}\left[F \mid \sigma(X_s)\right]$$

holds $\mathbb{P}$-a.s.

(4) For all s with $s \in [0,T]$, the equality

$$\mathbb{E}[FG] = \mathbb{E}\left[F\mathbb{E}\left[G \mid \sigma(X_s)\right]\right]$$

holds for all bounded real-valued random variables G and F such that G is $\mathcal{F}_s$-measurable and F is $\mathcal{F}_T^s$-measurable.

(5) For all $s \in [0,T]$, $A \in \mathcal{F}_s$, and $B \in \mathcal{F}_T^s$, the equality

$$\mathbb{P}\left[A \cap B \mid \sigma(X_s)\right] = \mathbb{P}\left[A \mid \sigma(X_s)\right]\mathbb{P}\left[B \mid \sigma(X_s)\right]$$

holds $\mathbb{P}$-a.s.

1.3 Transition Functions and Backward Transition Functions

This section is devoted to non-homogeneous transition functions. We will first introduce a forward transition probability function, or simply a transition probability function.

Definition 1.2 A transition probability function $P(r,x;s,A)$, where $0 \leq r < s \leq T$, $x \in E$, and $A \in \mathcal{E}$, is a nonnegative function for which the following conditions hold:

(1) For fixed r, s, and A, P is a nonnegative Borel measurable function on E.
(2) For fixed r, s, and x, P is a Borel measure on $\mathcal{E}$.
(3) $P(r,x;s,E) = 1$ for all r, s, and x.
(4) $P(r,x;s,A) = \int_E P(r,x;u,dy)P(u,y;s,A)$ for all $r < u < s$, and A.

A function P satisfying Condition (3) in Definition 1.2 is called normal, or conservative. Condition (4) is the Chapman-Kolmogorov equation for transition functions. In applications, a transition function P describes the time evolution of a random system. The number $P(r,x;s,A)$ can be interpreted as the probability of the following event: The random system located at $x \in E$ at time r hits the target $A \subset E$ at time s.

The next definition concerns backward transition probability functions.

Definition 1.3 A backward transition probability function $\widetilde{P}(\tau, A; t, y)$, where $0 \leq \tau \leq t \leq T$, $y \in E$, and $A \in \mathcal{E}$, is a nonnegative function for which the following conditions hold:

(1) For fixed τ, A, and t, $\widetilde{P}$ is a Borel function on E.
(2) For fixed τ, t, and y, $\widetilde{P}$ is a Borel measure on $\mathcal{E}$.
(3) The normality condition $\widetilde{P}(\tau, E; t, y) = 1$ holds for all τ, t, and y.
(4) The Chapman-Kolmogorov equation, that is,

$$\widetilde{P}(\tau, A; t, y) = \int_E \widetilde{P}(\tau, A; \lambda, x) \widetilde{P}(\lambda, dx; t, y),$$

holds for all $\tau < \lambda < t$, $A \in \mathcal{E}$, and $y \in E$.

There is a simple relation between forward and backward transition probability functions in the case of a finite time-interval $[0, T]$. Here we need the time reversal operation $t \mapsto T - t$, $t \in [0, T]$. It is easy to see that $\widetilde{P}$ is a backward transition probability function if and only if

$$P(\tau, x; t, A) = \widetilde{P}(T - t, A; t - \tau, x) \tag{1.3}$$

is a transition probability function.

If a function P satisfies conditions (1), (2), and (4) in Definition 1.2, but does not satisfy the normality condition, then it is called a transition function. Similarly, a function $\widetilde{P}$ satisfying conditions (1), (2), and (4) in Definition 1.3 is called a backward transition function. If the condition

$$P(\tau, x; t, E) < 1 \tag{1.4}$$

holds instead of condition (3) in Definition 1.2, then P is called a transition subprobability function. Similarly, if

$$\widetilde{P}(\tau, x; t, E) < 1, \tag{1.5}$$

then $\widetilde{P}$ is called a backward transition subprobability function. If P is such that (1.4) holds, then one can define a new state space $E^{\triangle} = E \cup \{\triangle\}$ where $\triangle$ is an extra point. If E is a compact space, then $\triangle$ is attached to E as an isolated point. If E is not compact, then the topology of $E^{\triangle}$ is that of a one-point compactification of E. The Borel σ-algebra of $E^{\triangle}$ will be denoted by $\mathcal{E}^\triangle$. Put

$$P^\triangle(\tau, x; t, A) = \begin{cases} 1, & \text{if } x = \triangle \text{ and } \triangle \in A \\ 0, & \text{if } x = \triangle \text{ and } \triangle \notin A \\ P(\tau, x; t, A), & \text{if } x \in E \text{ and } \triangle \notin A \\ 1 - P(\tau, x; t, E), & \text{if } x \in E \text{ and } A = \{\triangle\}. \end{cases} \quad (1.6)$$

Lemma 1.4 *Let P be a transition function satisfying (1.4), and let $P^\triangle$ be defined by (1.6). Then $P^\triangle$ is a transition probability function; moreover, the functions $P^\triangle$ and P coincide on E.*

Proof. It is clear that only conditions (3) and (4) in Definition 1.2 need to be checked for the function $P^\triangle$. For $x \in E$, (1.6) implies

$$\begin{aligned} P^\triangle(\tau, x; t, E^\triangle) &= P(\tau, x; t, E) + P^\triangle(\tau, x; t, \triangle) \\ &= P(\tau, x; t, E) + 1 - P(\tau, x; t, E) = 1. \end{aligned}$$

If $x = \triangle$, then (1.6) gives $P^\triangle(\tau, \triangle; t, E^\triangle) = 1$. Therefore, the function $P^\triangle$ is normal.

Our next goal is to prove that the function $P^\triangle$ satisfies the Chapman-Kolmogorov equation. For $x \neq \triangle$ and $A \in \mathcal{E}$, this fact follows from the Chapman-Kolmogorov equation for P. Let $x \in E^\triangle$, $\tau < r < t$, $A \in \mathcal{E}^\triangle$, $\triangle \in A$, and put $\widetilde{A} = A \backslash \triangle$. Then

$$\begin{aligned} &\int P^\triangle(\tau, x; r, dz) P^\triangle(r, z; t, A) \\ &= \int_E P^\triangle(\tau, x; r, dz) P^\triangle(r, z; t, \widetilde{A}) + \int_E P^\triangle(\tau, x; r, dz) P^\triangle(r, z; t, \{\triangle\}) \\ &\quad + \int_{\{\triangle\}} P^\triangle(\tau, x; r, dz) P^\triangle(r, z; t, \widetilde{A}) \\ &\quad + \int_{\{\triangle\}} P^\triangle(\tau, x; r, dz) P^\triangle(r, z; t, \{\triangle\}). \end{aligned} \quad (1.7)$$

If $x \in E$, then (1.7), (1.6), and the Chapman-Kolmogorov equation for P give

$$\int P^{\triangle}(\tau,x;r,dz)P^{\triangle}(r,z;t,A)$$
$$= P(\tau,x;t,\widetilde{A}) + \int_E P(\tau,x;r,dz)(1 - P(r,z;t,E)) + 1 - P(\tau,x;r,E)$$
$$= P(\tau,x;t,\widetilde{A}) + 1 - P(\tau,x;t,E)$$
$$= P^{\triangle}(\tau,x;t,\widetilde{A}) + P^{\triangle}(\tau,x;t,\{\triangle\}) = P^{\triangle}(\tau,x;t,A). \tag{1.8}$$

It is easy to see that if $x \in E$ and $\triangle \in A$, then Lemma 1.4 follows from (1.8). A similar reasoning can be used in the case where $x = \triangle$ and $\triangle \in A$, and in the case where $x = \triangle$ and $\triangle \notin A$. □

A nonnegative Borel measure m on $(E, \mathcal{E})$ will be fixed throughout the book. The measure m is called the reference measure. We will write dx instead of $m(dx)$ and assume that $0 < m(A) < \infty$ for any compact subset A of E with nonempty interior. It is said that a transition function P possesses a density p, provided that there exists a nonnegative function $p(r,x;s,y)$ such that for all $0 \leq r < s \leq T$, the function $(x,y) \to p(r,x;s,y)$ is $\mathcal{E} \otimes \mathcal{E}$-measurable, and the condition

$$P(r,x;s,A) = \int_A p(r,x;s,y)dy$$

holds for all $A \in \mathcal{E}$. The Chapman-Kolmogorov equation for transition densities is

$$p(\tau,x;t,y) = \int_E p(\tau,x;r,z)p(r,z;t,y)dz$$

for all $0 \leq \tau < r < t$ and $m \times m$ almost all $(x,y) \in E \times E$.

1.4 Markov Processes Associated with Transition Functions

Let E be a locally compact space such as in Section 1.1, and let P be a transition probability function. Our first goal is to construct a filtered measurable space $(\Omega, \mathcal{F}, \mathcal{F}_t^\tau)$, a family of probability measures $\mathbb{P}_{\tau,x}$, $x \in E$, $\tau \in [0,T]$, on the space $(\Omega, \mathcal{F}_T^\tau)$, and a Markov process X_t, $t \in [0,T]$, such

that
$$\mathcal{F}_t^\tau = \sigma\left(X_s : \tau \leq s \leq t\right), \quad 0 \leq \tau \leq t \leq T \tag{1.9}$$
and
$$\mathbb{P}_{\tau,x}(X_t \in A) = P(\tau, x; t, A), \quad 0 \leq \tau \leq t \leq T, \ A \in \mathcal{E}. \tag{1.10}$$

We begin with the construction of what is called the standard realization of such a process. Let $\Omega = E^{[0,T]}$ be the path space consisting of all functions mapping $[0,T]$ into E. The space Ω is equipped with the cylinder σ-algebra $\mathcal{F}$, which is the smallest σ-algebra containing all sets of the form $\{\omega \in \Omega : \omega(t_1) \in A_1, \ldots, \omega(t_k) \in A_k\}$, $0 \leq t_1 < \cdots < t_k \leq T$ and $A_i \in \mathcal{E}$ for all $1 \leq i \leq k$. Such sets are called finite-dimensional cylinders. The process X_t is defined on the space Ω by $X_t(\omega) = \omega(t)$. The σ-algebra $\mathcal{F}_t^\tau$ is defined by formula (1.9). The symbol $\mathbb{P}_{\tau,x}$, where $0 \leq \tau \leq T$ and $x \in E$, stands for the probability measure on $\mathcal{F}_T^\tau$ such that

$$\begin{aligned}
&\mathbb{P}_{\tau,x}\left(X_\tau \in A_0, X_{t_1} \in A_1, \cdots, X_{t_k} \in A_k\right) \\
&= \mathbb{P}_{\tau,x}\left(\omega \in \Omega : \omega(\tau) \in A_0, \omega(t_1) \in A_1, \cdots, \omega(t_k) \in A_k\right) \\
&= \chi_{A_0}(x) \int_{A_1} P(\tau, x; t_1, dx_1) \int_{A_2} P(t_1, x_1; t_2, dx_2) \cdots \\
&\quad \int_{A_k} P(t_{k-1}, x_{k-1}; t_k, dx_k)
\end{aligned} \tag{1.11}$$

for all $\tau < t_1 < t_2 < \cdots < t_k \leq T$ and $A_i \in \mathcal{E}$, $1 \leq i \leq k$. Such a measure exists by Kolmogorov's extension theorem (see Appendix A). The process X_t is a Markov process with respect to the family of measures $\mathbb{P}_{\tau,x}$, that is,

$$\mathbb{E}_{\tau,x}\left[f(X_t)|\mathcal{F}_s^\tau\right] = \mathbb{E}_{s,X_s} f(X_t) \ \mathbb{P}_{\tau,x}\text{-a.s.} \tag{1.12}$$

for all $\tau \leq s \leq t$, and all bounded Borel functions f on E. Recall that the symbol $\mathbb{E}_{\tau,x}$ in (1.12) stands for the expectation with respect to the measure $\mathbb{P}_{\tau,x}$, and the left-hand side of (1.12) is the conditional expectation of $f(X_t)$ with respect to the σ-algebra $\mathcal{F}_s^\tau$. Taking the conditional expectation with respect to the σ-algebra $\sigma(X_s)$ on both sides of equality (1.12) and using the $\sigma(X_s)$-measurability of the expression on the right-hand side of (1.12), we see that condition (1.1) holds for all measures $\mathbb{P}_{\tau,x}$ and all t, s with $\tau \leq s < t \leq T$. In a sense, the Markov property in (1.12) means that the past history of the process X_t does not affect predictions concerning the future of X_t. It is not hard to see that condition (1.10) holds for the process X_t. The expression on the left-hand side of (1.10) is called the marginal distribution of X_t, while the expression on the right-hand side of (1.11) is called the finite-dimensional distribution of X_t. It is not hard to see that for a Markov process, the finite-dimensional distributions can be recovered from the marginal ones.

Next, we discuss general non-homogeneous stochastic processes $(X_t, \mathcal{G}_t^\tau, \mathbb{P}_{\tau,x})$ defined on a general sample space $(\Omega, \mathcal{G})$ equipped with a two-parameter filtration $\mathcal{G}_t^\tau$, $0 \le \tau \le t \le T$. It is assumed that $\mathcal{F}_t^\tau \subset \mathcal{G}_t^\tau$, $0 \le \tau \le t \le T$. If this condition holds, then the process X_t is adapted to the two-parameter filtration $\mathcal{G}_t^\tau$ (see Definition 1.9 below). It is also assumed that for all $\tau \in [0, T]$ and $x \in E$, $\mathbb{P}_{\tau,x}$ is a probability measure on the σ-algebra $\mathcal{G}_T^\tau$.

Definition 1.4 A non-homogeneous stochastic process $(X_t, \mathcal{G}_t^\tau, \mathbb{P}_{\tau,x})$ is called a Markov process if

$$\mathbb{E}_{\tau,x}\left[f(X_t)|\mathcal{G}_s^\tau\right] = \mathbb{E}_{s,X_s} f(X_t) \quad \mathbb{P}_{\tau,x}\text{-a.s.} \tag{1.13}$$

for all $\tau \le s \le t$, and all bounded Borel functions f on E.

In the definition of the standard realization of the process X_t, the distribution of the state variable X_τ with respect to the measure $\mathbb{P}_{\tau,x}$ is δ_x, the Dirac measure at x. Let μ be a probability measure on $(E, \mathcal{E})$. Then, replacing (1.11) with

$$\mathbb{P}_{\tau,\mu}\left(\omega \in \Omega : \omega(\tau) \in A_0, \omega(t_1) \in A_1, \cdots, \omega(t_{k-1}) \in A_{k-1}, \omega(t_k) \in A_k\right)$$
$$= \int_{A_0} d\mu(x) \int_{A_1} P(\tau, x; t_1, dx_1) \int_{A_2} P(t_1, x_1; t_2, dx_2) \cdots$$
$$\int_{A_k} P(t_{k-1}, x_{k-1}; t_k, dx_k), \tag{1.14}$$

we get a stochastic process with the initial distribution at $t = \tau$ equal to μ.

Any two Markov processes associated with the same transition function P are called stochastically equivalent. Such processes may be defined on different sample spaces.

Next, we will consider backward transition probability functions. Let $\widetilde{P}$ be such a function, and let $t \in (0, T]$ and $y \in E$. Define a family of finite-dimensional distributions on the path space $\Omega = E^{[0,T]}$ equipped with the cylinder σ-algebra $\mathcal{F}$ by the following formula:

$$\widetilde{\mathbb{P}}^{t,y}\left(\omega \in \Omega : \omega(t_1) \in A_1, \cdots, \omega(t_{k-1}) \in A_{k-1}, \omega(t_k) \in A_k, \omega(t) \in A_{k+1}\right)$$
$$= \int_{A_1 \times \cdots \times A_k} \widetilde{P}(t_1, dx_1; t_2, x_2) \cdots \widetilde{P}(t_{k-1}, dx_{k-1}; t_k, x_k)$$
$$\widetilde{P}(t_k, dx_k; t, y) \chi_{A_{k+1}}(y), \qquad (1.15)$$

where $0 \le t_1 < t_2 < \cdots < t_k < t \le T$ and $A_i \in \mathcal{E}$, $1 \le i \le k+1$. Here we use Kolmogorov's extension theorem to establish the existence of the measure $\widetilde{\mathbb{P}}^{t,y}$. For all $t \in [0,T]$ and $y \in E$, the measure $\widetilde{\mathbb{P}}^{t,y}$ is defined on the σ-algebra $\mathcal{F}_t^0$. Consider the standard realization $X_s(\omega) = \omega(s)$ on the path space $(\Omega, \mathcal{F})$. Then the process X_s has $\widetilde{P}$ as its backward transition function, that is,

$$\widetilde{P}(\tau, A; t, y) = \widetilde{\mathbb{P}}^{t,y}[X_\tau \in A]$$

for all $0 \le \tau < t \le T$ and $A \in \mathcal{E}$. Moreover, the backward Markov property holds for X_s. This means that

$$\widetilde{\mathbb{E}}^{t,y}\left[f(X_\tau) \mid \mathcal{F}_t^s\right] = \widetilde{\mathbb{E}}^{t,y}\left[f(X_\tau) \mid \sigma(X_s)\right] = \widetilde{\mathbb{E}}^{s,X_s} f(X_\tau)$$

$\widetilde{\mathbb{P}}^{t,y}$-a.s. for all bounded Borel functions f and all $0 \le \tau < s < t \le T$. The process X_s has the terminal distribution at $s = t$ equal to the Dirac measure δ_y. In the case of a prescribed terminal distribution ν at $s = t$, the finite-dimensional distributions satisfy

$$\widetilde{\mathbb{P}}^{t,\nu}\left(\omega \in \Omega : \omega(t_1) \in A_1, \cdots, \omega(t_{k-1}) \in A_{k-1}, \omega(t_k) \in A_k, \omega(t) \in A_{k+1}\right)$$
$$= \int_{A_1 \times \cdots \times A_{k+1}} \widetilde{P}(t_1, dx_1; t_2, x_2) \cdots \widetilde{P}(t_{k-1}, dx_{k-1}; t_k, x_k)$$
$$\widetilde{P}(t_k, dx_k; t, y) d\nu(y). \qquad (1.16)$$

Since the relation described in (1.3) is a one-to-one correspondence between forward and backward transition functions, a backward transition probability function $\widetilde{P}$ generates a backward Markov process $\widetilde{X}^t$, $0 \le t \le T$, with respect to the family of σ-algebras $\widetilde{\mathcal{F}}_\tau^t = \sigma\left(\widetilde{X}_r : t \ge r \ge \tau\right) = \mathcal{F}_{T-\tau}^{T-t}$ and the family of measures $\widetilde{\mathbb{P}}^{t,x}$. This means that

$$\widetilde{\mathbb{E}}^{t,x}\left[f\left(\widetilde{X}^\tau\right) \mid \widetilde{\mathcal{F}}_s^t\right] = \widetilde{\mathbb{E}}^{t,x}\left[f\left(\widetilde{X}^\tau\right) \mid \widetilde{X}^s\right] = \widetilde{\mathbb{E}}^{s,\widetilde{X}^s}\left[f\left(\widetilde{X}^\tau\right)\right] \quad \widetilde{\mathbb{P}}^{t,x}\text{-a.s.}$$

for all $0 \le \tau \le s \le t \le T$ and all bounded Borel functions f on E. It is easy to see that the backward Markov property for the time reversed process $\widetilde{X}^s = X_{T-s}$ follows from the Markov property for the process X_t. Therefore, if X_t is a Markov process with transition function P, then the process X_{T-t} is a backward Markov process with backward transition function $\widetilde{P}$. In general, any property of Markov processes can be reformulated for backward Markov processes. We simply let time run backward from T to 0. For instance, if $\widetilde{P}$ is a backward transition function satisfying the subnormality condition, then we can use the construction in Section 1.3 to extend $\widetilde{P}$ to a backward transition probability function on the space $E^\triangle$.

1.5 Space-Time Processes

A transition function P is called time-homogeneous if

$$P(\tau, x; t, A) = P((\tau + h) \wedge T, x; (t + h) \wedge T, A) \tag{1.17}$$

for all $0 \leq \tau < t \leq T$, $h > 0$, $x \in E$, and $A \in \mathcal{E}$. The values of a time-homogeneous transition function depend only on the time span $t - \tau$ between the initial and final moments. We will write $P(t-\tau, x, A)$ instead of $P(\tau, x; t, A)$ in the case of a time-homogeneous transition function P. The minimum $(\tau + h) \wedge T$ appears in formula (1.17) because the parameters τ and t vary in a bounded interval.

Definition 1.5 Let $(\Omega, \mathcal{F})$ be a measurable space. A family of measurable mappings $\vartheta_s : \Omega \to \Omega$, $s \geq 0$, such that

$$\vartheta_t \circ \vartheta_s = \vartheta_{t+s}, \quad \vartheta_0 = I \tag{1.18}$$

for all $t \geq 0$ and $s \geq 0$, is called a family of time shift operators. The symbol I in (1.18) stands for the identity mapping on Ω.

If a stochastic process X_t with state space $(E, \mathcal{E})$ is given on a probability space $(\Omega, \mathcal{F})$, then it is natural to expect the process X_t to be related to the time shift operators θ_s as follows:

$$X_t \circ \vartheta_s = X_{(t+s) \wedge T} \tag{1.19}$$

for all $t \in [0, T]$ and $s \in [0, T]$. Equality (1.19) is often used in the theory of time-homogeneous stochastic processes. For some sample spaces, it is clear how to define the family of time shift operators $\{\vartheta_s\}$. For instance, if $\Omega = E^{[0,T]}$ and the process X_t is given by $X_t(\omega) = \omega(t)$, then we can define the time shift operators by $\vartheta_s(\omega)(t) = \omega((t + s) \wedge T)$.

Let P be a transition probability function, not necessarily time-homogeneous, and let $(X_t, \mathcal{F}_t^\tau, \mathbb{P}_{\tau,x})$ be a Markov process on $(\Omega, \mathcal{F})$ associated with P. Our next goal is to define the space-time process $\widehat{X}_t = (t, X_t)$, $t \in [0, T]$. It is clear that the state space of this process must be the space $[0, T] \times E$. However, it is not clear what is an optimal choice of the sample space for the space-time process and what is the transition function of the process $\widehat{X}_t$. It is natural to expect that the sample space for the process $\widehat{X}_t$ has to be the space $[0, T] \times \Omega$. We will show below that this is the case under certain restrictions.

Let B be a Borel set in $[0, T] \times E$. For every $s \in [0, T]$, the symbol $(B)_s$ will stand for the slice of the set B at level s, defined by $(B)_s = \{x \in E : (s, x) \in B\}$. Consider the following function

$$\widehat{P}(t, (\tau, x), B) = P\left(\tau, x; (\tau + t) \wedge T, (B)_{(\tau+t)\wedge T}\right), \qquad (1.20)$$

where $(\tau, x) \in [0, T] \times E$, $t \in [0, T]$, and $B \in \mathcal{B}_{[0,T] \times E}$. Equality (1.20) can be rewritten in the following form:

$$\widehat{P}(t, (\tau, x), dsdy) = P(\tau, x; s, dy) d\delta_{(\tau+t)\wedge T}(s),$$

where δ is the Dirac measure. If the transition probability function P has a density p, then formula (1.20) becomes

$$\begin{aligned}\widehat{P}(t, (\tau, x), B) &= \int_{(B)_{(\tau+t)\wedge T}} p(\tau, x; (\tau + t) \wedge T, y) dy \\ &= \int_\tau^t \int_B p(\tau, x; s, y) dy d\delta_{(\tau+t)\wedge T}(s). \end{aligned} \qquad (1.21)$$

The time variable in the definition of $\widehat{P}$ is t and the space variables are $(\tau, x) \in [0, T] \times E$ and $B \in \mathcal{B}_{[0,T] \times E}$.

Theorem 1.1 *The function $\widehat{P}$ given by (1.20) is a time-homogeneous transition probability function.*

Proof. It follows from (1.20) that the function $\widehat{P}$ is normal. Next, we will show that the function $\widehat{P}$ satisfies the Chapman-Kolmogorov equation, which in the case of time-homogeneous transition functions has the following form:

$$\widehat{P}((t_1 + t_2) \wedge T, (\tau, x), B) = \iint_{[0,T] \times E} \widehat{P}(t_1, (\tau, x), dsdy) \widehat{P}(t_2, (s, y), B) \qquad (1.22)$$

for all $(\tau, x) \in [0, T] \times E$, $B \in \mathcal{B}_{[0,T] \times E}$, $0 \le t_1 \le T$, and $0 \le t_2 \le T$. By using the Chapman-Kolmogorov equation for P and equality (1.20) twice, we see that

$$\widehat{P}((t_1 + t_2) \wedge T, (\tau, x), B) = P\left(\tau, x; (\tau + t_1 + t_2) \wedge T, (B)_{(\tau + t_1 + t_2) \wedge T}\right)$$
$$= \int_E P(\tau, x; (\tau + t_1) \wedge T, dy)$$
$$P\left((\tau + t_1) \wedge T, y; (\tau + t_1 + t_2) \wedge T, (B)_{(\tau + t_1 + t_2) \wedge T}\right)$$
$$= \iint_{[0,T] \times E} P(\tau, x; s, dy) P\left(s, y; (s + t_2) \wedge T, (B)_{(s + t_2) \wedge T}\right) d\delta_{(\tau + t_1) \wedge T}(s)$$
$$= \iint_{[0,T] \times E} \widehat{P}(t_1, (\tau, x), dsdy) \widehat{P}(t_2, (s, y), B).$$

This establishes (1.22). Therefore, the function $\widehat{P}$ defined by (1.20) is a transition probability function. □

Note that even if the transition function P has a density p, the measure $B \mapsto \widehat{P}(t, (\tau, x), B)$ on $\mathcal{B}_{[0,T] \times E}$ is singular with respect to the measure $dtdm$. The function $\widehat{P}$ will play the role of the transition function of the space-time process $\widehat{X}_t$.

There are several possible choices of the sample space $\widehat{\Omega}$ for the space-time process $\widehat{X}_t$. For instance, one can choose the full path space, that is, the space

$$\widehat{\Omega} = ([0, T] \times E)^{[0, T]}, \qquad (1.23)$$

to be the sample space of the space-time process $\widehat{X}_t$. In this case, the space-time process is defined by

$$\widehat{X}_t(\varphi, \omega) = (\varphi(t), \omega(t))$$

where $(\varphi, \omega) \in \widehat{\Omega}$, and the time shift operators on the space $\widehat{\Omega}$ are given by

$$\widehat{\vartheta}_s(\varphi, \omega)(t) = (\varphi((t + s) \wedge T), \omega((t + s) \wedge T)). \qquad (1.24)$$

Since $\widehat{P}$ is a transition probability function, the Kolmogorov extension theorem implies the existence of a family of measures $\widehat{\mathbb{P}}_{(\tau, x)}$ indexed by

$(\tau, x) \in [0, T] \times E$ such that

$$\widehat{\mathbb{P}}_{(\tau,x)} \left[\widehat{X}_{t_1} \in A_1, \cdots, \widehat{X}_{t_k} \in A_k \right]$$
$$= \widehat{\mathbb{P}}_{(\tau,x)} \left[(\varphi(t_1), \omega(t_1)) \in A_1, \cdots, (\varphi(t_k), \omega(t_k)) \in A_k \right]$$
$$= \mathbb{P}_{\tau,x} \left[X_{(\tau+t_1) \wedge T} \in (A_1)_{\varphi((\tau+t_1) \wedge T)}, \cdots, \right.$$
$$\left. X_{(\tau+t_k) \wedge T} \in (A_k)_{\varphi((\tau+t_k) \wedge T)} \right] \qquad (1.25)$$

for all $0 \leq t_1 < t_2 < \cdots < t_k \leq T$ and $A_i \in \mathcal{B}_{[0,T] \times E}$, $1 \leq i \leq k$. This construction results in a time-homogeneous Markov process $\left(\widehat{X}_t, \widehat{\mathcal{F}}_t^\tau, \widehat{\mathbb{P}}_{(\tau,x)} \right)$ on the space $\widehat{\Omega}$. The process $\widehat{X}_t$ is our first version of the space-time process. However, it may happen so that the first component $t \mapsto \varphi(t)$ of an element (φ, ω) of the sample space $\widehat{\Omega}$ is a non-measurable function with respect to the Lebesgue σ-algebra on the interval $[0, T]$. This makes the process $\widehat{X}_t$ practically useless. Our next goal is to restrict the process $\widehat{X}_t$ to an appropriate subset of the sample space $\widehat{\Omega}$.

Suppose that the original process X_t is defined on a subset Ω^* of the space $E^{[0,T]}$ such that for all $\tau \in [0, T]$ and $x \in E$,

$$\mathbb{P}^*_{\tau,x} \left(\vartheta_\tau^{-1}(\Omega^*) \right) = 1. \qquad (1.26)$$

In formula (1.26), the symbol $\mathbb{P}^*_{\tau,x}$ stands for the outer measure on the space $E^{[0,T]}$ generated by the measure $\mathbb{P}_{\tau,x}$, $\{\vartheta_\tau\}$ denotes the family of time shift operators on $E^{[0,T]}$ given by $\vartheta_\tau(\omega)(t) = \omega((\tau + t) \wedge T)$, and $\vartheta_\tau^{-1}(A)$ denotes the inverse image of the set A under the mapping ϑ_τ. Let $\widehat{\Omega}^*$ be the subspace of the space $\widehat{\Omega}$ consisting of all pairs $(\varphi_c, \omega) \in \widehat{\Omega}$ such that $\varphi_c(t) = (c + t) \wedge T$, $t \in [0, T]$, $c \geq 0$, and $\omega \in \Omega^*$. This means that we choose linear functions $t \mapsto (c + t) \wedge T$ as the first components of the elements of $\widehat{\Omega}^*$ and the paths ω from Ω^* as their second components. Note that we can identify the space $\widehat{\Omega}^*$ with the space $[0, T] \times \Omega^*$, using the mapping $j : \widehat{\Omega}^* \to [0, T] \times \Omega^*$ where $j(\varphi_c, \omega) = (c, \omega)$ with $c \in [0, T]$ and $\omega \in \Omega^*$.

Lemma 1.5 *The following equality holds for all $(\tau, x) \in [0, T] \times E$:*

$$\widehat{\mathbb{P}}^*_{(\tau,x)} \left(\widehat{\Omega}^* \right) = 1. \qquad (1.27)$$

Proof. Given $0 \le t_1 < \cdots < t_k \le T$ and $A_i \in \mathcal{B}_{[0,T] \times E}$, $1 \le i \le k$, define a cylinder in $\widehat{\Omega}$ by

$$C = \left\{ \widehat{X}_{t_1} \in A_1, \cdots, \widehat{X}_{t_k} \in A_k \right\}$$
$$= \left\{ (\varphi, \omega) : \omega(t_1) \in (A_1)_{\varphi(t_1)}, \cdots \omega(t_k) \in (A_k)_{\varphi(t_k)} \right\}. \quad (1.28)$$

For every $\tau \in [0, T]$, (1.28) gives

$$\widehat{\vartheta}_\tau^{-1}(C) = \left\{ (\varphi, \omega) : \omega((\tau + t_1) \wedge T) \in (A_1)_{\varphi((\tau + t_1) \wedge T)}, \right.$$
$$\left. \cdots, \omega((\tau + t_k) \wedge T) \in (A_1)_{\varphi((\tau + t_k) \wedge T)} \right\}. \quad (1.29)$$

Let $\pi_0 : E^{[0,T]} \mapsto \widehat{\Omega}$ be the following mapping: $\pi_0(\omega) = (\varphi_0, \omega)$ where $\omega \in E^{[0,T]}$ and $\varphi_0(t) = t$ for all $t \in [0, T]$. Then it follows from (1.29) that

$$\pi_0^{-1}\left(\widehat{\vartheta}_\tau^{-1}(C)\right) = \left\{ \omega : \omega((\tau + t_1) \wedge T) \in (A_1)_{(\tau + t_1) \wedge T}, \right.$$
$$\left. \cdots, \omega((\tau + t_k) \wedge T) \in (A_1)_{(\tau + t_k) \wedge T} \right\}. \quad (1.30)$$

Suppose that the set $\widehat{\Omega}^*$ is covered by a countable family $\{C_i\}$ of cylinders such as in (1.28). Then the family of cylinders $\left\{ \pi_0^{-1}\left(\widehat{\vartheta}_\tau^{-1}(C_i)\right) \right\}$ defined in (1.30) covers the set

$$\pi_0^{-1}\left(\widehat{\vartheta}_\tau^{-1}\left(\widehat{\Omega}^*\right)\right) = \Omega^*.$$

Using (1.25) and (1.26), we get

$$\sum_{i=1}^\infty \widehat{\mathbb{P}}_{(\tau,x)}(C_i) = \sum_{i=1}^\infty \mathbb{P}_{\tau,x}\left(\pi_0^{-1}\left(\widehat{\vartheta}_\tau^{-1}(C_i)\right)\right) \ge 1. \quad (1.31)$$

Now it is clear that equality (1.27) follows from (1.31).
This completes the proof of Lemma 1.5. $\square$

It follows from equality (1.27) that one can restrict the probability space structure from $\left(\widehat{\Omega}, \widehat{\mathcal{F}}_T^0, \widehat{\mathbb{P}}_{(\tau,x)}\right)$ to the set $\widehat{\Omega}^*$. The resulting probability space will be denoted by $\left(\widehat{\Omega}^*, \widehat{\mathcal{F}}_T^0, \widehat{\mathbb{P}}_{(\tau,x)}\right)$. Summarizing what has already been accomplished, we see that the space-time process $\widehat{X}_t$ can be defined on the probability space $\left(\widehat{\Omega}^*, \widehat{\mathcal{F}}_T^0, \widehat{\mathbb{P}}_{(\tau,x)}\right)$ as follows:

$$\widehat{X}_t(\varphi_c, \omega) = (\varphi_c(t), \omega(t)).$$

Here $\omega \in \Omega^*$, $c \geq 0$, and φ_c is the function on $[0, T]$ given by $\varphi_c(t) = (c + t) \wedge T$. Using the identification j of the spaces $\widehat{\Omega}^*$ and $[0, T] \times \Omega^*$, we see that the space-time process $\widehat{X}_t$ can be defined on the space $[0, T] \times \Omega^*$ by

$$\widehat{X}_t(c, \omega) = ((c + t) \wedge T, \omega(t)).$$

The state space of this process is the space $\left([0, T] \times E, \mathcal{B}_{[0,T] \times E}\right)$.

The space-time process $\widehat{X}_t$ can also be defined for a general Markov process $(X_t, \mathcal{F}_t^\tau, \mathbb{P}_{\tau,x})$ on the sample space Ω provided that there exists a family $\{\vartheta_s\}$ of time shift operators on Ω satisfying condition (1.19). In this case, the space-time process $\widehat{X}_t$ is defined on the sample space $\widehat{\Omega} = \{\varphi_c\}_{0 \leq c \leq T} \times \Omega$ by the formula

$$\widehat{X}_t(\varphi_c, \omega) = (\varphi_c(t), X_t(\omega)).$$

The family of time shift operators on the space $\widehat{\Omega}$ is given by

$$\widehat{\vartheta}_s(\varphi_c, \omega)(t) = \left(\varphi_{(c+s) \wedge T}(t), \vartheta_s(\omega)(t)\right).$$

It is not difficult to check that $\widehat{X}_t \circ \widehat{\vartheta}_\tau = \widehat{X}_{(\tau+t) \wedge T}$. As before, we can identify the space $\widehat{\Omega}$ with the space $[0, T] \times \Omega$, by using the mapping $j : \widehat{\Omega} \to [0, T] \times \Omega$ defined by $j(\varphi_c, \omega) = (c, \omega)$. Taking into account this identification, we see that

$$\widehat{X}_t(c, \omega) = ((c + t) \wedge T, X_t(\omega)),$$

and the time shift operators ϑ_s can be written as follows:

$$\widehat{\vartheta}_s(c, \omega) = ((c + s) \wedge T, \vartheta_s(\omega)).$$

The state space of the space-time process $\widehat{X}_t$ is $[0, T] \times E$. Our next goal is to define the measure $\widehat{\mathbb{P}}_{(\tau,x)}$ on the σ-algebra $\widehat{\mathcal{F}}_t^\tau$. For the cylinders C defined in (1.28), we can use formula (1.25). For a general set $A \in \widehat{\mathcal{F}}_T^\tau$, we extend this formula to

$$\widehat{\mathbb{P}}_{(\tau,x)}(A) = \mathbb{P}_{\tau,x}\left(\pi_0^{-1}\left(\widehat{\vartheta}_\tau^{-1}(A)\right)\right),$$

where $\pi_0 : \Omega \to \widehat{\Omega}$ is defined by $\pi_0(\omega) = (\phi_0, \omega)$.

Note that the time shift operator $\widehat{\vartheta}_\tau$ is $\widehat{\mathcal{F}}_{(\tau+t) \wedge T}^{(\tau+s) \wedge T} / \widehat{\mathcal{F}}_t^s$-measurable. Moreover, since the equality $\widehat{X}_t \circ \pi_0 = (t, X_t)$ holds, the mapping π_0 is $\mathcal{F}_t^s / \widehat{\mathcal{F}}_t^s$-measurable. It also follows that for any bounded $\widehat{\mathcal{F}}_T^0$-measurable random

variable F,
$$\widehat{\mathbb{E}}_{(\tau,x)}[F] = \mathbb{E}_{\tau,x}\left[F \circ \widehat{\vartheta}_\tau \circ \pi_0\right].$$

The space-time process $\widehat{X}_t$ is a Markov process. The Markov property of the process $\widehat{X}_t$ can be formulated as follows:
$$\widehat{\mathbb{E}}_{(\tau,x)}\left[f\left(\widehat{X}_{(t+s)\wedge T}\right) \mid \widehat{\mathcal{F}}_s^0\right] = \widehat{\mathbb{E}}_{(s,X_s)}\left[f\left(\widehat{X}_t\right)\right]$$
for all $t \in [0,T]$, $s \in [0,T]$, $x \in E$, and all bounded Borel functions f on the space $[0,T] \times E$. An equivalent formulation of the Markov property of the process $\widehat{X}_t$ is
$$\widehat{\mathbb{E}}_{(\tau,x)}\left[f\left(\widehat{X}_t\right) \circ \vartheta_s \mid \widehat{\mathcal{F}}_s\right] = \widehat{\mathbb{E}}_{(s,X_s)}\left[f\left(\widehat{X}_t\right)\right].$$

Remark 1.2 The space-time process $\widehat{X}_t$ associated with the given Markov process X_t is not simply the process $t \mapsto (t, X_t)$. For instance, if the space-time process $\widehat{X}_t$ is defined on the sample space $[0,T] \times \Omega$, then
$$\widehat{X}_t(c,\omega) = ((c+t) \wedge T, X_t(\omega))$$
where $t \in [0,T]$ and $(c,\omega) \in [0,T] \times \Omega$. For $c = 0$, we get $\widehat{X}_t(0,\omega) = (t, X_t(\omega))$.

Our next goal is to discuss space-time processes associated with backward transition probability functions. Let $\widetilde{P}(\tau, A; t, x)$ be such a function, and put
$$\widehat{P}(\tau, (t,x), B) = \widetilde{P}\left((t-\tau) \vee 0, (B)_{(t-\tau)\vee 0}; t, x\right). \quad (1.32)$$

Here $\tau \in [0,T]$ plays the role of the time-variable, whereas $(t,x) \in [0,T] \times E$ and $B \in \mathcal{B}_{[0,T]\times E}$ are the space variables. The equality in (1.32) can be rewritten as follows:
$$\widehat{P}(\tau, (t,x), dsdy) = \widetilde{P}(s, dy; t, x)\, d\delta_{(t-\tau)\vee 0}(s). \quad (1.33)$$

Theorem 1.2 Let $\widetilde{P}(\tau, A; t, x)$ be a backward transition probability function. Then the function $\widehat{P}$ defined by (1.32) is a time-homogeneous transition probability function.

Proof. It follows from equality (1.32) that $\widehat{P}$ is normal. Next let $\tau_1 \in [0,T]$, $\tau_2 \in [0,T]$, $(t,x) \in [0,T]\times E$, and $B \in \mathcal{B}_{[0,T]\times E}$. Then, using formulas

(1.32), (1.33), and the Chapman-Kolmogorov equation for the function $\widetilde{P}$, we obtain

$$\widehat{P}((\tau_1 + \tau_2) \wedge T, (t, x), B)$$
$$= \widetilde{P}\left((t - (\tau_1 + \tau_2) \wedge T) \vee 0, (B)_{(t-(\tau_1+\tau_2)\wedge T)\vee 0}; t, x\right)$$
$$= \int_E \widetilde{P}\left((t - (\tau_1 + \tau_2) \wedge T) \vee 0, (B)_{(t-(\tau_1+\tau_2)\wedge T)\vee 0}; (t - \tau_1) \vee 0, y\right)$$
$$\widetilde{P}((t - \tau_1) \vee 0, dy; t, x)$$
$$= \int\int_{[0,T]\times E} \widetilde{P}\left((s - \tau_2) \vee 0, (B)_{(s-\tau_2)\vee 0}; s, y\right) \widetilde{P}(s, dy; t, x) d\delta_{(t-\tau_1)\vee 0}(s)$$
$$= \int\int_{[0,T]\times E} \widehat{P}(\tau_1, (t, x), dsdy) \widehat{P}(\tau_2, (s, y), B). \tag{1.34}$$

In (1.34), we used the equality

$$((t - \tau_1) \vee 0 - \tau_2) \vee 0 = (t - (\tau_1 + \tau_2) \wedge T) \vee 0.$$

It is clear that (1.34) implies the Chapman-Kolmogorov equation for $\widehat{P}$.

This completes the proof of Theorem 1.2. □

Theorem 1.2 allows us to define the space-time process $\widehat{X}_\tau$ associated with the backward transition probability function $\widetilde{P}$. Arguing as in the case of transition probability functions, we first choose the space $\widehat{\Omega}$ defined by (1.23) as the sample space of the space-time process $\widehat{X}_\tau(\varphi, \omega) = (\varphi(\tau), \omega(\tau))$. Since $\widehat{P}$ is a time-homogeneous transition probability function, Kolmogorov's extension theorem implies the existence of a family of measures $\widehat{\mathbb{P}}^{(t,x)}$ indexed by the elements (t, x) of the state space $[0, T] \times E$ such that

$$\widehat{\mathbb{P}}^{(t,x)}\left[\widehat{X}_{\tau_1} \in A_1, \cdots, \widehat{X}_{\tau_k} \in A_k\right]$$
$$= \widehat{\mathbb{P}}^{(t,x)}\left[(\varphi(\tau_1), \omega(\tau_1)) \in A_1, \cdots, (\varphi(\tau_k), \omega(\tau_k)) \in A_k\right]$$
$$= \widetilde{\mathbb{P}}^{t,x}\left(X_{(t-\tau_1)\vee 0} \in (A_1)_{\varphi((t-\tau_1)\vee 0)}, \cdots, X_{(t-\tau_k)\vee 0} \in (A_k)_{\varphi((t-\tau_k)\vee 0)}\right). \tag{1.35}$$

In (1.35), $\widetilde{\mathbb{P}}^{t,x}$ is the family of measures defined by (1.15). The Markov process $\left(\widehat{X}_\tau, \widehat{\mathcal{F}}_\tau, \widehat{\mathbb{P}}^{(t,x)}\right)$ on the sample space $\widehat{\Omega}$ is our first version of the space-time process associated with the backward transition probability function $\widetilde{P}$. The following family of time shift operators on $\widehat{\Omega}$ can be used in this

case:
$$\widehat{\vartheta}_s(\varphi,\omega) = (\varphi((\tau-s)\vee 0), \omega((\tau-s)\vee 0)), \quad s \geq 0. \tag{1.36}$$

The operators $\widehat{\vartheta}_s$ are backward shifts by s with respect to the time variable τ. The time shift operators $\widehat{\vartheta}_s$ are connected with the space-time process $\widehat{X}_\tau$ as follows:
$$\widehat{X}_\tau \circ \widehat{\vartheta}_s = \widehat{X}_{(\tau-s)\vee 0} \tag{1.37}$$

for all $s \geq 0$ and $\tau \in [0,T]$. It is also possible to define space-time processes on smaller sample spaces as it has already been done in the case of transition probability functions. For instance, let $\widetilde{P}$ be a backward transition probability function, and let $\left(\widetilde{X}_\tau, \mathcal{F}_t^\tau, \widetilde{\mathbb{P}}^{t,x}\right)$ be a backward Markov process defined on the sample space $\widetilde{\Omega}$ and with $\widetilde{P}$ as its transition function. Assume that there exists a family of time shift operators $\widetilde{\vartheta}_s$ on the sample space $\widetilde{\Omega}$ such that
$$\widetilde{X}_\tau \circ \widehat{\vartheta}_s = \widetilde{X}_{(\tau-s)\vee 0} \tag{1.38}$$

for all $s \geq 0$ and $\tau \in [0,T]$. Then the space-time process $\left(\widehat{X}_\tau, \widehat{\mathcal{F}}_\tau, \widehat{\mathbb{P}}^{(t,x)}\right)$ can be defined on the sample space $[0,T] \times \widetilde{\Omega}$ by
$$\widehat{X}_\tau(c,\widetilde{\omega}) = \left((\tau-c)\vee 0, \widetilde{X}_\tau(\widetilde{\omega})\right)$$

where $(c,\widetilde{\omega}) \in [0,T] \times \widetilde{\Omega}$ and $\tau \in [0,T]$. The time shift operators $\widehat{\vartheta}_s$ on the sample space $[0,T] \times \widetilde{\Omega}$ of the space-time process are defined by
$$\widehat{\vartheta}_s(c,\widetilde{\omega}) = \left((\tau-c)\vee 0, \widetilde{\vartheta}_s(\widetilde{\omega})\right)$$

where $s \geq 0$ and $(c,\widetilde{\omega}) \in [0,T] \times \widetilde{\Omega}$. It is clear that condition (1.37) holds for $\widehat{X}_\tau$ and $\widehat{\vartheta}_s$.

1.6 Classes of Stochastic Processes

In this section we introduce and discuss various classes of stochastic processes. Let $(X_t, \mathcal{F}_t^\tau, \mathbb{P}_{\tau,x})$ be a stochastic process on $(\Omega, \mathcal{F})$ with state space $(E, \mathcal{E})$. A sample path of the process $(X_t, \mathcal{F}_t^\tau, \mathbb{P}_{\tau,x})$ corresponding to $\omega \in \Omega$ is the function $s \mapsto X_s(\omega)$ defined on the interval $[0,T]$. For a given $\omega \in \Omega$, the sample path $s \mapsto X_s(\omega)$ is often called the realization of ω, or a realization of the process X_t.

Definition 1.6

(1) A process $(X_t, \mathcal{F}_t^\tau, \mathbb{P}_{\tau,x})$ is called right-continuous if its sample paths are right-continuous functions on the interval $[0, T)$.
(2) A process $(X_t, \mathcal{F}_t^\tau, \mathbb{P}_{\tau,x})$ is called left-continuous if its sample paths are left-continuous functions on the interval $(0, T]$.
(3) It is said that a process $(X_t, \mathcal{F}_t^\tau, \mathbb{P}_{\tau,x})$ is right-continuous and has left limits if its sample paths are right-continuous functions on the interval $[0, T)$ and have left limits on the interval $(0, T]$.
(4) It is said that a process $(X_t, \mathcal{F}_t^\tau, \mathbb{P}_{\tau,x})$ is continuous if its sample paths are continuous functions on the interval $[0, T]$.

If the process $(X_t, \mathcal{F}_t^\tau, \mathbb{P}_{\tau,x})$ is right-continuous and has left limits, then the following notation will be used:

$$\lim_{s \uparrow t} X_s(\omega) = X_{s-}(\omega). \tag{1.39}$$

Since the process X_t is right-continuous, we also have

$$\lim_{s \downarrow t} X_s(\omega) = X_s(\omega). \tag{1.40}$$

Definition 1.7 A process $(X_t, \mathcal{F}_t^\tau, \mathbb{P}_{\tau,x})$ is called stochastically continuous if for all $x \in E$, $\tau \in [0, T]$, and $\varepsilon > 0$,

$$\lim_{t-s \downarrow 0; \tau \leq s \leq t \leq T} \mathbb{P}_{\tau,x}(\rho(X_s, X_t) > \varepsilon) = 0.$$

Recall that the symbol ρ in Definition 1.7 stands for the distance on $E \times E$. For $\varepsilon > 0$ and $y \in E$, put

$$G_\varepsilon(y) = \{x \in E : \rho(x, y) > \varepsilon\}. \tag{1.41}$$

Then an equivalent condition for the stochastic continuity of the process X_t is as follows: for all $x \in E$, $\tau \in [0, T]$, and $\varepsilon > 0$,

$$\lim_{t-s \downarrow 0; \tau \leq s \leq t \leq T} \int_E P(s, y; t, G_\varepsilon(y)) P(\tau, x; s, dy) = 0. \tag{1.42}$$

Definition 1.8 It is said that a process $(X_t, \mathcal{F}_t^\tau, \mathbb{P}_{\tau,x})$ is strongly stochastically continuous provided that for all $\varepsilon > 0$,

$$\lim_{t-s \downarrow 0; s \leq t \leq T} \sup_{(\tau,x) \in [0,s] \times E} \mathbb{P}_{\tau,x}(\rho(X_s, X_t) > \varepsilon) = 0.$$

An equivalent condition for the strong stochastic continuity of the process X_t is as follows: for all $\varepsilon > 0$,

$$\lim_{t-s \downarrow 0; s \leq t \leq T} \sup_{(\tau,x) \in [0,s] \times E} \int_E P(s,y;t,G_\varepsilon(y)) P(\tau,x;s,dy) = 0. \quad (1.43)$$

Definition 1.9 Let $(\Omega, \mathcal{F}, \mathbb{P})$ be a probability space equipped with a filtration $\mathcal{F}_t$, $0 \leq t \leq T$.

(1) A real-valued stochastic process Z_t, $0 \leq t < T$, on Ω is called adapted to the filtration $\mathcal{F}_t$ provided that Z_t is $\mathcal{F}_t/\mathcal{B}_\mathbb{R}$-measurable for all t with $0 \leq t < T$.
(2) A real-valued stochastic process Z_t, $0 \leq t < T$, is called an $\mathcal{F}_t$-martingale with respect to the measure $\mathbb{P}$ if for all $0 \leq t < T$, the inequality $\mathbb{E}|Z_t| < \infty$ holds, and for all s and t with $0 \leq s < t < T$,

$$\mathbb{E}(Z_t \mid \mathcal{F}_s) = Z_s$$

$\mathbb{P}$-a.s.
(3) A real-valued $\mathcal{F}_t$-adapted stochastic process Z_t, $0 \leq t < T$, is called an $\mathcal{F}_t$-supermartingale with respect to the measure $\mathbb{P}$ if for all $0 \leq t < T$, the inequality $\mathbb{E}|Z_t| < \infty$ holds, and for all s and t with $0 \leq s < t < T$,

$$\mathbb{E}(Z_t \mid \mathcal{F}_s) \leq Z_s$$

$\mathbb{P}$-a.s.
(4) A real-valued $\mathcal{F}_t$-adapted stochastic process Z_t, $0 \leq t < T$, is called an $\mathcal{F}_t$-submartingale with respect to the measure $\mathbb{P}$ provided that the process $(-Z_t)$ is an $\mathcal{F}_t$-supermartingale.

Definition 1.10 Let $(\Omega, \mathcal{G})$ be a sample space equipped with a two-parameter filtration $\mathcal{G}_t^\tau$, $0 \leq \tau \leq t \leq T$, and let X_t be a stochastic process on a filtered probability space $(\Omega, \mathcal{F}, \mathcal{G}_t^\tau)$. The process X_t is called adapted to the filtration $\mathcal{G}_t^\tau$ if for all $0 \leq \tau \leq t \leq T$, $\mathcal{F}_t^\tau \subset \mathcal{G}_t^\tau$ where $\mathcal{F}_t^\tau = \sigma(X_s : \tau \leq s \leq T)$, $0 \leq \tau \leq t \leq T$.

The next result is called the Martingale Convergence Theorem (see [Yeh (1995); Revuz and Yor (1991); Doob (2001)] for more information).

Theorem 1.3 *The following assertions hold:*

(1) Let Z_t, $0 \leq t < T$, be a right-continuous $\mathcal{F}_t$-martingale with respect to

the measure $\mathbb{P}$, and suppose that

$$\sup_{t\in[0,T)} \mathbb{E}\,|Z_t| < \infty.$$

Then the limit $Z_T = \lim_{t\uparrow T} Z_t$ exists and is finite $\mathbb{P}$-a.s.

(2) Let Z_t, $0 \leq t < T$, be a right-continuous $\mathcal{F}_t$-martingale with respect to the measure $\mathbb{P}$, and suppose that the family of random variables Z_t, $t \in [0,T)$, is uniformly integrable. Then $Z_T \in L^1$, and Z_t tends to X_T in L^1. Moreover, for all $t \in [0,T]$,

$$Z_t = \mathbb{E}\left[Z_T \mid \mathcal{F}_t\right]$$

$\mathbb{P}$-a.s.

The next assertion concerns martingales associated with transition probability functions.

Lemma 1.6 *Let P be a transition probability function, and let $(X_t, \mathcal{F}_t^\tau, \mathbb{P}_{\tau,x})$ be a Markov process with P as its transition function. Then for all $0 \leq \tau < t \leq T$, $x \in E$, and $A \in \mathcal{E}$, the process $s \mapsto P(s, X_s; t, A)$ is an $\mathcal{F}_s^\tau$-martingale on the interval $[\tau, t)$ with respect to the family of measures $\mathbb{P}_{\tau,x}$. Moreover, if P has a density p, then for all $0 \leq \tau < t \leq T$, $x \in E$, and $y \in E$, the process $s \mapsto p(s, X_s; t, y)$ is an $\mathcal{F}_s^\tau$-martingale on the interval $[\tau, t)$ with respect to the family of measures $\mathbb{P}_{\tau,x}$.*

Proof. Using the Markov property and the Chapman-Kolmogorov equation, we see that for all $\tau \leq r < s < t$,

$$\mathbb{E}_{\tau,x}\left[P(s, X_s; t, A) \mid \mathcal{F}_r^\tau\right] = \mathbb{E}_{r,X_r}\left[P(s, X_s; t, A)\right]$$
$$= \int_E P(r, X_r; s, dz) P(s, z; t, A) = P(r, X_r; t, A)$$

$\mathbb{P}_{\tau,x}$-a.s. This proves the first part of Lemma 1.6. The proof of the second part is similar.

This completes the proof of Lemma 1.6. □

1.7 Completions of σ-Algebras

This section is devoted to completions of σ-algebras with respect to families of measures.

Definition 1.11 A measure space $(X, \mathcal{F}, \mu)$ is called complete if
$$B \in \mathcal{F}, \ \mu(B) = 0, \ A \subset B \Longrightarrow A \in \mathcal{F}.$$

Let $(X, \mathcal{F}, \mu)$ be a measure space, and define a family of sets by
$$\mathcal{N} = \{A : \text{there exists } B \in \mathcal{F} \text{ such that } A \subset B \text{ and } \mu(B) = 0\}.$$

Definition 1.12 The completion $\mathcal{F}^\mu$ of the σ-algebra $\mathcal{F}$ with respect to the measure μ is the σ-algebra given by
$$\mathcal{F}^\mu = \sigma(\mathcal{F}, \mathcal{N}).$$

A σ-algebra $\mathcal{F}$ satisfying $\mathcal{F} = \mathcal{F}^\mu$ is called μ-complete.

The σ-algebra $\mathcal{F}^\mu$ is the smallest μ-complete σ-algebra containing the σ-algebra $\mathcal{F}$. The next lemma provides several equivalent conditions for $A \in \mathcal{F}^\mu$. In the formulation of Lemma 1.7 below, the symbol $A \triangle A'$ stands for the symmetric difference of the sets A and A'. It is defined by $A \triangle A' = (A \backslash A') \cup (A' \backslash A)$.

Lemma 1.7 Let $(X, \mathcal{F}, \mu)$ be a measure space, and let A be a subset of the set X. Then the following are equivalent:

(a) There exist sets $\widetilde{A} \in \mathcal{F}$ and $B \in \mathcal{F}$ such that $\widetilde{A} \subset A$, $A \backslash \widetilde{A} \subset B$, and $\mu(B) = 0$.
(b) There exist sets $A_1 \in \mathcal{F}$ and $A_2 \in \mathcal{F}$ such that $A_1 \subset A \subset A_2$ and $\mu(A_2 \backslash A_1) = 0$.
(c) There exist sets $A' \in \mathcal{F}$ and $B' \in \mathcal{F}$ such that $A \triangle A' \subset B'$ and $\mu(B') = 0$.

Proof.
(a)$\Rightarrow$(b). If there exist $\widetilde{A}$ and B such as in (a), then we set $A_1 = \widetilde{A}$ and $A_2 = \widetilde{A} \cup B$. It is clear that $A_1 \in \mathcal{F}$, $A_2 \in \mathcal{F}$, $A_1 \subset A \subset A_2$, and $\mu(A_2 \backslash A_1) = 0$.

(b)$\Rightarrow$(a). If there exist A_1 and A_2 such as in (b), then we set $\widetilde{A} = A_1$ and $B = A_2 \backslash A_1$. It is clear that $\widetilde{A} \in \mathcal{F}$, $B \in \mathcal{F}$, $A \backslash \widetilde{A} \subset A_2 \backslash A_1 = B$, and $\mu(B) = 0$.

(b)$\Rightarrow$(c). If there exist A_1 and A_2 such as in (b), then we put $A' = A_1$ and $B = A_2 \backslash A_1$. It follows that $A' \in \mathcal{F}$, $B \in \mathcal{F}$, $A \triangle A' \subset B$, and $\mu(B) = 0$.

(c)$\Rightarrow$(b). If there exist $A' \in \mathcal{F}$ and $B \in \mathcal{F}$ such as in (3), then we set $A_1 = A' \backslash B$ and $A_2 = A' \cup B$. It follows that $A_1 \in \mathcal{F}$ and $A_2 \in \mathcal{F}$. Moreover, $A_1 \subset A$. Indeed, if $x \in A_1$, then $x \in A'$ and $x \notin B$. Suppose that $x \notin A$. Then $x \in A' \backslash A \subset A \triangle A' \subset B$, which is a contradiction.

Hence $x \in A_1$, and the inclusion $A_1 \subset A$ holds. Next, we will prove that $A \subset A_2$. Let $x \in A$ and $x \notin A_2$. Then $x \notin A'$ and $x \notin B$. Therefore, $x \in A\backslash A' \subset A\triangle A' \subset B$, which is a contradiction. It follows that $A \subset A_2$. We also have $A_2\backslash A_1 \subset B$, and hence $\mu(A_2\backslash A_1) = 0$. □

Remark 1.3 It is not difficult to prove that the family $\mathcal{A}$ consisting of all sets A satisfying any of the equivalent conditions in Lemma 1.7, is a σ-algebra. Using condition (a) in Lemma 1.7, we can show that $\mathcal{A} = \mathcal{F}^\mu$. Indeed, since the inclusions $\mathcal{F} \subset \mathcal{A}$ and $\mathcal{N} \subset \mathcal{A}$ hold, we have $\mathcal{F}^\mu \subset \mathcal{A}$. Conversely, if $A \in \mathcal{A}$, then condition (a) in Lemma 1.7 holds for the set A, and hence $A \in \mathcal{F}^\mu$.

Definition 1.13 Let $(X, \mathcal{F})$ be a measurable space, and let V be a family of measures on $\mathcal{F}$. The completion $\mathcal{F}^V$ of the σ-algebra $\mathcal{F}$ with respect to the family V is defined as follows:

$$\mathcal{F}^V = \bigcap_{\mu \in V} \mathcal{F}^\mu.$$

A σ-algebra $\mathcal{F}$ satisfying $\mathcal{F} = \mathcal{F}^V$ is called V-complete.

Let $(X, \mathcal{F}, \mu)$ be a measure space. Then the measure μ can be extended to a measure μ_0 on $\mathcal{F}^\mu$ by setting $\mu_0(A) = \mu(A_1)$, where $A \in \mathcal{F}^\mu$ and A_1 is a set such as in part (b) of Lemma 1.7. It is not difficult to see that the number $\mu_0(A)$ does not depend on the choice of the set A_1 in (b). We will often use the same symbol μ for the extension μ_0 of μ. The measure space $(X, \mathcal{F}^\mu, \mu)$ is called the completion of the measure space $(X, \mathcal{F}, \mu)$ with respect to the measure μ. If V is a family of measures on $\mathcal{F}$, then every measure $\mu \in V$ can be extended to the σ-algebra $\mathcal{F}^V$ as above, and we will use the same symbol V for the family $\{\mu_0 : \mu \in V\}$ consisting of the extensions of measures $\mu \in V$. It is not hard to see that the σ-algebra $\mathcal{F}^V$ is V-complete.

Next, we will discuss completions of filtrations generated by Markov processes. Let P be a transition probability function, and let X_t be a Markov process on $(\Omega, \mathcal{F})$ associated with P. The process X_t generates the following families of σ-algebras:

$$\mathcal{F}_t^\tau = \sigma\left(X_s : \tau \leq s \leq t\right) \quad \text{and} \quad \mathcal{F}_{t+}^\tau = \bigcap_{s: t < s \leq T} \mathcal{F}_s^\tau \qquad (1.44)$$

where $0 \leq \tau \leq t \leq T$. Both families $\mathcal{F}_t^\tau$ and $\mathcal{F}_{t+}^\tau$ are increasing in t and decreasing in τ. For every $\tau \in [0, T]$, consider the family of measures on the σ-algebra $\mathcal{F}_T^\tau$ given by

$$V_\tau = \{\mathbb{P}_{s,x} : 0 \leq s \leq \tau,\ x \in E\}. \tag{1.45}$$

Suppose that $\{\mathcal{G}_t^\tau\}$, $0 \leq \tau \leq t \leq T$, is a family of σ-algebras such that it is increasing in t, decreasing in τ, and satisfies the condition

$$\mathcal{G}_t^\tau \subset [\mathcal{F}_T^\tau]^{V_\tau} \tag{1.46}$$

for all $0 \leq \tau \leq t \leq T$. In (1.46), V_τ is the family of measures defined by (1.45). For the sake of simplicity, we will denote the completion $[\mathcal{G}_t^\tau]^{V_\tau}$ of the σ-algebra $\mathcal{G}_t^\tau$ with respect to the family of measures V_τ by the symbol $\bar{\mathcal{G}}_t^\tau$. Since the families of σ-algebras $\{\mathcal{F}_t^\tau\}$ and $\{\mathcal{F}_{t+}^\tau\}$ satisfy condition (1.46), the families $\{\bar{\mathcal{F}}_t^\tau\}$ and $\{\bar{\mathcal{F}}_{t+}^\tau\}$ are well-defined.

Lemma 1.8 *The following assertions hold:*

(a) A set A belongs to the σ-algebra $\bar{\mathcal{F}}_t^\tau$ if and only if for all $(s,x) \in [0,\tau] \times E$ there exist sets $A_{s,x} \in \mathcal{F}_t^\tau$, $A'_{s,x} \in [\mathcal{F}_T^\tau]^{\mathbb{P}_{s,x}}$, and $A''_{s,x} \in [\mathcal{F}_T^\tau]^{\mathbb{P}_{s,x}}$ such that

$$A \cup A'_{s,x} = A_{s,x} \cup A''_{s,x} \quad and \quad \mathbb{P}_{s,x}\left(A'_{s,x}\right) = \mathbb{P}_{s,x}\left(A''_{s,x}\right) = 0.$$

(b) A set $\widetilde{A}$ belongs to the σ-algebra $\bar{\mathcal{F}}_{t+}^\tau$ if and only if for all $(s,x) \in [0,\tau] \times E$ there exist sets $\widetilde{A}_{s,x} \in \mathcal{F}_{t+}^\tau$, $\widetilde{A}'_{s,x} \in [\mathcal{F}_T^\tau]^{\mathbb{P}_{s,x}}$, and $\widetilde{A}''_{s,x} \in [\mathcal{F}_T^\tau]^{\mathbb{P}_{s,x}}$ such that

$$\widetilde{A} \cup \widetilde{A}'_{s,x} = \widetilde{A}_{s,x} \cup \widetilde{A}''_{s,x} \quad and \quad \mathbb{P}_{s,x}\left(\widetilde{A}'_{s,x}\right) = \mathbb{P}_{s,x}\left(\widetilde{A}''_{s,x}\right) = 0.$$

Proof. Let $(X, \mathcal{F}, \mu)$ be a measure space, and let $\mathcal{G}$ be a σ-algebra satisfying $\mathcal{G} \subset \mathcal{F}^\mu$. Let A be a set for which there exists $A' \in \mathcal{G}$ such that $A \triangle A' \in \mathcal{F}^\mu$ and $\mu(A \triangle A') = 0$. Then there exist $A'' \in \mathcal{G}$, $A_1 \in \mathcal{F}^\mu$, and $A_2 \in \mathcal{F}^\mu$ such that $\mu(A_1) = 0$, $\mu(A_2) = 0$, and $A \cup A_1 = A'' \cup A_2$. Indeed, it is sufficient to take $A'' = A'$, $A_1 = A' \backslash A$, and $A_2 = A \backslash A'$. Then $A_1 \subset A \triangle A'$, $A_2 \subset A \triangle A'$, and hence $A_1 \in \mathcal{F}^\mu$ and $A_2 \in \mathcal{F}^\mu$ with $\mu(A_1) + \mu(A_2) = 0$.

Now let A be a set such that there exist sets A'', A_1, and A_2 for which the conditions formulated above hold. Set $A' = A''$. Since $A \triangle A' = A \triangle A'' \subset A_1 \cup A_2$, we have $A \triangle A' \in \mathcal{F}^\mu$ and $\mu(A \triangle A') = 0$. Next, arguing as above, we see that Lemma 1.8 holds. $\square$

Lemma 1.9 *Let P be a transition probability function, and let X_t be a corresponding Markov process on $(\Omega, \mathcal{F})$. Then for all $0 \leq \tau \leq t \leq T$,*

$$\bar{\mathcal{F}}_t^\tau \subset \bar{\mathcal{F}}_{t+}^\tau \tag{1.47}$$

and

$$\bar{\mathcal{F}}_{t+}^\tau = \bigcap_{u:t<u\leq T} \bar{\mathcal{F}}_u^\tau. \tag{1.48}$$

Proof. The inclusion in (1.47) is straightforward. Next, we will prove the equality in (1.48). Let $\widetilde{A} \in \bar{\mathcal{F}}_{t+}^\tau$, and let $\widetilde{A}_{s,x} \in \bigcap_{u:t<u\leq T} \mathcal{F}_u^\tau$, $\widetilde{A}'_{s,x} \in [\mathcal{F}_T^\tau]^{\mathbb{P}_{s,x}}$, and $\widetilde{A}''_{s,x} \in [\mathcal{F}_T^\tau]^{\mathbb{P}_{s,x}}$ be the sets from part (b) of Lemma 1.8. Then it follows from part (a) of Lemma 1.8 that $\widetilde{A} \in \bigcap_{u:t<u\leq T} \bar{\mathcal{F}}_u^\tau$. Therefore, the inclusion

$$\bar{\mathcal{F}}_{t+}^\tau \subset \bigcap_{u:t<u\leq T} \bar{\mathcal{F}}_u^\tau \tag{1.49}$$

holds.

Let $A \in \bigcap_{u:t<u\leq T} \bar{\mathcal{F}}_u^\tau$. Then by part (a) of Lemma 1.8, for all $u \in (t,T]$, $s \in [0,\tau]$, and $x \in E$, there exist sets $A_{u,s,x} \in \mathcal{F}_u^\tau$, $A'_{u,s,x} \in [\mathcal{F}_T^\tau]^{\mathbb{P}_{s,x}}$, and $A''_{u,s,x} \in [\mathcal{F}_T^\tau]^{\mathbb{P}_{s,x}}$ such that $A \cup A'_{u,s,x} = A_{u,s,x} \cup A''_{u,s,x}$ and $\mathbb{P}_{s,x}(A'_{u,s,x}) = \mathbb{P}_{s,x}(A''_{u,s,x}) = 0$. Fix a sequence of numbers $u_n \in (t,T]$ such that $u_n \downarrow t$ as $n \to \infty$, and put

$$A_{s,x} = \limsup_{n\to\infty} A_{u_n,s,x}, \; A'_{s,x} = \limsup_{n\to\infty} A'_{u_n,s,x}, \; \text{and} \; A''_{s,x} = \limsup_{n\to\infty} A''_{u_n,s,x}.$$

Since

$$A \cup A'_{u_n,s,x} = A_{u_n,s,x} \cup A''_{u_n,s,x}$$

for all $n \geq 1$, we see that

$$A \cup A'_{s,x} = A_{s,x} \cup A''_{s,x}.$$

Therefore, $A_{s,x} \in \mathcal{F}_{t+}^\tau$, $A'_{s,x} \in [\mathcal{F}_T^\tau]^{\mathbb{P}_{s,x}}$, $A''_{s,x} \in [\mathcal{F}_T^\tau]^{\mathbb{P}_{s,x}}$, and $\mathbb{P}_{s,x}(A'_{s,x}) = \mathbb{P}_{s,x}(A''_{s,x}) = 0$. It follows from part (b) of Lemma 1.8 that $A \in \bar{\mathcal{F}}_{t+}^\tau$. Moreover,

$$\bigcap_{u:t<u\leq T} \bar{\mathcal{F}}_u^\tau \subset \bar{\mathcal{F}}_{t+}^\tau. \tag{1.50}$$

Now we see that (1.48) follows from (1.49) and (1.50).

This completes the proof of Lemma 1.9. □

1.8 Path Properties of Stochastic Processes: Separability and Progressive Measurability

It will be established in this section and in Section 1.9 that under certain restrictions on a transition probability function P, there exists a Markov process associated with P and with prescribed measurability or continuity properties of its sample paths.

Let P be a transition probability function, and let $(X_t, \mathcal{F}_t^\tau, \mathbb{P}_{\tau,x})$ be a Markov process on $(\Omega, \mathcal{F})$ with P as its transition function. If $\widetilde{X}_t$ is a modification of X_t, then the process $\widetilde{X}_t$ is not necessarily adapted to the filtration $\mathcal{F}_t^\tau$. However, if we replace the family $\mathcal{F}_t^\tau$ by its completion $\bar{\mathcal{F}}_t^\tau$, then it is not hard to prove that the process $(X_t, \bar{\mathcal{F}}_t^\tau, \mathbb{P}_{\tau,x})$ is a Markov process, and any modification Y_t of X_t is an $\bar{\mathcal{F}}_t^\tau$-adapted process. We will always assume that filtrations are complete when dealing with modifications of stochastic processes (see Subsection 1.1.5).

Let J be a countable dense subset of the interval $[a, b]$, and let g be an E-valued function defined on $[a, b]$. The function g is called minimally continuous with respect to the set J if for every $t \in [a, b]$ there exists a sequence $t_k \in J$ such that $t_k \to t$ and $g(t_k) \to g(t)$ as $k \to \infty$. Similarly, the function g is called minimally right-continuous with respect to J if for every $t \in [a, b)$ there exists a sequence $t_k \in J$ such that $t_k \downarrow t$ and $g(t_k) \to g(t)$ as $k \to \infty$.

The next definitions concern the separability and measurability properties of Markov processes.

Definition 1.14 A Markov process X_t is called separable if there exists a countable dense set $J \subset [0, T]$ such that the sample paths $t \mapsto X_t(\omega)$ are minimally continuous with respect to J $\mathbb{P}_{\tau,x}$-almost surely for all $(\tau, x) \in [0, T] \times E$. Similarly, a Markov process X_t is called separable from the right if there exists a countable dense subset $J \subset [0, T]$ such that the sample paths $t \mapsto X_t(\omega)$ are minimally right-continuous with respect to J $\mathbb{P}_{\tau,x}$-almost surely for all $(\tau, x) \in [0, T] \times E$.

Definition 1.15

(a) A stochastic process X_t on a measurable space $(\Omega, \mathcal{F})$ with state space $(E, \mathcal{E})$ is called a measurable process if the function $(s, \omega) \mapsto X_s(\omega)$ is $\mathcal{B}_{[0,T]} \otimes \mathcal{F}/\mathcal{E}$-measurable, where $\mathcal{B}_{[0,T]}$ stands for the Borel σ-algebra of the interval $[0, T]$.

(b) Let X_t be a stochastic process on a measurable space $(\Omega, \mathcal{F})$ with state space $(E, \mathcal{E})$, and let $\mathcal{F}_t$ be a filtration such that X_t is $\mathcal{F}_t$-adapted. The process X_t is called $\mathcal{F}_t$-progressively measurable if for every $t \in [0,T]$, the restriction of the function $(s,\omega) \mapsto X_s(\omega)$ to the set $[0,t] \times \Omega$ is $\mathcal{B}_{[0,t]} \otimes \mathcal{F}_t/\mathcal{E}$-measurable.

Definition 1.16 Let P be a transition probability function, and let $(X_t, \mathcal{F}_t^\tau, \mathbb{P}_{\tau,x})$ be a corresponding Markov process with state space $(E,\mathcal{E})$. The process X_t is called $\mathcal{F}_t^\tau$-progressively measurable if for every τ and t with $0 \leq \tau < t \leq T$, the restriction of the function $(s,\omega) \mapsto X_s(\omega)$ to the set $[\tau,t] \times \Omega$ is $\mathcal{B}_{[\tau,t]} \otimes \mathcal{F}_t^\tau/\mathcal{E}$-measurable.

The next result states that measurable processes always have progressively measurable modifications.

Theorem 1.4 *Let X_t be a measurable stochastic process on a probability space $(\Omega, \mathcal{F}, \mathbb{P})$ and with $(E,\mathcal{E})$ as its state space. Let $\mathcal{F}_t$ be a filtration such that the process X_t is $\mathcal{F}_t$-adapted. Then there exists an $\mathcal{F}_t$-progressively measurable modification Y_t of the process X_t.*

Proof. Any measurable process can be identified with a $\mathcal{B}_{[0,T]} \otimes \mathcal{F}/\mathcal{E}$-measurable function $(t,\omega) \mapsto X_t(\omega)$. Denote by $\widetilde{\mathcal{M}}_E$ the space of classes of equivalence of $\mathcal{F}/\mathcal{E}$-measurable functions from the space Ω into the space E, equipped with the metric

$$d(f,g) = \inf_{\epsilon>0}\left(\epsilon + \mathbb{P}\left[\omega : \rho(f(\omega),g(\omega)) > \epsilon\right]\right).$$

Then the convergence in the metric topology of the space $\widetilde{\mathcal{M}}_E$ is equivalent to the convergence in probability. Moreover, if $f \in \widetilde{\mathcal{M}}_E$ and $f_n \in \widetilde{\mathcal{M}}_E$ are such that

$$\sum_{n=1}^{\infty} d(f,f_n) < \infty, \tag{1.51}$$

then $f_n(\omega) \to f(\omega)$ $\mathbb{P}$-almost surely on Ω. Any $\mathcal{B}_{[0,T]} \otimes \mathcal{F}/\mathcal{E}$-measurable function f generates a function $\widehat{f} : [0,T] \to \widetilde{\mathcal{M}}_E$ defined as follows: for $t \in [0,T]$, $\widehat{f}(t)$ is the class of equivalence in $\widetilde{\mathcal{M}}_E$ containing the function $\omega \mapsto f(t,\omega)$.

A simple function is a function from the space $[0,T] \times \Omega$ into the space E assuming only finitely many values, each on a $\mathcal{B}_{[0,T]} \otimes \mathcal{F}$-measurable set. A simple product-space function is a simple function such that each value is assumed on a set that can be represented as a finite disjoint union of direct products of sets from $\mathcal{B}_{[0,T]}$ and $\mathcal{F}$. An elementary measurable process Y_t is a stochastic process on the space $(\Omega, \mathcal{F}, \mathbb{P})$ such that there exist a partition

$\{A_k : k \geq 1\}$ of the interval $[0,T]$ into Borel measurable sets and a sequence $\{f_k : k \geq 1\}$ of $\mathcal{F}/\mathcal{E}$-measurable mappings of the space Ω into the space E such that $Y_t = f_k$ for all $t \in A_k$.

Denote by D_E the class consisting of all $\mathcal{B}_{[0,T]} \otimes \mathcal{F}/\mathcal{E}$-measurable functions f, for which the function $\widehat{f} : [0,T] \to \widetilde{\mathcal{M}}_E$ is $\mathcal{B}_{[0,T]}/\mathcal{B}_{\widetilde{\mathcal{M}}_E}$-measurable and has a separable range. Here $\mathcal{B}_{\widetilde{\mathcal{M}}_E}$ denotes the Borel σ-algebra of the space $\widetilde{\mathcal{M}}_E$. An equivalent definition of the class D_E is as follows. A $\mathcal{B}_{[0,T]} \otimes \mathcal{F}/\mathcal{E}$-measurable function f belongs to the class D_E if the function $\widehat{f}$ can be approximated by a sequence of elementary measurable processes in the sense of pointwise convergence on Ω uniformly in $t \in [0,T]$.

Lemma 1.10 *The class D_E coincides with the class of all $\mathcal{B}_{[0,T]} \otimes \mathcal{F}/\mathcal{E}$-measurable functions.*

Proof. Let us first prove the lemma in the case where $E = \mathbb{R}$. It is not difficult to see that the class $D_\mathbb{R}$ is closed under pointwise convergence of sequences of functions and contains all simple product-space functions. By the monotone class theorem for functions, Lemma 1.10 holds for $E = \mathbb{R}$.

Our next goal is to prove Lemma 1.10 for any finite subset of $\mathbb{R}$. Let $\mathbb{R}_0 = \{c_1, c_2, \cdots, c_n\}$ be a finite subset of $\mathbb{R}$ equipped with the metric inherited from the space $\mathbb{R}$. Next, using Lemma 1.10 for $E = \mathbb{R}$, we see that if $f : [0,T] \times \Omega \mapsto \mathbb{R}_0$ is a $\mathcal{B}_{[0,T]} \otimes \mathcal{F}/\mathcal{B}_{\mathbb{R}_0}$-measurable function and $i \in \mathbb{N}$, then the range $\left\{\widehat{f}(t) : 0 \leq t \leq T\right\}$ of the function $\widehat{f}$ can be covered by a countable disjoint family A_k^i, $k \in \mathbb{N}$, of Borel subsets of the space $\widetilde{\mathcal{M}}_\mathbb{R}$ so that $\widehat{f}^{-1}\left(A_k^i\right) \in \mathcal{B}_{[0,T]}$, and the diameter of any set A_k^i is less than $\frac{1}{i}$. Therefore, there exists a sequence f_i of elementary measurable processes such that $\widehat{f}_i(t) \in \mathcal{M}_{\mathbb{R}_0}$ for all $t \in [0,T]$, and moreover

$$\sup_{t \in [0,T]} d\left(f(t,\cdot) - f_i(t,\cdot)\right) < \frac{1}{i} \tag{1.52}$$

for all $i \in \mathbb{N}$. It follows from the second definition of the class $D_{\mathbb{R}_0}$ that $f \in D_{\mathbb{R}_0}$. This establishes Lemma 1.10 for $E = \mathbb{R}_0$.

Next, we will pass from a finite subset $\mathbb{R}_0 = \{c_1, c_2, \cdots, c_n\}$ of the space $\mathbb{R}$ to a finite subset $E_0 = \{x_1, x_2, \cdots, x_n\}$ of the given state space E. Let $g : [0,T] \times \Omega \mapsto E_0$ be a $\mathcal{B}_{[0,T]} \otimes \mathcal{F}/\mathcal{B}_{E_0}$-measurable function. Then for every $1 \leq j \leq n$, we have $g(t,\omega) = x_j$ on a set $B_j \in \mathcal{B}_{[0,T]} \otimes \mathcal{F}$. The sets B_j may be empty. It is also true that the nonempty sets B_j are disjoint and cover $[0,T] \times \Omega$. Let us consider a $\mathcal{B}_{[0,T]} \otimes \mathcal{F}/\mathcal{B}_{\mathbb{R}_0}$-measurable function defined by $f(t,\omega) = c_j$ on the set B_j with $1 \leq j \leq n$. By the previous part of the proof, there exists a sequence f_i of elementary measurable processes such that $\widehat{f_i}(t) \in \mathcal{M}_{\mathbb{R}_0}$ and inequality (1.52) holds. Replacing c_j by x_j in the function f_i, we get a function g_i. Taking into account (1.52) and the fact that $\rho(x_m, x_k) \leq c \mid c_m - c_k \mid$ for $1 \leq m \leq k \leq n$, where $c > 0$ is a finite constant, we see that

$$\lim_{i \to \infty} \sup_{t \in [0,T]} d(g(t,\cdot), g_i(t,\cdot)) = 0. \tag{1.53}$$

Since g_i is an elementary process and (1.53) holds, we have $g \in D_{E_0}$. It follows that any simple function $s : [0,T] \times \Omega \to E$ belongs to the class D_E. Now let f be a $\mathcal{B}_{[0,T]} \otimes \mathcal{F}/\mathcal{E}$-measurable function. Then the function f can be approximated pointwise by simple functions, and since the class D_E is closed under pointwise convergence, we have $f \in D_E$.

This completes the proof of Lemma 1.10. $\square$

Let us continue the proof of Theorem 1.4. By Lemma 1.10, the class D_E coincides with the class of all $\mathcal{B}_{[0,T]} \otimes \mathcal{F}/\mathcal{E}$-measurable functions. Approximating $\mathcal{B}_{[0,T]}/\mathcal{B}_{\widetilde{\mathcal{M}}}$-measurable functions with separable range by simple functions, we see that for any measurable process X_t, there exists a sequence Y_t^n of elementary measurable processes such that

$$d(X_t, Y_t^n) \leq \frac{1}{2^{n+1}} \tag{1.54}$$

for all $t \in [0,T]$ and $n \geq 1$. Recall that to every elementary process Y_t^n there corresponds a partition $\{A_k^n : k \geq 1\}$ of the interval $[0,T]$ into Borel measurable sets and a sequence $\{f_k^n : k \geq 1\}$ of $\mathcal{F}/\mathcal{E}$-measurable mappings of the space Ω into the space E such that $Y_t^n = f_k^n$ for all $t \in A_k^n$. Fix $n \geq 1$ and $\delta \in (0,T)$. Our next goal is to modify the process Y_t^n as follows. Put $s_k^n = \inf\{t : t \in A_k^n\}$. If $s_k^n \in A_k^n$, then the new process Z_t^n is defined for $t \in A_k^n$ by $Z_t^n = X_{s_k^n}$. If $s_k^n \notin A_k^n$, then we fix $t_k^n \in A_k^n$ such that $t_k^n - s_k^n < \delta$, and put $Z_t^n = X_{t_k^n}$ for all $t \in A_k^n$. It is clear that the new processes Z_t^n are elementary measurable processes. Since X_s is an adapted process, it is easy to see that for every $t \in [0, T-\delta]$, the restriction of the function $(s, \omega) \mapsto Z_s^n(\omega)$ to the space $[0,t] \times \Omega$ is $\mathcal{B}_{[0,t]} \otimes \mathcal{F}_{t+\delta}/\mathcal{E}$-measurable. Moreover, inequality (1.54) implies

$$d(X_t, Z_t^n) \leq \frac{1}{2^n} \tag{1.55}$$

for all $t \in [0,T]$ and $n \geq 1$. By (1.51), $Z_t^n(\omega) \to X_t(\omega)$ as $n \to \infty$ for all $t \in [0,T]$ almost surely on Ω. Fix $x_0 \in E$, and put $Y_t(\omega) = \lim_{n \to \infty} Z_t^n(\omega)$

if the limit exists, and $Y_t(\omega) = x_0$ otherwise. Then the process Y_t is a modification of the process X_t. We will next show that the process Y_t is progressively measurable. It is clear from the definition of the process Y_t that it suffices to prove that every process Z_t^n is progressively measurable.

Since the process X_t is adapted, we see that for every $u \in [0, T - \delta]$, the restriction of the function $(s, \omega) \mapsto Z_s^n(\omega)$ to the space $[0, u] \times \Omega$ is $\mathcal{B}_{[0,u]} \otimes \mathcal{F}_{u+\delta}/\mathcal{E}$-measurable. Now let $t \in [0, T]$. Then using the previous assertion with $u_m = t - \frac{1}{m}$, $m \geq m_0$, we see that the restriction of the function $(s, \omega) \mapsto Z_s^n(\omega)$ to the space $[0, t) \times \Omega$ is $\mathcal{B}_{[0,t)} \otimes \mathcal{F}_t/\mathcal{E}$-measurable. Since the process X_t is adapted, we see that the process Z_t^n is progressively measurable for all $n \geq 1$.

This completes the proof of Theorem 1.4. □

Every sample path of a measurable process is a $\mathcal{B}_{[0,T]}/\mathcal{E}$ measurable function. The next assertion provides examples of progressively measurable stochastic processes.

Theorem 1.5 *Every left- or right-continuous process is progressively measurable.*

Proof. Let X be a right-continuous process. Fix τ and t with $0 \leq \tau < t \leq T$, and consider a sequence of partitions

$$\pi_k = \left\{\tau = s_1^{(k)} < s_2^{(k)} < \cdots < s_k^{(k)} = t\right\}$$

such that the mesh of the partition π_k tends to zero as $k \to \infty$. For every $k \geq 1$, define a simple process X^k on $[\tau, t]$ as follows: $X_s^k = X_{s_j^{(k)}}$, if $s \in [s_j^{(k)}, s_j^{(k+1)})$, and $X_t^k = X_t$. It is clear that the function $(s, \omega) \mapsto X_s^k(\omega)$ defined on $[\tau, t] \times \Omega$ is $\mathcal{B}_{[\tau,t]} \otimes \mathcal{F}_t^\tau/\mathcal{E}$-measurable. It follows from the right-continuity of the process X_t that

$$\lim_{k \to \infty} X_s^k(\omega) = X_s(\omega)$$

for all $s \in [\tau, t]$ and $\omega \in \Omega$. Hence, the function $(s, \omega) \mapsto X_s(\omega)$ defined on $[\tau, t] \times \Omega$ is $\mathcal{B}_{[\tau,t]} \otimes \mathcal{F}_t^\tau/\mathcal{E}$-measurable. This means that the process X_t is progressively measurable.

The proof of Theorem 1.5 is thus completed for right-continuous processes. The proof for left-continuous processes is similar. □

The next result concerns separable processes and stochastic equivalence.

Theorem 1.6 *Let P be a transition probability function satisfying condition (1.42). Then there exists a separable process $(X_t, \mathcal{F}_t^\tau, \mathbb{P}_{\tau,x})$ on $(\Omega, \mathcal{F})$*

with state space $(E, \mathcal{E})$ such that the transition function of X_t coincides with P.

Proof. The following lemma will be used in the proof of Theorem 1.6:

Lemma 1.11 *Let P be a transition probability function, and let X_t be a Markov process associated with P. Then for every $(\tau, x) \in [0, T] \times E$ there exists a separable $\mathbb{P}_{\tau,x}$-modification $\widehat{X}_s$, $\tau \leq s \leq T$, of the process X_s, $\tau \leq s \leq T$. The process $\widehat{X}_s$ depends on τ and x.*

Remark 1.4 Note that the stochastic continuity condition is not assumed in Lemma 1.11.

Proof. The next assertion is an important part of the proof of Lemma 1.11.

Lemma 1.12 *Let $B \in \mathcal{E}$, $\tau \in [0, T]$, and $x \in E$. Then, there exists a finite or countable sequence $t_k \in [\tau, T]$ depending on B for which $\mathbb{P}_{\tau,x}[N(t, B)] = 0$, $t \in [\tau, T]$, $x \in E$, where $N(t, B) = \{X_{t_k} \in B : k \geq 1, X_t \notin B\}$.*

Proof. We will use the method of mathematical induction in the proof of Lemma 1.12. Let t_1 be any number in $[\tau, T]$. If the numbers $t_1, t_2, \ldots, t_k$ have already been chosen, then we put

$$\gamma_k = \sup_{t \in [\tau, T]} \mathbb{P}_{\tau,x}\left[X_{t_j} \in B : 1 \leq j \leq k, X_t \notin B\right].$$

If $\gamma_k = 0$, then we are done. If $\gamma_k > 0$, then there exists $t_{k+1} \in [\tau, T]$ such that

$$\sup_{t \in [\tau, T]} \mathbb{P}_{\tau,x}\left[X_{t_j} \in B : 1 \leq j \leq k, X_{t_{k+1}} \notin B\right] \geq \frac{\gamma_k}{2}.$$

Put

$$N_k(B) = \left\{X_{t_j} \in B : 1 \leq j \leq k, X_{t_{k+1}} \notin B\right\}.$$

It is clear that the sets $N_k(B)$ are disjoint. Therefore,

$$1 \geq \sum_k \mathbb{P}_{\tau,x}[N_k(B)] \geq \frac{1}{2} \sum_k \gamma_k,$$

and hence, $\gamma_k \to 0$ as $k \to \infty$. Since $\mathbb{P}_{\tau,x}[N(t, B)] \leq \gamma_k$ for any $k \geq 1$, we have $\mathbb{P}_{\tau,x}[N(t, B)] = 0$ for all $t \in [\tau, T]$.

This completes the proof of Lemma 1.12. □

Now let B_i be a sequence of Borel sets in E. It follows from Lemma 1.12 that for every $i \geq 1$ there exists a sequence t_k^i, $k \geq 1$, such that

$$\mathbb{P}_{\tau,x}[N(t, B_i)] = 0 \tag{1.56}$$

for all $t \in [\tau, T]$. By enumerating the set $\{t_k^i : i \geq 1, k \geq 1\}$, we see that the sequence t_k can be chosen independently of i. It is not hard to prove that the sets $N(t, B_i)$ constructed for this sequence are subsets of the similar sets in (1.56). This means that (1.56) holds for the new sequence. Next we will show that more is true.

Lemma 1.13 *Let $\tau \in [0, T]$, $x \in E$, and let B_i be a sequence of Borel sets in E. Then there exists a sequence $t_k \in [\tau, T]$, $k \geq 1$, and for every $t \in [\tau, T]$ there exists a set $N(t) \in \mathcal{F}_T^\tau$ such that*

$$\mathbb{P}_{\tau,x}[N(t)] = 0, \tag{1.57}$$

and

$$N(t, B) \subset N(t) \tag{1.58}$$

for all $t \in [\tau, T]$ and for all sets B which can be represented as a countable intersection of elements of the family $\{B_i\}$.

Proof. Put $N(t) = \bigcup_i N(t, B_i)$ (here we use the sequence t_k constructed after equality (1.56)). Let $B = \bigcap_j B_{i_j}$. Then

$$N(t, B) \subset \bigcup_j \{X_{t_k} \in B, \, k \geq 1, \, X_t \notin B_{i_j}\}$$

$$\subset \bigcup_j \{X_{t_k} \in B_{i_j}, \, k \geq 1, \, X_t \notin B_{i_j}\} = \bigcup_j N(t, B_{i_j}) \subset N(t). \tag{1.59}$$

Now it is clear that (1.57) follows from (1.56), while (1.58) follows from (1.59). □

Let us return to the proof of Lemma 1.11. Denote by C_i the family of all open balls of rational radii centered at the points of a countable dense subset of E, and put $B_i = E \backslash C_i$. It is clear that the family of sets which are representable as countable intersections of the sets from the family $\{B_i\}$ contains the family of all closed subsets of E. Applying Lemma 1.13, we see that there exists a sequence $t_k \in [\tau, T]$, and for every $t \in [\tau, T]$ there exists a set $N(t)$ such that $\mathbb{P}_{\tau,x}[N(t)] = 0$ and $N(t, C) \subset N(t)$ for all closed subsets C of E and all $t \in [\tau, T]$.

Let Y_t be a separable process on Ω, and define a new process $\widehat{X}_t$ as follows: $\widehat{X}_t(\omega) = X_t(\omega)$ if $t \in \{t_k\}$ or $\omega \notin N(t)$, and $\widehat{X}_t(\omega) = Y_t(\omega)$ otherwise. Our next goal is to show that

$$\mathbb{P}_{\tau,x}\left[\widehat{X}_t = X_t\right] = 1 \tag{1.60}$$

for all $t \in [\tau, T]$ and also to prove that $\widehat{X}_t$ is a separable process. Indeed, if $t \in \{t_i\}$, then $\left\{\widehat{X}_t = X_t\right\} = \Omega$. If $t \notin \{t_i\}$, then $\left\{\widehat{X}_t \neq X_t\right\} \subset N(t)$. This gives (1.60).

Now we are ready to prove the separability of the process $\widehat{X}_t$. If $t = t_i$ for some $i \geq 1$, then $\widehat{X}_{t_i} = X_{t_i}$, and hence for all $\omega \in \Omega$, the corresponding sample path is minimally continuous with respect to $\{t_i\}$. If $t \notin \{t_k\}$ and $\omega \in N(t)$, then $\widehat{X}_t$ coincides with a $\{t_i\}$-separable process. If $t \notin \{t_i\}$ and $\omega \notin N(t)$, then $\widehat{X}_t(\omega) = X_t(\omega)$, and we proceed as follows. Suppose that $X_t(\omega)$ cannot be approximated by a subsequence of the sequence $X_{t_i}(\omega)$. Then there exists a ball C centered at $X_t(\omega)$ such that $X_{t_i}(\omega) \notin C$ for all $i \geq 1$. Moreover, for every $i \geq 1$, we have $X_{t_i}(\omega) \in B$ and $X_t(\omega) \notin B$ where $B = E \backslash C$. Hence, $\omega \in N(t, B) \subset N(t)$, which is a contradiction. Therefore, $X_t(\omega)$ can be approximated by a subsequence of the sequence $X_{t_i}(\omega)$, and this implies the separability of the process $\widehat{X}_t$.

This completes the proof of Lemma 1.11. □

Note that we have not yet employed the stochastic continuity condition in the proof of Theorem 1.6. This condition is needed to guarantee that any countable dense subset of $[\tau, T]$ can be used as a separability set.

Lemma 1.14 *Suppose that P is a transition probability function satisfying the stochastic continuity condition (1.8), and let X_t be a corresponding Markov process. Fix $(\tau, x) \in [0, T] \times E$. Then any countable dense subset of $[\tau, T]$ can be used as a separability set in Lemma 1.11.*

Proof. By Lemma 1.11, there exists a separable process $\widehat{X}_t$ on $[\tau, T]$ associated with a separability set $\{t_k\}$. Let $\{s_i\}$ be any countable dense subset of $[\tau, T]$. Next, we will show that for $\mathbb{P}_{\tau,x}$-almost all $\omega \in \Omega$ and all $k \geq 1$, the element $\widehat{X}_{t_k}(\omega)$ of E belongs to the set $A(\omega)$ consisting of all limit points of the set $\left\{\widehat{X}_{s_i}(\omega)\right\}$. By Fatou's Lemma and the stochastic continuity of the process $\widehat{X}_t$, we see that for every $k \geq 1$,

$$\mathbb{P}_{\tau,x}\left[\liminf_{i\to\infty} \rho\left(\widehat{X}_{t_k}, \widehat{X}_{s_i}\right) > 0\right]$$
$$\leq \lim_{n\to\infty} \mathbb{P}_{\tau,x}\left[\liminf_{i\to\infty} \rho\left(\widehat{X}_{t_k}, \widehat{X}_{s_i}\right) > \frac{1}{n}\right]$$
$$\leq \lim_{n\to\infty} \liminf_{i\to\infty} \mathbb{P}_{\tau,x}\left[\rho\left(\widehat{X}_{t_k}, \widehat{X}_{s_i}\right) > \frac{1}{n}\right] = 0. \qquad (1.61)$$

Condition (1.61) means that for $\mathbb{P}_{\tau,x}$-almost every $\omega \in \Omega$ and every $k \geq 1$, the element $\widehat{X}_{t_k}(\omega)$ of E belongs to the set $A(\omega)$. Therefore, the process $\widehat{X}_t$ is separable with respect to the set $\{s_i\}$.

This completes the proof of Lemma 1.14. $\qquad\square$

The next result (Lemma 1.15) will allow us to get rid of the dependence of the process $\widehat{X}_s$ in Lemma 1.11 on the variables τ and x. The conditions in Lemma 1.15 are as follows. A family of stochastic processes $X_t^{(\tau,x)}$ parameterized by $(\tau, x) \in [0, T] \times E$ is given, and it is known that the sample paths of these processes possess a certain property. Our goal is to construct a single non-homogeneous process X_t from the processes $X_t^{(\tau,x)}$ so that the sample paths of X_t possess the same property.

Lemma 1.15 *Let P be a transition probability function and suppose that for every pair $(\tau, x) \in [0, T] \times E$, a stochastic process $\widehat{X}_t^{(\tau,x)}$, $0 \leq t \leq T$, is given on $(\Omega, \mathcal{F})$. Suppose also that*

$$\widehat{\mathbb{P}}_{\tau,x}\left[\widehat{X}_t^{(\tau,x)} \in B\right] = P(\tau, x; t, B) \qquad (1.62)$$

for all t with $\tau \leq t \leq T$ and all $B \in \mathcal{E}$. Let F be a class of E-valued functions defined on $[0, T]$, and assume that the sample paths of all the processes $\widehat{X}_t^{(\tau,x)}$ belong to F. Then there exists a Markov process X_t such that its sample paths belong to the class F and P is its transition function.

Proof. Consider a new sample space $\widehat{\Omega} = [0, T] \times E \times \Omega$. This space will be equipped with the σ-algebra

$$\widehat{\mathcal{F}} = \left\{\widehat{A} \subset \widehat{\Omega} : \widehat{A}_{\tau,x} \in \mathcal{F} \text{ for all } (\tau, x) \in [0, T] \times E\right\}.$$

Here $\widehat{A}_{\tau,x} = \{\omega : (\tau, x, \omega) \in \widehat{A}\}$. Define a stochastic process on $\widehat{\Omega}$ by $\widehat{X}_t(\tau, x, \omega) = X_t^{(\tau,x)}(\omega)$ and consider the family of σ-algebras given by $\widehat{\mathcal{F}}_t^\tau = \sigma\left(\widehat{X}_s : \tau \leq s \leq t\right)$. The family of probability measures $\widehat{\mathbb{P}}_{\tau,x}$ is defined as follows: $\widehat{\mathbb{P}}_{\tau,x}\left[\widehat{A}\right] = \mathbb{P}_{\tau,x}\left[\widehat{A}_{\tau,x}\right]$ for every $\widehat{A} \in \widehat{\mathcal{F}}$. It is not hard to

see from the definition of the process $\widehat{X}_t$ that its sample paths belong to
F. It remains to prove that $\widehat{X}_t$ is associated with P. Let $B \in \mathcal{E}$ and $\tau \leq t$.
Then

$$\widehat{\mathbb{P}}_{\tau,x}\left[\widehat{X}_t \in B\right] = \widehat{\mathbb{P}}_{\tau,x}\left[\widehat{\omega} = (u,y,\omega) \in \widehat{\Omega} : X_t^{(\tau,x)}(\omega) \in B\right]$$
$$= \mathbb{P}_{\tau,x}\left[\omega : X_t^{(\tau,x)}(\omega) \in B\right] = \mathbb{P}_{\tau,x}\left[\omega : X_t(\omega) \in B\right]. \quad (1.63)$$

It follows from (1.62) and (1.63) that P is the transition function of the process $\widehat{X}_t$.

This completes the proof of Lemma 1.15. □

Finally, we are ready to finish the proof of Theorem 1.6. Let P be a given transition probability function satisfying the stochastic continuity condition, and let Y_t be a Markov process on $(\Omega, \mathcal{F})$ with transition function P. By Lemma 1.11, for every (τ, x) there exists a separable process $\widehat{Y}_t^{(\tau,x)}$, $\tau \leq t \leq T$, that is stochastically equivalent to the process Y_t, $\tau \leq t \leq T$. Put

$$\widehat{X}_t^{(\tau,x)} = \begin{cases} \widehat{Y}_t^{(\tau,x)}, & \text{if } \tau \leq t \\ x, & \text{if } 0 \leq t < \tau. \end{cases}$$

Then it is clear that the process $\widehat{X}_t^{(\tau,x)}$ is separable for every τ and x. Since the stochastic continuity condition holds, Lemma 1.14 and the definition of the process $\widehat{X}_t^{(\tau,x)}$ imply that any countable dense subset of $[0,T]$ serves as a separability set for the processes $\widehat{X}_t^{(\tau,x)}$. Moreover, condition (1.62) holds. Let us fix a separability set $\{t_k\}$, and define a class of functions F as follows. The class F contains all functions on $[0,T]$ with values in E which are minimally continuous with respect to $\{t_k\}$. Now it is clear that we can apply Lemma 1.15, and get a separable process X_t with transition function P.

This completes the proof of Theorem 1.6. □

The next theorem concerns the existence of a progressively measurable Markov process associated with a given transition probability function P.

Theorem 1.7 *Let P be a transition probability function satisfying the strong stochastic continuity condition:*

$$\lim_{t-s\downarrow 0; s\leq t\leq T} \sup_{(\tau,x)\in[0,s]\times E} \int_E P(s,y;t,G_\epsilon(y)) P(\tau,x;s,dy) = 0$$

for all $\varepsilon > 0$. *Then there exists a progressively measurable process* $(X_t, \mathcal{F}_t^\tau, \mathbb{P}_{\tau,x})$ *on the space* $(\Omega, \mathcal{F})$ *with P as its transition function.*

Proof. The strong continuity condition was discussed in Section 1.6 (see (1.43) and Definition 1.8). Let X_t be a strongly stochastically continuous Markov process with P as its transition function and assume that $\mathcal{F}_t^\tau$ is the filtration $\sigma(X_s : \tau \leq s \leq t)$ augmented by the family of sets

$$\{B \in \mathcal{F} : \mathbb{P}_{\tau,x}[B] = 0 \text{ for some } (\tau, x) \in [0, T] \times E\}.$$

Consider a sequence of partitions $0 = t_0^n < t_1^n < \cdots < t_{m_n}^n = T$ such that $\max_{j:1\leq j\leq m_n} (t_j^n - t_{j-1}^n) \to 0$ as $n \to \infty$, and define a sequence of stochastic processes by

$$X_s^n = \begin{cases} X_{t_{j-1}^n}, & \text{if } t_{j-1}^n \leq s < t_j^n \text{ with } 1 \leq j \leq m_n, \\ X_T, & \text{if } s = T. \end{cases}$$

It is clear that every process X^n is $\mathcal{F}_t^s$-progressively measurable. It follows from the definition of X^n and Definition 1.8 that

$$\lim_{n\to\infty} \sup_{(\tau,x)\in[0,T]\times E} \sup_{s:0\leq s\leq T} \mathbb{P}_{\tau,x}\left[\rho(X_s^n, X_s) \geq \epsilon\right] = 0 \quad (1.64)$$

for all $\epsilon > 0$. Next, using (1.64) and reasoning as in the standard proof of the fact that the convergence in measure of a sequence of functions implies the existence of an almost everywhere convergent subsequence, we see that there exists a sequence $i_n \uparrow \infty$ such that for every $s \in [0,T]$, $X_s^{i_n}$ converges to X_s $\mathbb{P}_{\tau,x}$-a.s. for all $(\tau, x) \in [0,T] \times E$. Indeed, from (1.64) we see that there exists a sequence $i_n \uparrow \infty$ such that

$$\mathbb{P}_{\tau,x}\left[\rho(X_s^{i_n}, X_s) \geq \frac{1}{2^n}\right] \leq \frac{1}{2^n}$$

for all $s \in [0,T]$, $\tau \in [0,T]$, and $x \in E$. Therefore,

$$\mathbb{P}_{\tau,x}\left[\bigcap_{j\geq 1}\bigcup_{n\geq j}\left\{\rho(X_s^{i_n}, X_s) \geq \frac{1}{2^n}\right\}\right] = 0$$

for all $s \in [0,T]$, $\tau \in [0,T]$, and $x \in E$. It follows that for every $s \in [0,T]$, the sequence $X_s^{i_n}$ converges $\mathbb{P}_{\tau,x}$-a.s. to X_s for every $\tau \in [0,T]$ and $x \in E$.

Without loss of generality, we can assume that $i_n = n$ for all $n \geq 1$. Let $\mathcal{A}$ be the set of all $(s, \omega) \in [\tau, T] \times \Omega$ such that $\lim_{n\to\infty} X_s^n(\omega)$ exists.

It is clear that the set $\mathcal{A}$ is $(\mathcal{B}_{[0,T]} \otimes \mathcal{F})$-measurable. Moreover, for every $s \in [0,T]$,

$$\mathbb{P}_{\tau,x} \{\omega : (s, \omega) \in \mathcal{A}\} = 1 \qquad (1.65)$$

for all $(\tau, x) \in [0,T] \times E$. Indeed, if $s \in [0,T]$ and $(\tau, x) \in [0,T] \times E$, then the set $\{\omega : (s, \omega) \in \mathcal{A}\}$ contains an $\mathcal{F}$-measurable set $A_s^{(\tau,x)}$ such that $\mathbb{P}_{\tau,x}(A_s(\tau, x)) = 1$. This follows from the fact that if $s \in [0,T]$, then X_s^n converges to X_s $\mathbb{P}_{\tau,x}$-a.s. for every $(\tau, x) \in [0,T] \times E$.

Define a new process $\widehat{X}_s$ on Ω by $\widehat{X}_s(\omega) = \lim_{n \to \infty} X_s^n(\omega)$, if $(s, \omega) \in \mathcal{A}$, and $\widehat{X}_s(\omega) = x_0$ for all $(s, \omega) \notin \mathcal{A}$, where x_0 is a fixed point in E. Since for all $(\tau, x) \in [0,T] \times E$, the process $\widehat{X}_t$ is an $\mathbb{P}_{\tau,x}$-modification of the process X_t, it is clear that $\mathbb{P}_{\tau,x}\left[\widehat{X}_t \in B\right] = P(\tau, x; t, B)$ for all $0 \leq \tau < t \leq T$ and all $B \in \mathcal{E}$. Moreover, since the processes X^n are $\mathcal{F}_t^s$-progressively measurable and (1.65) holds, the process $\widehat{X}_t$ is $\widehat{\mathcal{F}}_t^s$-progressively measurable.

This completes the proof of Theorem 1.7. □

1.9 Path Properties of Stochastic Processes: One-Sided Continuity and Continuity

In this section we continue our exploration of the properties of paths of Markov processes. Specifically, we will study the processes with continuous sample paths and the processes for which the sample paths have only jump discontinuities.

Theorem 1.8 *Let P be a transition probability function, and let $\widetilde{X}_t$ be a Markov process with transition function P. Suppose that for all $(\tau, x) \in [0, T) \times E$ the following condition holds:*

$$\lim_{t-s \downarrow 0; \tau \leq s < t \leq T} \mathbb{P}_{\tau,x}\text{-ess sup} \, P\left(s, \widetilde{X}_s; t, G_\epsilon\left(\widetilde{X}_s\right)\right) = 0 \qquad (1.66)$$

for all $\epsilon > 0$, where $G_\epsilon(y)$ is defined by (1.41). The essential supremum in (1.66) is taken with respect to the measure $\mathbb{P}_{\tau,x}$. Then there exists a Markov process $(X_t, \mathcal{F}_t^\tau, \mathbb{P}_{\tau,x})$ on $(\Omega, \mathcal{F})$ such that X_t is right-continuous, has left limits, and the transition function of X_t coincides with P.

Corollary 1.1 *Let P be a transition probability function satisfying the following condition:*

$$\lim_{t-s \to 0+} \sup_{y \in E} P(s, y; t, G_\epsilon(y)) = 0 \qquad (1.67)$$

for all $\epsilon > 0$, where $G_\epsilon(y)$ is defined by (1.41). Then there exists a Markov process $(X_t, \mathcal{F}_t^\tau, \mathbb{P}_{\tau,x})$ on $(\Omega, \mathcal{F})$ such that X_t is right-continuous, has left limits, and the transition function of X_t coincides with P.

Remark 1.5 It can be shown that condition (1.67) implies condition (1.42). Indeed, if a transition probability function P satisfies condition (1.67), then we have

$$\int_E P(s, y; t, G_\epsilon(y)) P(\tau, x; s, dy) \le \sup_{y \in E} P(s, y; t, G_\epsilon(y)),$$

and hence condition (1.42) holds.

Proof. The structure of the proof of Theorem 1.8 is similar to that of Theorem 1.6. We start with the following lemma.

Lemma 1.16 *Let P be a transition probability function, and let $\widetilde{X}_t$ be a Markov process with transition function P. Suppose that condition (1.66) holds for a fixed pair $(\tau, x) \in [0, T) \times E$. Then there exists a $\mathbb{P}_{\tau,x}$-modification $\widehat{X}_t$ of the process $\widetilde{X}_t$, $\tau \le t \le T$, which is right-continuous and has left limits. The process $\widehat{X}_t$ depends on τ and x.*

Proof. Without loss of generality, we may assume that $\widetilde{X}_t$ is a separable process on $(\Omega, \mathcal{F})$ (see Lemma 1.11). Fix τ and x. The following random variables will be used in the proof:

$$\varphi(\epsilon, r, s, t) = \mathbb{P}_{\tau,x}\left[\rho(X_s, X_t) \ge \epsilon \mid \mathcal{F}_r^\tau\right] \tag{1.68}$$

where $\tau \le r \le s \le t \le T$ and $\epsilon > 0$. Denote

$$\alpha(\epsilon, \delta) = \sup_{(s,t): 0 < t-s \le \delta} \mathbb{P}_{\tau,x}\text{-ess sup}_\omega \varphi(\epsilon, s, s, t) \tag{1.69}$$

where $0 < \delta < T - \tau$ and $\epsilon > 0$. Using the properties of conditional expectations, we get

$$\varphi(\epsilon, r, s, t) \le \alpha(\epsilon, t - s) \quad \mathbb{P}_{\tau,x} \text{ a.s.} \tag{1.70}$$

Indeed,

$$\varphi(\epsilon, r, s, t) = \mathbb{E}_{\tau,x}\left[\mathbb{P}_{\tau,x}\left[\rho(X_s, X_t) \ge \epsilon \mid \mathcal{F}_s^\tau\right] \mid \mathcal{F}_r^\tau\right] \le \alpha(\epsilon, t - s).$$

Moreover, using the Markov property, we obtain

$$\alpha(\epsilon, \delta) \le \sup_{(s,t): 0 < t-s \le \delta} \mathbb{P}_{\tau,x}\text{-ess sup}_\omega P(s, X_s; t, G_\epsilon(X_s)). \tag{1.71}$$

Indeed, since for $0 \leq \tau \leq r < s \leq t \leq T$ we have $\mathcal{F}_r^\tau \subset \mathcal{F}_s^\tau$, it follows that

$$\alpha(\epsilon,\delta) \leq \sup_{(s,t):0<t-s\leq\delta} \operatorname{ess\,sup}_\omega \mathbb{P}_{s,X_s}\left[\rho(X_s,X_t) \geq \epsilon\right]$$

$$= \sup_{(s,t):0<t-s\leq\delta} \mathbb{P}_{\tau,x}\text{-ess\,sup}_\omega P\left(s,X_s;t,G_\epsilon(X_s)\right).$$

By condition (1.66) for (τ,x), we see that

$$\lim_{\delta\downarrow 0}\alpha(\epsilon,\delta) = 0 \tag{1.72}$$

for every $\epsilon > 0$.

In order to continue the proof of Lemma 1.16, we will need the following definition.

Definition 1.17 Let I be a subset of $[\tau,T]$ and f be a function defined on I and taking values in E. Let $\epsilon > 0$ and $k \geq 1$. It is said that the function f has at least k ϵ-oscillations on the set I provided that there exists a finite subset $J = \{s_1 < s_2 < \cdots < s_k < s_{k+1}\}$ of the set I such that $\rho\left(f(s_i),f(s_{i+1})\right) \geq \epsilon$ for all $1 \leq i \leq k$.

Denote by $Z_{k,\epsilon}$ the class of all functions having at least k ϵ-oscillations on I. We will say that the function f has a finite number of ϵ-oscillations on I provided that there exists $k \in \mathbb{N}$ such that $f \notin Z_{k,\epsilon}$.

Lemma 1.17 *For every subinterval $I = [a,b]$ of the interval $[\tau,T]$, the class of functions on I having a finite number of ϵ-oscillations for every $\epsilon > 0$ coincides with the class of functions on I with no discontinuities of the second kind.*

Proof. Suppose that the left limit of f does not exist at a point $t \in (a,b]$. Then

$$\zeta = \inf_{s:a<s<t}\sup_{r:s<r<t}\rho\left(f(s),f(r)\right) > 0. \tag{1.73}$$

Now let ϵ be such that $0 < \epsilon < \zeta$. Then it follows from (1.73) that f has an infinite number of ϵ-oscillations on I. The proof of this fact in the case of the right-hand limit is similar.

Conversely, let us assume that there exists $\epsilon > 0$ such that f has an infinite number of ϵ-oscillations on I. Then there exists a sequence $I_k = \{t_1^k < t_2^k < \cdots < t_{k+1}^k\}$, $k \geq 2$, of finite subsets of $[a,b]$ such that $\rho\left(f\left(t_j^k\right),f\left(t_{j+1}^k\right)\right) \geq \epsilon$ for all $1 \leq j \leq k$. This implies the existence of a sequence $\left(t_{i_k-1}^k, t_{i_k}^k, t_{i_k+1}^k\right)$ such that $t_{i_k+1}^k - t_{i_k-1}^k \to 0$ as $k \to \infty$. By passing to a subsequence, we may assume that $t_{i_k}^k$ has a one-sided limit t. With no loss of generality, we may assume that $t_{i_k}^k \uparrow t$. This is a contradiction, since $t_{i_k}^k \uparrow t$, $t_{i_k-1}^k \to t$, and $\rho\left(f\left(t_{i_k-1}^k\right), f\left(t_{i_k}^k\right)\right) \geq \epsilon$ for all $k \geq 2$.

This completes the proof of Lemma 1.17. □

Let us continue the proof of Lemma 1.16. For every $\epsilon > 0$, $k \geq 1$, and any Borel subset H of the interval $[\tau, T]$, define the following events:

$$\Gamma_k(\epsilon, H) = \{\omega : \text{the function } t \mapsto X_t(\omega) \text{ has at least } k \text{ } \epsilon\text{-oscillations on } H\} \tag{1.74}$$

and

$$\Gamma_\infty(\epsilon, H) = \bigcap_{k \geq 1} \Gamma_k(\epsilon, H). \tag{1.75}$$

Our goal is to prove that

$$\mathbb{P}_{\tau,x}\left[\Gamma_\infty(\epsilon, [\tau, T])\right] = 0 \tag{1.76}$$

for every $\epsilon > 0$. Then

$$\mathbb{P}_{\tau,x}\left[\bigcup_{\epsilon > 0} \Gamma_\infty(\epsilon, [\tau, T])\right] = 0,$$

and hence, by Lemma 1.17,

$$\mathbb{P}_{\tau,x}\{\omega : \text{ the function } t \mapsto X_t(\omega) \text{ has no discontinuities of the second} $$
$$\text{kind on } [\tau, T]\} = 1. \tag{1.77}$$

Let $I = \{t_1 < t_2 < \cdots < t_n\}$ be a finite subset of the interval $[\tau, T]$. Put

$$\Phi_k(\epsilon, I) = \mathbb{P}_{\tau,x}\left[\Gamma_k(\epsilon, I) \mid \mathcal{F}_{t_1}^\tau\right] \tag{1.78}$$

and

$$\beta_k(\epsilon, I) = \operatorname*{ess\,sup}_{\omega} \Phi_k(\epsilon, I)(\omega). \tag{1.79}$$

Lemma 1.18 *For every $k \geq 1$, the following estimate holds:*

$$\beta_k(\epsilon, I) \leq \left(2\alpha\left(\frac{\epsilon}{4}, t_n - t_1\right)\right)^k. \tag{1.80}$$

Proof. For any i with $1 \leq i \leq n$, denote by $A_1^i(\epsilon, I)$ the event consisting of all $\omega \in \Gamma_1(\epsilon, I)$ such that $\rho(X_{t_j}, X_{t_m}) \geq \epsilon$ for some j and m with $i \leq j < m \leq n$. In addition, let us denote by $B_{k-1}^i(\epsilon, I)$ the event consisting of all $\omega \in \Gamma_{k-1}(\epsilon, I)$ such that ω has at least $k - 1$ ϵ-oscillations on the set $(t_1, \ldots, t_i)$, but the number of ϵ-oscillations on the set $(t_1, \ldots, t_i)$ is less than $k-1$. Then the events $B_{k-1}^i(\epsilon, I)$ are disjoint and $\mathcal{F}_{t_i}^\tau$-measurable, $\bigcup_{i=1}^n B_{k-1}^i(\epsilon, I) = \Gamma_{k-1}(\epsilon, I)$, and moreover,

$$\Gamma_k(\epsilon, I) \subset \bigcup_{i=1}^n \left[B_{k-1}^i(\epsilon, I) \cap A_1^i(\epsilon, I)\right].$$

For every $1 \leq i \leq n$, put $I_i = \{t_m : i \leq m \leq n\}$. It follows that

$$\begin{aligned}
\Phi_k(\epsilon, I) &\leq \sum_{i=1}^n \mathbb{E}_{\tau,x}\left[\mathbb{E}_{\tau,x}\left[\chi_{A_1^i(\epsilon,I)}\chi_{B_{k-1}^i(\epsilon,I)} \mid \mathcal{F}_{t_i}^\tau\right] \mid \mathcal{F}_{t_1}^\tau\right] \\
&= \sum_{i=1}^n \mathbb{E}_{\tau,x}\left[\chi_{B_{k-1}^i(\epsilon,I)}\mathbb{E}_{\tau,x}\left[\chi_{A_1^i(\epsilon,I)} \mid \mathcal{F}_{t_i}^\tau\right] \mid \mathcal{F}_{t_1}^\tau\right] \\
&\leq \sum_{i=1}^n \beta_{k-1}(\epsilon, I_i)\mathbb{P}_{\tau,x}\left[B_{k-1}^i(\epsilon, I) \mid \mathcal{F}_{t_1}^\tau\right] \\
&= \max_{i: 1 \leq i \leq n} \beta_{k-1}(\epsilon, I_i)\mathbb{E}_{\tau,x}\left[\Gamma_{k-1}(\epsilon, I) \mid \mathcal{F}_{t_1}^\tau\right] \\
&= \max_{i: 1 \leq i \leq n} \beta_{k-1}(\epsilon, I_i)\Phi_1(\epsilon, I). \tag{1.81}
\end{aligned}$$

Now we see that in order to prove (1.80), it suffices to show that

$$\beta_1(\epsilon, J) \leq 2\alpha\left(\frac{\epsilon}{4}, s_m - s_1\right), \tag{1.82}$$

for any set $J = \{s_1, \ldots, s_m\}$ with $\tau \leq s_1 < \cdots < s_m < T$.

Define the following events:

$$K_i(\epsilon, J) = \left\{\rho\left(X_{s_1}, X_{s_j}\right) < \frac{\epsilon}{2} \text{ for all } 1 \leq j \leq i-1, \text{ and } \rho\left(X_{s_1}, X_{s_i}\right) > \frac{\epsilon}{2}\right\}$$

for all $2 \leq i \leq m$, and

$$L_i(\epsilon, J) = \left\{\rho\left(X_{s_i}, X_{s_m}\right) \geq \frac{\epsilon}{4}\right\}$$

for all $1 \leq i \leq m$. It is clear that the events $K_i(\epsilon, J)$ are disjoint and $\mathcal{F}_{s_i}^\tau$-measurable. Moreover,

$$\Gamma_1(\epsilon, J) \subset L_1(\epsilon, J) \cup \bigcup_{i=2}^m \left(K_i(\epsilon, J) \cap L_i(\epsilon, J)\right). \tag{1.83}$$

Indeed, using the triangle inequality, we obtain

$$\Gamma_1(\epsilon, J) \subset \bigcup_{i=2}^m K_i(\epsilon, J).$$

Let $\omega \in \Gamma_1(\epsilon, J)$. Then $\omega \in K_i(\epsilon, J)$ for some i with $2 \leq i \leq m$. Here i depends on ω. If $\omega \notin L_i(\epsilon, J)$, then it follows from

$$\rho\left(X_{s_1}(\omega), X_{s_m}(\omega)\right) \geq \rho\left(X_{s_1}(\omega), X_{s_i}(\omega)\right) - \rho\left(X_{s_i}(\omega), X_{s_n}(\omega)\right) \geq \frac{\epsilon}{4}$$

that $\omega \in L_1(\epsilon, J)$. This gives (1.83).

It is not hard to see from (1.83) that

$$\Phi_1(\epsilon, J) \leq \mathbb{P}_{\tau,x}\left[L_1(\epsilon, J) \mid \mathcal{F}_{s_1}^\tau\right] + \sum_{i=2}^m \mathbb{P}_{\tau,x}\left[K_i(\epsilon, J) \cap L_i(\epsilon, J) \mid \mathcal{F}_{s_1}^\tau\right]$$

$$\leq \mathbb{P}_{\tau,x}\left[L_1(\epsilon, J) \mid \mathcal{F}_{s_1}^\tau\right] + \sum_{i=2}^m \mathbb{E}_{\tau,x}\left[\mathbb{E}_{\tau,x}\left[\chi_{K_i(\epsilon,J)}\chi_{L_i(\epsilon,J)} \mid \mathcal{F}_{s_i}^\tau\right] \mid \mathcal{F}_{s_1}^\tau\right]$$

$$\leq \mathbb{P}_{\tau,x}\left[L_1(\epsilon, J) \mid \mathcal{F}_{s_1}^\tau\right] + \sum_{i=2}^m \mathbb{E}_{\tau,x}\left[\chi_{K_i(\epsilon,J)}\mathbb{E}_{\tau,x}\left[\chi_{L_i(\epsilon,J)} \mid \mathcal{F}_{s_i}^\tau\right] \mid \mathcal{F}_{s_1}^\tau\right].$$

(1.84)

Therefore, taking the essential supremum in (1.84), we get

$$\beta_1(\epsilon, J) \leq \alpha\left(\frac{\epsilon}{4}, s_m - s_1\right) + \alpha\left(\frac{\epsilon}{4}, s_m - s_1\right) \sum_{i=2}^m \mathbb{P}_{\tau,x}\left[K_i(\epsilon, J) \mid \mathcal{F}_{s_1}^\tau\right]$$

$$\leq 2\alpha\left(\frac{\epsilon}{4}, s_m - s_1\right).$$

Now it is clear that (1.82) holds. Moreover, (1.82) and (1.81) imply (1.80). This completes the proof of Lemma 1.18. □

Next, we will prove conditions (1.76) and (1.77). Fix $\epsilon > 0$. Since (1.72) holds, there exists $\delta > 0$ such that

$$\alpha\left(\frac{\epsilon}{4}, \delta\right) < \frac{1}{2}. \tag{1.85}$$

Given $\epsilon > 0$, fix a number δ for which (1.85) holds. Let $[a, b]$ be any subinterval of $[\tau, T]$ such that $b - a \leq \delta$. Then (1.80) gives

$$\lim_{k \to \infty} \sup \{\beta_k(\epsilon, I) : I \text{ finite and } I \subset [a, b]\} = 0.$$

Let us subdivide the interval $[\tau, T]$ into a finite family $\{\triangle_j\}$, $1 \le j \le m$, of intervals of length at most δ. Recall that we assumed the separability of the process X_t. Moreover, any countable dense subset $J = \{s_\ell : \ell \ge 1\}$ of $[\tau, T]$ may serve as a separability set for X_t on $[\tau, T]$. Fix such a set J and put $J_n = \{s_\ell : 1 \le \ell \le n\}$, $J_{n,j} = J_n \cap \triangle_j$, and $a_{n,j} = \min\{s : s \in J_{n,j}\}$. Then for every $k \ge 1$, we have

$$\mathbb{P}_{\tau,x}\left[\Gamma_\infty\left(\epsilon, J_{n,j}\right)\right] \le \mathbb{P}_{\tau,x}\left[\Gamma_k\left(\epsilon, J_{n,j}\right)\right] = \mathbb{E}_{\tau,x}\left[\mathbb{P}_{\tau,x}\left[\Gamma_k\left(\epsilon, J_{n,j}\right) \mid \mathcal{F}^\tau_{a_{n,j}}\right]\right]$$

$$\le \beta_k\left(\epsilon, J_{n,j}\right) \le \left\{2\alpha\left(\frac{\epsilon}{4}, \delta\right)\right\}^k.$$

It follows from (1.85) that for every $n \ge 1$ and j with $1 \le j \le m$, $\mathbb{P}_{\tau,x}\left[\Gamma_\infty\left(\epsilon, J_{n,j}\right)\right] = 0$. It is easy to see that

$$\Gamma_\infty\left(\epsilon, J \cap \triangle_j\right) = \cup_{n \ge 1} \Gamma_\infty\left(\epsilon, J_{n,j}\right).$$

Therefore, $\mathbb{P}_{\tau,x}\left[\Gamma_\infty\left(\epsilon, J \cap \triangle_j\right)\right] = 0$ for all j with $1 \le j \le m$. Moreover,

$$\Gamma_\infty\left(\epsilon, J\right) = \bigcup_{1 \le j \le m} \Gamma_\infty\left(\epsilon, J \cap \triangle_j\right),$$

implies $\mathbb{P}_{\tau,x}\left[\Gamma_\infty\left(\epsilon, J\right)\right] = 0$. Since J is a separability set for X_t on the interval $[\tau, T]$, we see that (1.76) and (1.77) hold.

Denote by $\widetilde{\Omega}$ the event consisting of all $\omega \in \Omega$ for which $t \mapsto X_t(\omega)$ has no discontinuities of the second kind on $[\tau, T]$, and define a stochastic process by $\widetilde{X}_t(\omega) = X_t(\omega)$ if $\omega \in \widetilde{\Omega}$, and by $\widetilde{X}_t(\omega) = x$ if $\omega \in \Omega \backslash \widetilde{\Omega}$. It is clear that the process $\widetilde{X}_t$ is separable, and that its sample paths have no discontinuities of the second kind. Moreover, $\widetilde{X}_t$ is stochastically equivalent to X_t on $[\tau, T]$. It follows from the properties of the process X_t that the limit $\widehat{X}_t(\omega) = \lim_{n \to \infty} X_{t+\frac{1}{n}}(\omega)$ exists for all $t \in [\tau, T)$ and $\omega \in \Omega$. The process $\widehat{X}_t$ is right-continuous on $[0, T)$ and has left limits on $(\tau, T]$. Moreover, for a fixed $t \in [\tau, T)$, we have

$$\left\{\widehat{X}_t \ne X_t\right\} = \bigcup_{m=1}^{\infty} \left\{\rho(X_t, \widehat{X}_t) > \frac{1}{m}\right\}.$$

It follows that

$$\mathbb{P}_{\tau,x}\left[\widehat{X}_t \neq X_t\right] \leq \lim_{m\to\infty} \mathbb{P}_{\tau,x}\left[\rho(X_t,\widehat{X}_t) > \frac{1}{m}\right]$$

$$= \lim_{m\to\infty} \mathbb{P}_{\tau,x}\left[\rho\left(X_t, \lim_{n\to\infty} X_{t+\frac{1}{n}}\right) > \frac{1}{m}\right]$$

$$\leq \lim_{m\to\infty}\lim_{n\to\infty} \mathbb{P}_{\tau,x}\left[\rho\left(X_t, X_{t+\frac{1}{n}}\right) > \frac{1}{m}\right].$$

Now using the stochastic continuity of the process X_t, we get $\mathbb{P}_{\tau,x}\left[\widehat{X}_t \neq X_t\right] = 0$. Therefore, the process $\widehat{X}_t$, $[\tau, T]$, is a modification of the process X_t, $[\tau, T]$, and it follows that the process $\widehat{X}_t$ has P as its transition function.

This completes the proof of Lemma 1.16. □

Finally, using Lemma 1.16 and Lemma 1.15 for the class F consisting of all right-continuous functions on $[0,T]$ with left limits, we see that Theorem 1.8 holds. □

The next result concerns the continuity of the sample paths.

Theorem 1.9 *Let P be a transition probability function, and let $\widetilde{X}_t$ be a Markov process with transition function P. Suppose that for all $(\tau, x) \in [0,T) \times E$ the following condition holds:*

$$\lim_{t-s\downarrow 0;\tau\leq s<t\leq T} \frac{1}{t-s}\mathbb{P}_{\tau,x}\text{-ess sup}\, P\left(s, \widetilde{X}_s; t, G_\epsilon(\widetilde{X}_s)\right) = 0 \quad (1.86)$$

for all $\epsilon > 0$. Then there exists a continuous process $(X_t, \mathcal{F}_t^\tau, \mathbb{P}_{\tau,x})$ with P as its transition function.

The next corollary follows from Theorem 1.9.

Corollary 1.2 *Let P be a transition probability function satisfying the following condition:*

$$\lim_{t-s\downarrow 0;0\leq s<t\leq T} \sup_{y\in E} \frac{P(s,y;t,G_\epsilon(y))}{t-s} = 0 \quad (1.87)$$

for all $\epsilon > 0$. Then there exists a continuous process $(X_t, \mathcal{F}_t^\tau, \mathbb{P}_{\tau,x})$ with P as its transition function.

Proof. The following lemma will be used in the proof of Theorem 1.9:

Lemma 1.19 *Let P be a transition probability function, and let $\widetilde{X}_t$ be a Markov process with transition function P. Suppose that for a given pair $(\tau, x) \in [0, T) \times E$,*

$$\lim_{t-s\downarrow 0; \tau \le s < t \le T} \frac{1}{t-s} \mathbb{P}_{\tau,x}\text{-ess sup } P\left(s, \widetilde{X}_s; t, G_\epsilon(\widetilde{X}_s)\right) = 0 \qquad (1.88)$$

for all $\epsilon > 0$. Then there exists a continuous $\mathbb{P}_{\tau,x}$-modification $\widehat{X}_t$ of the process $\widetilde{X}_t$. The process $\widehat{X}_t$ depends on τ and x.

Proof. By Lemma 1.16, there exists a $\mathbb{P}_{\tau,x}$-modification Y_t of the process $\widetilde{X}_t$ which is right-continuous and has left limits. Let

$$\pi_i = \left\{\tau = t_0^i < t_1^i < \cdots < t_{n-1}^i < t_{n_i}^i = T\right\}, \quad i \ge 1$$

be a sequence of partitions of the interval $[\tau, T]$ such that π_{i+1} is a refinement of π_i for all $i \ge 1$. Let us also assume that

$$\delta_i = \max\left\{t_{k+1}^i - t_k^i : 0 \le k \le n_i - 1\right\}$$

tends to 0 as i tends to ∞. Note that condition (1.88) still holds if we replace $\widetilde{X}_t$ by Y_t. Put

$$z(\epsilon, \delta) = \sup_{(s,t): 0 < t-s \le \delta; \tau \le s < t} \mathbb{P}_{\tau,x}\text{-ess sup } P(s, Y_s; t, G_\epsilon(Y_s))$$

and

$$\widetilde{z}(\epsilon, \delta) = \sup_{(s,t): 0 < t-s \le \delta; \tau \le s < t} \mathbb{P}_{\tau,x}\text{-ess sup } \frac{P(s, Y_s; t, G_\epsilon(Y_s))}{t-s}$$

for all $\epsilon > 0$ and $\delta > 0$. Then, using (1.68)–(1.71), we obtain

$$\sum_{k=0}^{n_i-1} \mathbb{P}_{\tau,x}\left[\rho\left(Y_{t_k^i}, Y_{t_{k+1}^i}\right) \ge \epsilon\right]$$

$$= \sum_{k=0}^{n_i-1} \mathbb{E}_{\tau,x}\left[\mathbb{P}_{\tau,x}\left[\rho\left(Y_{t_k^i}, Y_{t_{k+1}^i}\right) \ge \epsilon \mid \mathcal{F}_{t_k^i}^\tau\right]\right] \le \sum_{k=0}^{n_i-1} \alpha\left(\epsilon, t_{k+1}^i - t_k^i\right)$$

$$\le \sum_{k=0}^{n_i-1} z\left(\epsilon, t_{k+1}^i - t_k^i\right) = \sum_{k=0}^{n_i-1} \left(t_{k+1}^i - t_k^i\right) \frac{z\left(\epsilon, t_{k+1}^i - t_k^i\right)}{t_{k+1}^i - t_k^i}$$

$$\le \sum_{k=0}^{n_i-1} \left(t_{k+1}^i - t_k^i\right) \widetilde{z}\left(\epsilon, t_{k+1}^i - t_k^i\right) \le (T-\tau)\widetilde{z}(\epsilon, \delta_i) \le T\widetilde{z}(\epsilon, \delta_i).$$

By (1.86), $\widetilde{z}(\epsilon,\delta_i) \to 0$ as $i \to \infty$. Therefore

$$\lim_{i\to\infty} \sum_{k=0}^{n_i-1} \mathbb{P}_{\tau,x}\left[\rho\left(Y_{t_k^i}, Y_{t_{k+1}^i}\right) \geq \epsilon\right] = 0. \tag{1.89}$$

For every $i \geq 1$ and $\epsilon > 0$, put

$$B_i(\epsilon) = \left\{\omega \in \Omega : \max_{k:0\leq k<n_i} \rho\left(Y_{t_k^i}(\omega), Y_{t_{k+1}^i}(\omega)\right) \geq \epsilon\right\},$$

and

$$C_j(\epsilon) = \bigcup_{i=j}^{\infty} B_i(\epsilon).$$

It is easy to see, using (1.89) and passing to a subsequence of the sequence π_i, that without loss of generality we may assume that

$$\lim_{j\to\infty} \mathbb{P}_{\tau,x}[C_j(\epsilon)] = 0.$$

The sequence of events $C_j(\epsilon)$ is decreasing. Hence, for $C(\epsilon) = \bigcap_{j\geq 1} C_j(\epsilon)$, we have $\mathbb{P}_{\tau,x}[C(\epsilon)] = 0$. The complement $D(\epsilon)$ of the event $C(\epsilon)$ can be described as follows: $\omega \in D(\epsilon)$ if and only if there exists $j > 1$ depending on ω and such that

$$\max_{k:0\leq k<n_i} \rho\left(Y_{t_k^i}(\omega), Y_{t_{k+1}^i}(\omega)\right) < \epsilon \tag{1.90}$$

for all $i \geq j$. Recall that the process Y_t is right-continuous and has left limits. For all $t \in (\tau, T]$, put $Y_{t-}(\omega) = \lim_{s\uparrow t} Y_s(\omega)$. Our next goal is to show that if $\omega \in D(\epsilon)$, then

$$\rho\left(Y_{t-}(\omega), Y_t(\omega)\right) < 2\epsilon \tag{1.91}$$

for all $t \in (\tau, T]$. Indeed, let $\omega \in D(\epsilon)$, and let j be the integer corresponding to ω in the definition of $D(\epsilon)$. Assume that

$$\rho\left(Y_{s-}(\omega), Y_s(\omega)\right) \geq 2\epsilon$$

for some $s \in (\tau, T]$. It is not hard to see that this inequality contradicts the inequality in (1.90). Hence, (1.91) is satisfied. Moreover, if $\omega \in \bigcap_{\epsilon>0} D(\epsilon)$, then (1.91) holds for any $\epsilon > 0$. This establishes the continuity condition for all $\omega \in \bigcap_{\epsilon>0} D(\epsilon)$ and all $t \in (\tau, T]$. Since the process X_t is right-continuous at $t = \tau$ and $\mathbb{P}_{\tau,x}\left[\cap_{\epsilon>0} D(\epsilon)\right] = 1$, there exists a continuous modification $\widehat{X}_t$ of the process Y_t.

This completes the proof of Lemma 1.19. □

It is not hard to see that Lemma 1.15 with the class F consisting of all continuous functions on $[0, T]$ implies Theorem 1.9. □

Remark 1.6 Note that if τ and x in Theorem 1.9 are fixed, and if X_t is a Markov process with transition function P, then there exists a continuous $\mathbb{P}_{\tau,x}$-modification $\widehat{X}_t$ of the process X_t.

Remark 1.7 Let P be a transition probability function, and let $\widetilde{X}_t$ be a Markov process associated with P. Fix τ with $0 \leq \tau < T$ and $x \in E$, and let $\triangle$ be a closed subinterval of the interval $[\tau, T]$. Suppose that

$$\lim_{t-s\downarrow 0; s,t\in\triangle} \mathbb{P}_{\tau,x}\text{- ess sup} \, P\left(s, \widetilde{X}_s; t, G_\epsilon(\widetilde{X}_s)\right) = 0$$

for all $\epsilon > 0$. Then, arguing as in the proof of Lemma 1.16, we see that there exists a $\mathbb{P}_{\tau,x}$-modification $\widehat{X}_t$ of the process $\widetilde{X}_t$, which is right-continuous and has left limits on the interval $\triangle$.

Similarly, if P is a transition probability function such that

$$\lim_{t-s\downarrow 0; s,t\in\triangle} \frac{1}{t-s} \mathbb{P}_{\tau,x}\text{- ess sup} \, P\left(s, \widetilde{X}_s; t, G_\epsilon(\widetilde{X}_s)\right) = 0$$

for all $\epsilon > 0$, then there exists a $\mathbb{P}_{\tau,x}$-modification $\widehat{X}_t$ of the process $\widetilde{X}_t$, which is continuous on the interval $\triangle$.

Next, we will formulate two results concerning the path properties of supermartingales (Theorems 1.10 and 1.11 below). The proofs of these results will be omitted, and we refer the reader to Section 2.9 in [Yeh (1995)] for more information.

Theorem 1.10 *Let Z_t be a supermartingale on a filtered probability space $(\Omega, \mathcal{F}, \mathcal{F}_t, \mathbb{P})$ where $0 \leq t \leq T$. Denote by $\mathbb{Q}$ the set of all rational numbers in $[0, T]$. Then there exists a set $\Lambda \in \mathcal{F}$ such that $\mathbb{P}(\Lambda) = 0$, and the limits*

$$\lim_{r\downarrow t, r\in\mathbb{Q}} Z_r(\omega) \quad \text{and} \quad \lim_{r\uparrow t, r\in\mathbb{Q}} Z_r(\omega)$$

exist for all $\omega \in \Omega\backslash\Lambda$.

The set Λ in the formulation of Theorem 1.10 can be chosen independently of $t \in [0, T]$.

Theorem 1.11 *Let Z_t be a supermartingale on a filtered probability space $(\Omega, \mathcal{F}, \mathcal{F}_t, \mathbb{P})$, and assume that $\mathcal{F}_t$ is an augmented right-continuous filtration. Then the function $t \mapsto \mathbb{E}[Z_t]$ is right-continuous if and only if there exists a modification $\widetilde{Z}_t$ of the process Z_t, which is right-continuous and has left limits.*

It is clear that a martingale is also a supermartingale, and for a martingale Z_t, the function $t \mapsto \mathbb{E}[Z_t]$ is equal to a constant. Hence, Theorem 1.11 implies that every martingale on a filtered probability space satisfying the conditions in Theorem 1.11 has a modification that is right-continuous and has left limits.

1.10 Reciprocal Transition Functions and Reciprocal Processes

Let X_t be a Markov process with transition function P. Then the time reversed process $\widehat{X}_t = X_{T-t}$ is a backward Markov process with backward transition function $\widetilde{P}$ given by

$$\widetilde{P}(\tau, A; t, y) = P(T - t, y; T - \tau, A)$$

(see Section 1.4). If the motion of a random system described by the process $\{X_s : \tau \le s \le T\}$ starts at $x \in E$ at time τ, one says that the process X_s is pinned at x at time τ. Similarly, if the backward Markov process $\{\widehat{X}_s : 0 \le s \le t\}$ starts at $y \in E$ at time t, one says that the process $\widehat{X}_s$ is pinned at y at time t. Our goal is to study Markov processes possessing both features described above, that is, we would like to understand whether it is possible to simultaneously pin the process X_s at $x \in E$ at time τ and at $y \in E$ at time t. This problem goes back to Schrödinger (see [Schrödinger (1931)]) and Bernstein (see [Bernstein (1932)]). More references can be found in Section 1.13.

Let $(\Omega, \mathcal{F}, \mathbb{P})$ be a probability space, and let X_t be a stochastic process on Ω with state space $(E, \mathcal{E})$. For all $\tau \le u \le v \le t$, put

$$\mathcal{F}_u^\tau \vee \mathcal{F}_t^v = \sigma\left(X_s : \tau \le s \le u \text{ or } v \le s \le t\right).$$

The family of σ-algebras $\{\mathcal{F}_u^\tau \vee \mathcal{F}_t^v\}$ will play an important role in the present section.

Definition 1.18 The process X_s, $s \in [\tau, t]$, is called a reciprocal Markov process on the interval $[\tau, t]$ provided that for all u, s, and v with $\tau \le u \le$

$s \leq v \leq t$ and all bounded Borel functions f on E, the equality

$$\mathbb{E}\left[f\left(X_s\right) \mid \mathcal{F}_u^\tau \vee \mathcal{F}_t^v\right] = \mathbb{E}\left[f\left(X_s\right) \mid \sigma(X_u, X_v)\right] \tag{1.92}$$

holds $\mathbb{P}$-a.s.

Condition (1.92) is called the two-sided Markov property of the process X_s. The σ-algebra $\sigma\left(X_u, X_v\right)$ in (1.92) is interpreted as the present information about the random system described by the process X_s, the σ-algebra $\mathcal{F}_u^\tau \vee \mathcal{F}_t^v$ contains the information about the system before the moment u in the past and the moment v in the future, and finally the σ-algebra $\mathcal{F}_v^u$ contains the information about the system after the moment u in the past and the moment v in the future.

The following lemma gives several equivalent conditions for the given stochastic process X_s to be a reciprocal Markov process on the interval $[\tau, t]$.

Lemma 1.20 *Let X_s be a stochastic process on $(\Omega, \mathcal{F}, \mathbb{P})$ with state space $(E, \mathcal{E})$. Then the following are equivalent:*

(1) Condition (1.92) holds.
(2) For all u and v with $\tau \leq u \leq v \leq t$, all s with $u < s < v$, and all finite subsets $\{r_1, \ldots, r_n\}$ of the set $[\tau, t] \backslash (u, v)$, the equality

$$\mathbb{E}\left[f(X_s) \mid \sigma\left(X_{r_1}, \ldots, X_{r_n}, X_u, X_v\right)\right] = \mathbb{E}\left[f(X_s) \mid \sigma(X_u, X_v)\right]$$

holds for any bounded Borel function f on E.
(3) For all u and v with $\tau \leq u \leq v \leq t$, the equality

$$\mathbb{E}\left[F \mid \mathcal{F}_u^\tau \vee \mathcal{F}_t^v\right] = \mathbb{E}\left[F \mid \sigma(X_u, X_v)\right]$$

holds for any bounded real-valued $\mathcal{F}_v^u$-measurable random variable F.
(4) For all u and v with $\tau \leq u \leq v \leq t$, the equality

$$\mathbb{E}\left[GF\right] = \mathbb{E}\left[G\mathbb{E}\left[F \mid \sigma\left(X_u, X_v\right)\right]\right]$$

holds for any bounded real-valued random variables G and F such that G is $\mathcal{F}_u^\tau \vee \mathcal{F}_t^v$-measurable and F is $\mathcal{F}_v^u$-measurable.
(5) For all u and v with $\tau \leq u \leq v \leq t$, the equality

$$\mathbb{P}\left[A \cap B \mid \sigma(X_u, X_v)\right] = \mathbb{P}\left[A \mid \sigma(X_u, X_v)\right] \mathbb{P}\left[B \mid \sigma(X_u, X_v)\right]$$

holds for all $A \in \mathcal{F}_u^\tau \vee \mathcal{F}_t^v$ and $B \in \mathcal{F}_v^u$.

Remark 1.8 Condition 5 in Lemma 1.20 states that the future and the past are conditionally independent if the present is known. Here the present, the past, and the future are interpreted in the sense of reciprocal processes. Condition 5 was used in [Jamison (1970); Jamison (1974)] as the definition of a reciprocal process.

Proof. $(1) \Longrightarrow (2)$.
This implication follows from the inclusions
$$\sigma(X_u, X_v) \subset \sigma(X_{r_1}, \ldots, X_{r_n}, X_u, X_v) \subset \mathcal{F}_u^\tau \vee \mathcal{F}_t^v.$$

$(2) \Longrightarrow (3)$.
Assume that condition (2) in Lemma 1.20 holds. In order to prove the implication $(2) \Longrightarrow (3)$, we will first show that the equality
$$\mathbb{E}\left[F \mid \sigma(X_{r_1}, \ldots, X_{r_n}, X_u, X_v)\right] = \mathbb{E}\left[F \mid \sigma(X_u, X_v)\right] \tag{1.93}$$
holds if the random variable F has a special form. Let
$$F = \prod_{j=1}^m f_j(X_{s_j}),$$
where f_j, $1 \leq j \leq m$, is a bounded Borel function on E and $u \leq s_1 < s_2 < \cdots < s_m \leq v$. We will prove equality (1.93) by induction with respect to m. Since we assumed that condition (2) holds, (1.93) is valid for $m = 1$. Next, suppose that (1.93) is true for $m \in \mathbb{N}$ and for all finite sets $\{s_j : 1 \leq j \leq m\}$ with $u \leq s_1 < s_2 < \cdots < s_m \leq v$. Let $\{s_1, \ldots, s_{m+1}\}$ be a subset of $[u, v]$ with $m+1$ elements. Without loss of generality, we can assume that $u \leq s_1 < s_2 < \cdots < s_{m+1} \leq v$. Put
$$\bar{\mathcal{F}} = \sigma(X_{r_1}, \ldots, X_{r_n}, X_u, X_v)$$
and
$$\mathcal{H} = \sigma(X_{r_1}, \ldots, X_{r_n}, X_{s_1}, \ldots, X_{s_m}, X_u, X_v).$$
Then the tower property of conditional expectations gives
$$\mathbb{E}\left[\prod_{j=1}^{m+1} f_j(X_{s_j}) \mid \bar{\mathcal{F}}\right] = \mathbb{E}\left[\left(\prod_{j=1}^m f_j(X_{s_j})\right) f_{m+1}(X_{s_{m+1}}) \mid \bar{\mathcal{F}}\right]$$
$$= \mathbb{E}\left[\prod_{j=1}^m f_j(X_{s_j}) \mathbb{E}\left[f_{m+1}(X_{s_{m+1}}) \mid \mathcal{H}\right] \mid \bar{\mathcal{F}}\right]. \tag{1.94}$$

It follows from condition (2) in Lemma 1.20 that

$$\mathbb{E}\left[f_{m+1}\left(X_{s_{m+1}}\right) \mid \mathcal{H}\right] = \mathbb{E}\left[f_{m+1}\left(X_{s_{m+1}}\right) \mid \sigma\left(X_{s_m}, X_v\right)\right]. \quad (1.95)$$

Therefore, (1.94) gives

$$\mathbb{E}\left[\prod_{j=1}^{m+1} f_j(X_{s_j}) \mid \bar{\mathcal{F}}\right]$$
$$= \mathbb{E}\left[\prod_{j=1}^{m} f_j(X_{s_j})\mathbb{E}\left[f_{m+1}\left(X_{s_{m+1}}\right) \mid \sigma\left(X_{s_m}, X_v\right)\right] \mid \bar{\mathcal{F}}\right]. \quad (1.96)$$

Our next goal is to prove that there exists a bounded Borel function g on $E \times E$ such that

$$\mathbb{E}\left[f_{m+1}\left(X_{s_{m+1}}\right) \mid \sigma\left(X_{s_m}, X_v\right)\right] = g\left(X_{s_m}, X_v\right) \quad (1.97)$$

$\mathbb{P}$-almost surely. Actually, we will prove a more general known assertion: If F is a bounded $\sigma\left(X_{s_m}, X_v\right)$-measurable random variable on Ω, then there exists a bounded Borel function g on $E \times E$ such that

$$F = g\left(X_{s_m}, X_v\right) \quad (1.98)$$

$\mathbb{P}$-almost surely. Denote by $\mathcal{H}$ the class of all real functions F on Ω for which there exists a bounded Borel function g on $E \times E$ such that equality (1.98) holds. The class $\mathcal{H}$ is a vector space containing the constant functions. Moreover, the product of any two functions from $\mathcal{H}$ belongs to $\mathcal{H}$, and any function of the form $\chi_A\left(X_{s_m}\right)\chi_B\left(X_v\right)$ with $A \in \mathcal{E}$ and $B \in \mathcal{E}$ belongs to $\mathcal{H}$. Next, let $h_n \in \mathcal{H}$, $n \in \mathbb{N}$, be an increasing uniformly bounded sequence of positive functions. Then

$$h_n = g_n\left(X_{s_m}, X_v\right), \ n \in \mathbb{N},$$

where g_n is a bounded Borel function on $E \times E$. It is easy to see that without loss of generality we can assume that the sequence g_n is increasing as $n \to \infty$ (otherwise, we can replace the sequence g_n by the sequence $\max_{1 \leq i \leq n} g_i$). Then,

$$\sup_n h_n = \left(\sup_n g_n\right)\left(X_{s_m}, X_v\right),$$

and thus $\sup_n h_n \in \mathcal{H}$. It follows from the monotone class theorem for functions (Theorem 5.2) that any bounded $\sigma(X_{s_m}, X_v)$-measurable random variable on Ω belongs to $\mathcal{H}$. Therefore, equality (1.97) holds.

It is clear that (1.96) and (1.97) imply that

$$\mathbb{E}\left[\prod_{j=1}^{m+1} f_j(X_{s_j}) \mid \bar{\mathcal{F}}\right] = \mathbb{E}\left[\prod_{j=1}^{m} f_j(X_{s_j}) g(X_{s_m}, X_v) \mid \bar{\mathcal{F}}\right] \quad (1.99)$$

$\mathbb{P}$-almost surely. Applying the induction hypothesis to the right-hand side of (1.99) and using (1.95) and the tower property of conditional expectations, we get

$$\mathbb{E}\left[\prod_{j=1}^{m+1} f_j(X_{s_j}) \mid \bar{\mathcal{F}}\right] = \mathbb{E}\left[\prod_{j=1}^{m} f_j(X_{s_j}) g(X_{s_m}, X_v) \mid \sigma(X_u, X_v)\right]$$

$$\mathbb{E}\left[\prod_{j=1}^{m} f_j(X_{s_j}) \mathbb{E}\left[f_{m+1}(X_{s_{m+1}}) \mid \sigma(X_{s_m}, X_v)\right] \mid \sigma(X_u, X_v)\right]$$

$$= \mathbb{E}\left[\prod_{j=1}^{m} f_j(X_{s_j}) \mathbb{E}\left[f_{m+1}(X_{s_{m+1}}) \mid \mathcal{H}\right] \mid \sigma(X_u, X_v)\right]$$

$$= \mathbb{E}\left[\mathbb{E}\left[\prod_{j=1}^{m+1} f_j(X_{s_j}) \mid \mathcal{H}\right] \mid \sigma(X_u, X_v)\right]$$

$$= \mathbb{E}\left[\prod_{j=1}^{m+1} f_j(X_{s_j}) \mid \sigma(X_u, X_v)\right]. \quad (1.100)$$

By the monotone class theorem and (1.100),

$$\mathbb{E}\left[F \mid \bar{\mathcal{F}}\right] = \mathbb{E}\left[F \mid \sigma(X_u, X_v)\right] \quad (1.101)$$

for any bounded $\mathcal{F}_v^u$-measurable random variable F. Next, applying the monotone class theorem again and using the definition of conditional expectations, we obtain

$$\mathbb{E}\left[F \mid \mathcal{F}_u^\tau \vee \mathcal{F}_t^v\right] = \mathbb{E}\left[F \mid \sigma(X_u, X_v)\right].$$

Therefore, condition (3) in Lemma 1.20 holds.

(3) $\implies$ (4).

This implication follows from the definition of conditional expectations, the formula

$$G = \int_0^\infty \chi_{\{G^+ \geq \lambda\}} d\lambda - \int_0^\infty \chi_{\{G^- \geq \lambda\}} d\lambda, \qquad (1.102)$$

and Tonelli's theorem.

(4) $\implies$ (5).

Suppose that condition (4) in Lemma 1.20 holds, and let $A \in \mathcal{F}_u^\tau \vee \mathcal{F}_t^v$, $B \in \mathcal{F}_v^u$, and $D \in \sigma(X_u, X_v)$. Then, using condition (4) with $G = \chi_A$ and $F = \chi_B \chi_D$, we obtain

$$\begin{aligned} \mathbb{E}\left[\chi_D \chi_A \chi_B\right] &= \mathbb{E}\left[\chi_A \mathbb{E}\left[\chi_D \chi_B \mid \sigma(X_u, X_v)\right]\right] \\ &= \mathbb{E}\left[\chi_D \chi_A \mathbb{E}\left[\chi_B \mid \sigma(X_u, X_v)\right]\right]. \end{aligned} \qquad (1.103)$$

Therefore,

$$\begin{aligned} \mathbb{E}\left[\chi_A \chi_B \mid \sigma(X_u, X_v)\right] &= \mathbb{E}\left[\chi_A \mathbb{E}\left[\chi_B \mid \sigma(X_u, X_v)\right] \mid \sigma(X_u, X_v)\right] \\ &= \mathbb{P}\left[A \mid \sigma(X_u, X_v)\right] \mathbb{P}\left[B \mid \sigma(X_u, X_v)\right], \end{aligned} \qquad (1.104)$$

and hence condition (5) in Lemma 1.20 holds.

(5) $\implies$ (1).

Suppose that condition (5) in Lemma 1.20 holds. Then, comparing (1.103) and (1.104), we see that (1.103) holds. Let $D = \Omega$ and $G = f(X_s)$. It follows from (2.110) that

$$\mathbb{E}\left[\chi_A f(X_s)\right] = \mathbb{E}\left[\chi_A \mathbb{E}\left[f(X_s) \mid \sigma(X_u, X_v)\right]\right].$$

Next, using the definition of conditional expectations, we get

$$\mathbb{E}\left[f(X_s) \mid \mathcal{F}_u^\tau \vee \mathcal{F}_t^v\right] = \mathbb{E}\left[f(X_s) \mid \sigma(X_u, X_v)\right]$$

for all s with $u < s < v$. Therefore, condition (1) in Lemma 1.20 holds.

This completes the proof of Lemma 1.20. $\square$

It has already been established in Section 1.2 that the Markov property is invariant under time-reversal (see Lemma 1.2 and Lemma 1.3). In fact, more is true.

Lemma 1.21 *Let $(\Omega, \mathcal{F}, \mathbb{P})$ be a probability space, and let X_t with $0 \leq t \leq T$ be a Markov process with respect to the measure $\mathbb{P}$. Then X_t is a reciprocal process with respect to $\mathbb{P}$.*

Proof. We will prove that equality (1.92) holds. Suppose that $0 \leq \tau < u < s < v < t \leq T$, Y is an $\mathcal{F}_u^\tau$-measurable random variable, and Z is an

$\mathcal{F}_t^v$-measurable random variable. Let f be a bounded Borel function on E. It follows from the properties of conditional expectations that

$$\mathbb{E}\left[Yf(X_s)Z \mid \sigma(X_u, X_v)\right]$$
$$= \mathbb{E}\left[\mathbb{E}\left[Yf(X_s)Z \mid \sigma(X_u, X_s, X_v)\right] \mid \sigma(X_u, X_v)\right]$$
$$= \mathbb{E}\left[f(X_s)\mathbb{E}\left[YZ \mid \sigma(X_u, X_s, X_v)\right] \mid \sigma(X_u, X_v)\right]. \quad (1.105)$$

Next, using the properties of conditional expectations and the Markov property of the processes X_r and X_{T-r} (see Lemma 1.2 and Lemma 1.3), we see that

$$\mathbb{E}\left[YZ \mid \sigma(X_u, X_s, X_v)\right]$$
$$= \mathbb{E}\left[Y\mathbb{E}\left[Z \mid \sigma(Y, X_u, X_s, X_v)\right] \mid \sigma(X_u, X_s, X_v)\right]$$
$$= \mathbb{E}\left[Y\mathbb{E}\left[Z \mid \sigma(X_v)\right] \mid \sigma(X_u, X_s, X_v)\right]$$
$$= \mathbb{E}\left[Z \mid \sigma(X_v)\right]\mathbb{E}\left[Y \mid \sigma(X_u, X_s, X_v)\right]$$
$$= \mathbb{E}\left[Z \mid \sigma(X_v)\right]\mathbb{E}\left[Y \mid \sigma(X_u)\right]$$
$$= \mathbb{E}\left[Z \mid \sigma(X_v)\right]\mathbb{E}\left[Y \mid \sigma(X_u, X_v)\right]$$
$$= \mathbb{E}\left[Y\mathbb{E}\left[Z \mid \sigma(X_v)\right] \mid \sigma(X_u, X_v)\right]$$
$$= \mathbb{E}\left[Y\mathbb{E}\left[Z \mid \sigma(Y, X_u, X_v)\right] \mid \sigma(X_u, X_v)\right]$$
$$= \mathbb{E}\left[\mathbb{E}\left[YZ \mid \sigma(Y, X_u, X_v)\right] \mid \sigma(X_u, X_v)\right]$$
$$= \mathbb{E}\left[YZ \mid \sigma(X_u, X_v)\right]. \quad (1.106)$$

Therefore, (1.106) and (1.105) give

$$\mathbb{E}\left[Yf(X_s)Z \mid \sigma(X_u, X_v)\right] = \mathbb{E}\left[Y\mathbb{E}\left[Z \mid \sigma(X_u, X_v)\right] \mid \sigma(X_u, X_v)\right]$$
$$= \mathbb{E}\left[Y \mid \sigma(X_u, X_v)\right]\mathbb{E}\left[Z \mid \sigma(X_u, X_v)\right]. \quad (1.107)$$

Now it is clear that (1.107) implies (1.92), and hence X_r is a reciprocal Markov process.

This completes the proof of Lemma 1.21. $\square$

The remaining part of the present section is devoted to reciprocal transition probability functions. Such a function describes the evolution of a random system as follows. The value of a reciprocal transition probability function is the probability of finding the system inside a set $B \in \mathcal{E}$ at an intermediate moment of time knowing the location of the system at the initial and the final moments of time.

Definition 1.19 A function $Q(\tau, x; s, B; t, y)$ where $0 \leq \tau < s < t \leq T$, $B \in \mathcal{E}$, $x \in E$, and $y \in E$, is called a reciprocal transition probability function if the following conditions are satisfied:

(i) The mapping $B \mapsto Q(\tau, x; s, B; t, y)$ is a probability measure on $\mathcal{E}$ for all $(x, y) \in E \times E$ and $0 \leq \tau < s < t \leq T$.
(ii) The function $(x, y) \mapsto Q(\tau, x; s, B; t, y)$ is $\mathcal{E} \otimes \mathcal{E}$-measurable for all $B \in \mathcal{E}$ and $0 \leq \tau < s < t \leq T$.
(iii) For all $(C, D) \in \mathcal{E} \times \mathcal{E}$, $(x, y) \in E \times E$, and $0 \leq \tau < u < v < t \leq T$, the following equality holds:

$$\int_D Q(\tau, x; v, dz; t, y) Q(\tau, x; u, C; v, z)$$
$$= \int_C Q(\tau, x; u, dw; t, y) Q(u, w; v, D; t, y). \quad (1.108)$$

The equation in (iii) is an analogue of the Chapman-Kolmogorov equation for transition probability functions. The following equalities follow from (1.108):

$$Q(\tau, x; v, D; t, y) = \int_E Q(\tau, x; u, dw; t, y) Q(u, w; v, D; t, y) \quad (1.109)$$

for all $0 \leq \tau < u < v \leq T$ and $D \in \mathcal{E}$; and

$$Q(\tau, x; u, C; t, y) = \int_E Q(\tau, x; u, C; v, z) Q(\tau, x; v, dz; t, y) \quad (1.110)$$

for all $0 \leq \tau < u < v \leq T$ and $C \in \mathcal{E}$. The equality in (1.108) can be interpreted as follows. Suppose we know that a random system with the transition law Q is at $x \in E$ at the initial moment τ and at $y \in E$ at the final moment t. Let u and v be such that $\tau < u < v < t$, and let $C \in \mathcal{E}$ and $D \in \mathcal{E}$. Then the probability of the system hitting the target C at the moment u and then the target D at the moment v in forward motion is equal to the probability of hitting the target D at the moment v and then the target C at the moment u in backward motion.

Lemma 1.22 Let $Q(\tau, x; s, A; t, y)$ be a reciprocal transition probability function. Then for all $(t, y) \in [0, T] \times E$, the function

$$P_{t,y}(\tau, x; s, A) = Q(\tau, x; s, A; t, y)$$

is a transition probability function, while for all $(\tau, x) \in [0, T] \times E$, the function

$$\widetilde{P}_{\tau,x}(s, A; t, y) = Q(\tau, x; s, A; t, y)$$

is a backward transition probability function.

Proof. Using condition (iii) in Definition 1.19 with $C = E$ and $D = A$, we get

$$\begin{aligned} P_{t,y}(\tau, x; s, A) &= Q(\tau, x; s, A; t, y) \\ &= \int_A Q(\tau, x; s, dz; t, y) \, Q(\tau, x; u, E; s, z) \\ &= \int_E Q(\tau, x; u, dw; t, y) \, Q(u, w; s, A; t, y) \\ &= \int_E P_{t,y}(\tau, x; u, dw) \, P_{t,y}(u, w; s, A). \end{aligned} \quad (1.111)$$

On the other hand, using condition (iii) in Definition 1.19 with $C = A$ and $D = E$, we obtain

$$\begin{aligned} \widetilde{P}_{\tau,x}(s, A; t, y) &= Q(\tau, x; s, A; t, y) \\ &= \int_A Q(\tau, x; s, dw; t, y) \, Q(s, w; v, E; t, y) \\ &= \int_E Q(\tau, x; v, dz; t, y) \, Q(\tau, x; s, A, v, z) \\ &= \int_E \widetilde{P}_{\tau,x}(v, dz; t, y)) \, \widetilde{P}_{\tau,x}(s, A; v, z). \end{aligned} \quad (1.112)$$

Now it is clear that Lemma 1.22 follows from (1.111) and (1.112). □

Let Q be a reciprocal transition probability function. Next, we will construct a stochastic process X_t associated with Q. Fix $(\tau, x) \in [0, T] \times E$ and $(t, y) \in [0, T] \times E$ with $0 \le \tau < t \le T$, put $\Omega = E^{[0,T]}$, and consider the process $X_s(\omega) = \omega(s)$ for all $\omega \in \Omega$ and $0 \le s \le T$. Let $\mathcal{F}$ be the σ-algebra generated by finite-dimensional cylinders. Then there exists a probability measure $\mathbb{P}_{(\tau,t),(x,y)}$ on $(\Omega, \mathcal{F})$ such that

$$\mathbb{P}_{(\tau,t),(x,y)} [X_\tau \in A_0, X_{s_1} \in A_1, \ldots, X_{s_n} \in A_n, X_t \in A_{n+1}]$$
$$= \chi_{A_0}(x)\chi_{A_{n+1}}(y) \int_{A_1} Q(\tau, x; s_1, dz_1; t, y)$$
$$\int_{A_2} \cdots \int_{A_n} Q(s_{n-1}, z_{n-1}; s_n, dz_n; t, y) \tag{1.113}$$

where $A_i \in \mathcal{E}$ for all $0 \leq i \leq n+1$ and $\tau < s_1 < \cdots < s_n < t$. The existence of the measure $\mathbb{P}_{(\tau,t),(x,y)}$ follows from the Kolmogorov extension theorem, since the expression on the right-hand side of (1.113) defines a projective system of measures. This fact is a consequence of the equalities in (1.109) and (1.110). More precisely, suppose that for some k with $1 \leq k < n$ we have $A_k = E$. Then (1.110) implies the equality

$$Q(s_{k-1}, z_{k-1}; s_{k+1}, C; t, y)$$
$$= \int_E Q(s_{k-1}, z_{k-1}; s_k, dz_k; t, y) Q(s_k, z_k; s_{k+1}, C; t, y), \quad C \in \mathcal{E}.$$

In other words, the number of integrations on the right-hand side of (1.113) can be reduced by one. If $k = n$, then applying (1.109) we arrive at the same conclusion. Now it is not hard to see that (1.113) defines a projective system of measures. A special case of (1.113) is

$$\mathbb{P}_{(\tau,t),(x,y)} [X_s \in A] = Q(\tau, x; s, A; t, y) \tag{1.114}$$

for all $0 \leq \tau < s < t \leq T$, $x \in E$, $y \in E$, and $A \in \mathcal{E}$. Formula (1.114) corresponds to the case $n = 1$.

The initial distribution of the process X_s, $\tau \leq s \leq t$, is the Dirac measure δ_x, while the final distribution of X_s is δ_y. Since the present state of the reciprocal process X_s, $\tau \leq s \leq t$, can be interpolated from the past and future information, the joint distribution of X_τ and X_t plays an important role. Next, we will construct a reciprocal process X_t with a prescribed joint distribution μ of X_τ and X_t. Let μ be a probability measure on $\mathcal{E} \otimes \mathcal{E}$. Then there exists a probability measure $\mathbb{P}_{(\tau,t),\mu}$ on

$(\Omega, \mathcal{F})$ for which

$$\mathbb{P}_{(\tau,t),\mu}[X_\tau \in A_0, X_{s_1} \in A_1, \ldots, X_{s_n} \in A_n, X_t \in A_{n+1}]$$
$$= \iint_{A_0 \times A_{n+1}} d\mu(x,y) \int_{A_1} Q(\tau, x; s_1, dz_1; t, y)$$
$$\int_{A_2} \cdots \int_{A_n} Q(s_{n-1}, z_{n-1}; s_n, dz_n; t, y) \tag{1.115}$$

where $A_i \in \mathcal{E}$ for all $0 \le i \le n+1$ and $\tau < s_1 < \cdots < s_n < t$. The existence of the measure $\mathbb{P}_{(\tau,t),\mu}$ follows from the Kolmogorov extension theorem. Note that our previous notation $\mathbb{P}_{(\tau,t),(x,y)}$ is nothing else but $\mathbb{P}_{(\tau,t),\delta_x \times \delta_y}$. We will call the measure μ the initial-final distribution of the process X_s on the interval $[\tau, t]$.

Theorem 1.12 *Let Q be a reciprocal transition probability function, and let the measures $\mathbb{P}_{(\tau,t),(x,y)}$ be defined by (1.113). Then the following conditions hold for the process X_s:*

(1) For all $0 \le \tau < t \le T$, $x \in E$, $y \in E$, $A \in \mathcal{E}$, and $B \in \mathcal{E}$,

$$\mathbb{P}_{(\tau,t),(x,y)}[X_\tau \in A, X_t \in B] = \delta_x \times \delta_y(A \times B).$$

(2) For all $0 \le \tau \le u < s < v \le t \le T$, $x \in E$, $y \in E$, and $A \in \mathcal{E}$,

$$\mathbb{P}_{(\tau,t),(x,y)}\left[X_s \in A \mid \mathcal{F}_u^\tau \vee \mathcal{F}_t^v\right] = Q(u, X_u; s, A; v, X_v).$$

$\mathbb{P}_{(\tau,t),(x,y)}$-*a.s.*

(3) The two-sided Markov property holds for the process X_s. This means that one (all) of the following equivalent conditions is valid $\mathbb{P}_{(\tau,t),(x,y)}$-a.s. for all $0 \le \tau \le u \le s \le v \le t \le T$, $\tau < t$, $x \in E$, $y \in E$, and all bounded Borel functions f on E:

$$\mathbb{E}_{(\tau,t),(x,y)}\left[f(X_s) \mid \mathcal{F}_u^\tau \vee \mathcal{F}_t^v\right] = \mathbb{E}_{(\tau,t),(x,y)}\left[f(X_s) \mid \sigma(X_u, X_v)\right],$$
$$\tag{1.116}$$
$$\mathbb{E}_{(\tau,t),(x,y)}\left[f(X_s) \mid \mathcal{F}_u^\tau \vee \mathcal{F}_t^v\right] = \mathbb{E}_{(u,v),(X_u,X_v)}\left[f(X_s)\right], \tag{1.117}$$
and
$$\mathbb{E}_{(\tau,t),(x,y)}\left[f(X_s) \mid \mathcal{F}_u^\tau \vee \mathcal{F}_t^v\right] = \int f(y) Q(u, X_u; s, dy; v, X_v). \tag{1.118}$$

(4) For all $0 \le \tau \le u \le v \le t \le T$, $\tau < t$, $x \in E$, $y \in E$, and all bounded $\mathcal{F}_v^u$-measurable random variables F, the following equivalent conditions

hold $\mathbb{P}_{(\tau,t),(x,y)}$-a.s.:

$$\mathbb{E}_{(\tau,t),(x,y)}\left[F \mid \mathcal{F}_u^\tau \vee \mathcal{F}_t^v\right] = \mathbb{E}_{(\tau,t),(x,y)}\left[F \mid \sigma(X_u, X_v)\right] \quad (1.119)$$

and

$$\mathbb{E}_{(\tau,t),(x,y)}\left[F \mid \mathcal{F}_u^\tau \vee \mathcal{F}_t^v\right] = \mathbb{E}_{(u,v),(X_u,X_v)}\left[F\right]. \quad (1.120)$$

Proof. Condition (1) in Theorem 1.12 is a corollary of (1.113). Our next goal is to prove that condition (2) holds. Fix $\tau < u < s < t$, and let $A_0, A_1, \ldots, A_{n+1}$ be an arbitrary finite family of Borel subsets of E. By the monotone class theorem and the definition of the conditional expectation, condition (2) follows from the following assertion: The equality

$$\mathbb{P}_{(\tau,t),(x,y)}\left[X_\tau \in A_0, X_{r_1} \in A_1, \ldots, X_{r_{k-1}} \in A_{k-1}, X_{r_k} \in A_k,\right.$$
$$\left. X_{r_{k+1}} \in A_{k+1}, \ldots, X_{r_n} \in A_n, X_t \in A_{n+1}\right]$$
$$= \mathbb{E}_{(\tau,t),(x,y)}\left[Q(u, X_u; s, A_k; v, X_v), X_\tau \in A_0, X_{r_1} \in A_1, \ldots,\right.$$
$$\left. X_{r_{k-1}} \in A_{k-1}, X_{r_{k+1}} \in A_{k+1}, \ldots, X_{r_n} \in A_n, X_t \in A_{n+1}\right]$$
$$(1.121)$$

holds for all finite subsets $\{r_1, \ldots, r_n\}$ of the set $(\tau, t) \setminus \{(u,s) \cup (s,v)\}$ such that

$$\tau < r_1 < \cdots < r_{k-1} = u < r_k = s < v = r_{k+1} < \cdots < r_n < t.$$

It is not hard to see that the equality in (1.121) is a consequence of (1.113).

We will next prove condition (3) in Theorem 1.12. Note that (1.102) with f instead of G and Tonelli's theorem imply equality (1.118). The fact that the expressions on the right-hand side of (1.117) and (1.118) are equal follows from (1.113). This shows that the function

$$(x,y) \mapsto \mathbb{E}_{(\tau,t),(x,y)}\left[f(X_s)\right]$$

is Borel measurable. Therefore, the right-hand side of (1.117), that is, the expression $\mathbb{E}_{(u,v),(X_u,X_v)}\left[f(X_s)\right]$, is measurable with respect to the σ-algebra $\sigma(X_u, X_v)$. Using the definition of the conditional expectation and formula (1.113), we see that the expressions on the right-hand side of (1.116) and (1.117) are equal. For any s with $u \leq s \leq v$, put $F = f(X_s)$. Then the equalities in part (4) of Theorem 1.12 can be obtained from part (3) of this theorem. Equality (1.119) follows from part (3) of Lemma 1.20. The fact that the expressions on the right-hand side of (1.119) and (1.120) coincide is a consequence of (1.113) and the monotone class theorem. More precisely, (1.113) implies that

$$\mathbb{E}_{(\tau,t),(x,y)}\left[F \mid \sigma\left(X_u, X_v\right)\right] = \mathbb{E}_{(u,v),(X_u,X_v)}[F] \qquad (1.122)$$

for $F = \chi_{\mathcal{A}}$, where

$$\mathcal{A} = \{X_\tau \in A_0, X_{r_1} \in A_1, \ldots, X_{r_{k-1}} \in A_{k-1}, X_{r_k} \in A_k,$$
$$X_{r_{k+1}} \in A_{k+1}, \ldots, X_{r_n} \in A_n, X_t \in A_{n+1}\}.$$

Here each A_j, $0 \leq j \leq n+1$, is a Borel subset of E and $\tau < r_1 < \cdots < r_{k-1} = u < r_k = s < v = r_{k+1} < \cdots r_n < t$. The monotone class theorem then yields equality (1.122) in its full generality.

This completes the proof of Theorem 1.12. □

The next theorem concerns general initial-final distributions. The proof of Theorem 1.13 is similar to that of Theorem 1.12.

Theorem 1.13 *Let Q be a reciprocal transition probability function, and let μ be a Borel probability measure on $\mathcal{E} \times \mathcal{E}$. Suppose that the measure $\mathbb{P}_{(\tau,t),\mu}$ is defined by (1.115). Then the following conditions hold for the reciprocal process X_s corresponding to Q:*

(1) For all $0 \leq \tau < t \leq T$, $A \in \mathcal{E}$, and $B \in \mathcal{E}$,

$$\mathbb{P}_{(\tau,t),\mu}\left[X_\tau \in A, X_t \in B\right] = \mu(A \times B).$$

(2) For all $0 \leq \tau \leq u < s < v \leq t \leq T$ and $A \in \mathcal{E}$,

$$\mathbb{P}_{(\tau,t),\mu}\left[X_s \in A \mid \mathcal{F}_u^\tau \vee \mathcal{F}_t^v\right] = Q\left(u, X_u; s, A; v, X_v\right) \ \mathbb{P}_{(\tau,t),\mu}\text{-a.s.}$$

(3) The two-sided Markov property with respect to the measure $\mathbb{P}_{(\tau,t),\mu}$ holds for the process X_s. This means that for all $0 \leq \tau \leq u \leq s \leq v \leq t \leq T$ with $\tau < t$,

$$\mathbb{E}_{(\tau,t),\mu}\left[f(X_s) \mid \mathcal{F}_u^\tau \vee \mathcal{F}_t^v\right] = \mathbb{E}_{(\tau,t),\mu}\left[f(X_s) \mid \sigma\left(X_u, X_v\right)\right] \ \mathbb{P}_{(\tau,t),\mu}\text{-a.s.}$$

The next part of the present section is devoted to the relations between Markov processes and reciprocal Markov processes. Let Q be a reciprocal transition probability function. It is said that Q possesses a density q if there exists a function $q(\tau, x; s, z; t, y)$ such that for all τ, s, and t with $0 \leq \tau < s < t \leq T$ and all $z \in E$, the function $(x, y) \mapsto q(\tau, x; s, z; t, y)$ is a Borel function on $E \times E$, and the condition

$$Q(\tau, x; s, A; t, y) = \int_A q(\tau, x; s, z; t, y)\, dm(z) \qquad (1.123)$$

holds. In (1.123), m is the reference measure (see Section 1.3).

The following construction goes back to Schrödinger (see [Schrödinger (1931)]). Let P be a transition function possessing everywhere positive density p. Here we do not assume that P is normal. Consider the so-called derived density

$$q(\tau, x; s, z; t, y) = \frac{p(\tau, x; s, z)p(s, z; t, y)}{p(\tau, x; t, y)}, \qquad (1.124)$$

where $0 \leq \tau < s < t \leq T$ and $x, y, z \in E$. Then it is easy to see that q is a density of a reciprocal transition probability function Q defined in (1.123). Schrödinger considered the density q in the case of a standard Brownian motion on the real line $\mathbb{R}$. In this case, the role of the transition density p is played by the one-dimensional standard Gaussian density p_1. Recall that the standard Gaussian density p_d on d-dimensional Euclidean space $\mathbb{R}^d$ is defined by

$$p_d(\tau, x; t, y) = \frac{1}{[2\pi(t-\tau)]^{d/2}} \exp\left\{-\frac{|x-y|^2}{2(t-\tau)}\right\}. \qquad (1.125)$$

The reciprocal transition probability function Q with derived density q given by (1.124) is associated with a reciprocal Markov process X_s satisfying the conditions in Theorem 1.12. The density q is nothing else but the conditional density of the process X_s subject to the conditions $X_\tau = x$ and $X_t = y$. Indeed, for any open neighborhood U of the point $y \in E$, we have

$$\mathbb{P}_{\tau,x}\left[X_s \in A \mid X_t \in U\right] = \frac{\mathbb{P}_{\tau,x}\left[X_s \in A, X_t \in U\right]}{\mathbb{P}_{\tau,x}\left[X_t \in U\right]}$$
$$= \frac{\int_{A \times U} p(\tau, x; s, z_1) p(s, z_1; t, z_2) dz_1 dz_2}{\int_U p(\tau, x; t, z_2) dz_2}.$$

By shrinking U to y, we get the following formula:

$$\mathbb{P}_{\tau,x}\left[X_s \in A \mid X_t = y\right] = \frac{\int_A p(\tau, x; s, z_1) p(s, z_1; t, y) dz_1}{p(\tau, x; t, y)}.$$

Let μ be a Borel probability measure on $E \times E$, and let X_s be the reciprocal Markov process on $[\tau, t]$ with derived density q and such that μ is its initial-final distribution; that is,

$$\mu(A \times B) = \mathbb{P}_{(\tau,t),\mu}\left[X_\tau \in A, X_t \in B\right] \qquad (1.126)$$

for all Borel subsets A and B of E. We will denote by μ_τ and μ_t the marginal distributions of μ. They are defined as follows: $\mu_\tau(A) = \mu(A \times E)$

and $\mu_t(B) = \mu(E \times B)$. For all u and v with $\tau \leq u < v \leq t$, the symbol $\mu_{u,v}$ will stand for the joint distribution of the random variables X_u and X_v with respect to the measure $\mathbb{P}_{(\tau,t),\mu}$.

Jamison posed and solved the following problem in [Jamison (1974)]: Determine under what restrictions on the initial-final distribution μ a reciprocal Markov process X_s on the interval $[\tau, t]$ associated with the derived density q is a Markov process with respect to the measure $\mathbb{P}_{(\tau,t),\mu}$.

Theorem 1.14 *Let $p(r, x; s, y)$ be a transition density, and let $q(\tau, x; s, z; t, y)$ be the derived density given by (1.124). Fix τ and t with $0 \leq \tau < t \leq T$, and let X_s be a reciprocal process on $[\tau, t]$ with transition density q and the initial-final distribution μ given by (1.126). Then the following are equivalent:*

(1) X_s is a Markov process with respect to the measure $\mathbb{P}_{(\tau,t),\mu}$.
(2) There exist σ-finite Borel measures ν_τ and ν_t on E such that

$$\mu(G) = \iint_G p(\tau, x; t, y) d\nu_\tau(x) d\nu_t(y) \qquad (1.127)$$

for all Borel measurable subsets G of $E \times E$.

Proof. (2) $\Longrightarrow$ (1). Suppose that there exist measures ν_τ and ν_t for which condition (1.127) holds. Let $\tau < s_1 < s_2 < \cdots < s_n < t$ and $A_i \in \mathcal{E}$ for all $1 \leq i \leq n$. Given $n \geq 1$, put

$$E \times E \times A_1 \times \cdots \times A_{n-1} = A'_{n-1}.$$

Then, using equalities (1.124) and (1.127) and making cancellations, we get

$$\mathbb{P}_{(\tau,t),\mu}[X_{s_i} \in A_i, 1 \leq i \leq n]$$
$$= \int_{A'_{n-1} \times A_n} q(\tau, x; s_1, z_1; t, y) \ldots q(s_{n-1}, z_{n-1}; s_n, z_n; t, y)$$
$$q(s_{n-1}, z_{n-1}; s_n, z_n; t, y) d\mu(x, y) dz_1 \ldots dz_{n-1} dz_n$$
$$= \int_{A'_{n-1} \times A_n} p(\tau, x; s_1, z_1) \ldots p(s_{n-1}, z_{n-1}; s_n, z_n)$$
$$p(s_n, z_n; t, y) d\nu_\tau(x) d\nu_t(y) dz_1 \ldots dz_{n-1} dz_n. \qquad (1.128)$$

Denote by f the following function:

$$f(z_{n-1}) = \frac{\int_{A_n \times E} p(s_{n-1}, z_{n-1}; s_n, z_n) p(s_n, z_n; t, y) dz_n d\nu_t(y)}{\int_E p(s_{n-1}, z_{n-1}; t, y) d\nu_t(y)}.$$

The function f depends on s_{n-1}, s_n, and A_n. It follows from (1.128) that

$$\mathbb{P}_{(\tau,t),\mu}[X_{s_i} \in A_i, 1 \le i \le n]$$
$$= \int_{A'_{n-1}} d\nu_\tau(x) d\nu_t(y) p(\tau, x; s_1, z_1)$$
$$\ldots p(s_{n-2}, z_{n-2}; s_{n-1}, z_{n-1}) p(s_{n-1}, z_{n-1}; t, y) f(z_{n-1}) dz_1 \ldots dz_{n-1}$$
$$= \int_{A'_{n-1}} d\mu(x,y) q(\tau, x; s_1, z_1; t, y)$$
$$\ldots q(s_{n-2}, z_{n-2}; s_{n-1}, z_{n-1}; t, y) f(z_1) dz_1 \ldots dz_{n-1}$$
$$= \mathbb{E}_{(\tau,t),\mu}\left[f\left(X_{s_{n-1}}\right), X_{s_i} \in A_i, 1 \le i \le n-1\right]. \tag{1.129}$$

Therefore, (1.129) gives

$$\mathbb{P}_{(\tau,t),\mu}\left[X_{s_n} \in A_n \mid \mathcal{F}^\tau_{s_{n-1}}\right] = f(X_{s_{n-1}})$$
$$= \mathbb{P}_{(\tau,t),\mu}\left[X_{s_n} \in A_n \mid \sigma(X_{s_{n-1}})\right]. \tag{1.130}$$

The cases where $s_1 = \tau$ and $s_n < t$; or $s_1 > \tau$ and $s_n = t$; or $s_1 = \tau$ and $s_n = t$ are similar. Now it is not hard to see that the Markov property holds for X_s.

(1) $\implies$ (2).

Let us suppose that the reciprocal process X_s is simultaneously a Markov process on $[\tau, t]$ with respect to the measure $\mathbb{P}_{(\tau,t),\mu}$. For all s with $\tau < s < t$ and all triples (A, B, C) of Borel subsets of E, put

$$I(s, A, B, C) = \mathbb{P}_{(\tau,t),\mu}[X_\tau \in A, X_s \in B, X_t \in C].$$

Then the following equalities hold:

$$I(s, A, B, C) = \int_{A \times C} d\mu(x, y) \int_B q(\tau, x; s, z; t, y) dz$$
$$= \int_{A \times C} d\mu(x, y) \int_B \frac{p(\tau, x; s, z) p(s, z; t, y)}{p(\tau, x; t, y)} dz. \tag{1.131}$$

On the other hand, since X_s is a Markov process, there exists the Markov transition $P^f(u, X_u; s, A)$. It follows that

$$I(s, A, B, C) = \int_A d\mu_\tau(x) \int_B P^f(\tau, x; s, dz) P^f(s, z; t, C). \tag{1.132}$$

Let u and s be such that $\tau \le u < s < t$. Then, since the process X_s is simultaneously reciprocal and Markov, and the Borel σ-algebra of E is countably generated, we have

$$P^f(u,x;s,B) = \int P^f(u,x;t,dy) \int_B q(u,x;s,z;t,y)dz \qquad (1.133)$$

for μ_u-almost all $x \in E$ and all $B \in \mathcal{E}$. Let us denote by A_0 the set of all $x \in E$ such that (1.133) holds for all $B \in \mathcal{E}$. Then $\mu_u(E \backslash A_0) = 0$.

Our next goal is to prove that for μ_u-almost all $x \in E$, the measure $B \mapsto P^f(u,x;s,B)$ and the reference measure m are mutually absolutely continuous. Indeed, let $m(B) = 0$. Then (1.133) gives $P^f(u,x;s,B) = 0$ for μ_u-almost all $x \in E$. On the other hand, if $x \in A_0$ and $P^f(u,x;s,B) = 0$, then (1.133) implies

$$\int_B dz \int P^f(u,x;t,dy)\, q(u,x;s,z;t,y) = 0.$$

Since p is a strictly positive function and $P^f(u,x;t,E) = 1$, we get $m(B) = 0$. This completes the proof of the mutual absolute continuity of the measures $B \mapsto P^f(u,x;s,B)$ and m. It follows from the reasoning above that

$$P^f(u,x;s,B) = \int_B p^f(u,x;s,z)dz \qquad (1.134)$$

for all $x \in A_0$ and $B \in \mathcal{E}$. In (1.134),

$$p^f(u,x;s,z) = \int_E P^f(u,x;t,dy)q(u,x;s,z;t,y).$$

We will next prove that the measure $D \mapsto P^f(u,z;t,D)$ is absolutely continuous with respect to the measure μ_t for μ_u-almost all $z \in E$. Indeed, we have

$$\mu_t(D) = \int d\mu_u(x) P^f(u,x;t,D) \qquad (1.135)$$

for all $D \in \mathcal{E}$. Suppose that $C \in \mathcal{E}$ is such that $\mu_t(C) = 0$. Then it follows from (1.135) with $D = C$ that

$$P^f(u,x;t,C) = 0 \text{ for } \mu_u\text{-almost all } x \in E. \qquad (1.136)$$

Using the Markov property, we see that for every $D \in \mathcal{E}$,

$$P^f(u,x;t,D) = \int P^f(u,x;s,dy)\, P^f(s,y;t,D) \qquad (1.137)$$

μ_u-almost everywhere on E. Let us denote by A_1 the set of those $x \in E$ for which (1.137) holds for all $D \in \mathcal{E}$. It follows from the separability of $\mathcal{E}$

that $\mu_u(E\backslash A_1) = 0$. Fix $x_1 \in A_0 \cap A_1$ such that (1.136) holds for x_1. Note that x_1 depends on the set C. Then (1.137) and (1.136) with $x = x_1$ show that

$$P^f(s, y; t, C) = 0 \tag{1.138}$$

for $P^f(u, x_1; s, \cdot)$-almost all $y \in E$. Since $x_1 \in A_0$, equality (1.138) holds for m-almost all $y \in E$. Now we see that for all $x_0 \in A_0 \cap A_1$, equality (1.138) holds for $P^f(u, x_0; s, \cdot)$-almost all $y \in E$. Then (1.137) with $D = C$ gives $P^f(u, x_0; t, C) = 0$. We have thus completed the proof of the fact that the measure $D \mapsto P^f(u, z; t, D)$ is absolutely continuous with respect to the measure μ_t for μ_u-almost all $z \in E$. Indeed, the equality $P^f(u, x_0; t, C) = 0$ holds for all C with $\mu_t(C) = 0$ for all x_0 belonging to the set $A_0 \cap A_1$. This set is independent of C and such that $\mu_u(E \backslash (A_0 \cap A_1)) = 0$. It follows that there exists a function $\overline{P}^f$ such that

$$P^f(u, z; t, D) = \int_D \overline{P}^f(u, z; t, y) \, d\mu_t(y) \tag{1.139}$$

for all $D \in \mathcal{E}$ and for μ_u-almost all $z \in E$. Now it is not hard to show that for every $u \in [\tau, t)$ there exists a function $(z, y) \mapsto \widetilde{P}^f(u, z; t, y)$ such that it is $\mathcal{E} \otimes \mathcal{E}$-measurable and coincides with the function $(z, y) \mapsto \overline{P}^f(u, z; t, y)$ $\mu_u \times \mu_t$-almost everywhere. Indeed, using (1.135) and (1.139), we get

$$\int_E d\mu_u(z) \int_E \overline{P}^f(u, z; t, y) \, d\mu_t(y) = \mu_t(E) = 1,$$

and hence, there exists a probability measure ν on $\mathcal{E} \otimes \mathcal{E}$ such that

$$\nu(A \times D) = \int_A d\mu_u(z) \int_D \overline{P}^f(u, z; t, y) \, d\mu_t(y).$$

This measure is defined as follows. For any set $G \in \mathcal{E} \otimes \mathcal{E}$,

$$\nu(G) = \int_E d\mu_u(z) \int_{G_z} \overline{P}^f(u, z; t, y) \, d\mu_t(y)$$

where $G_z = \{y \in E : (z, y) \in G\}$. It is easy to see that the measure ν is absolutely continuous with respect to the measure $\mu_u \times \mu_t$. Let us denote the Radon-Nikodym derivative $\dfrac{d\nu}{d(\mu_u \times \mu_t)}$ by $\widetilde{P}^f$. Then for all u with $\tau \leq u < t$, the function $(z, y) \mapsto \widetilde{P}^f(u, z; t, y)$ is $\mathcal{E} \otimes \mathcal{E}$-measurable, and moreover,

$$\int_A d\mu_u(z) \int_D \overline{P}^f(u, z; t, y) \, d\mu_t(y)$$

$$= \int_A d\mu_u(z) \int_D \widetilde{P}^f(u, z; t, y) \, d\mu_t(y) \tag{1.140}$$

for all Borel sets A and D in E. It follows from (1.140) and from the separability of the σ-algebra $\mathcal{E}$ that for every u with $\tau \leq u < t$,

$$\overline{P}^f(u, z; t, y) = \widetilde{P}^f(u, z; t, y)$$

for $\mu_u \times \mu_t$-almost all $(z, y) \in E \times E$.

Let us fix s with $\tau < s < t$. Then, using (1.131), (1.132), and (1.134), we get

$$\int_{A \times C} d\mu(x, y) \int_B \frac{p(\tau, x; s, z) p(s, z; t, y)}{p(\tau, x; t, y)} dz$$
$$= \int_A d\mu_\tau(x) \int_B p^f(\tau, x; s, z) dz P^f(s, z; t, C). \tag{1.141}$$

Now it would be natural to use equality (1.139) with $u = s$ in (1.141). However, we only know that equality (1.139) holds for μ_s-almost all $z \in E$. The following equality justifies the use of (1.139) with $u = s$ in (1.141):

$$\mu_s(B) = \int d\mu_\tau(x) P^f(\tau, x; s, B) = \int_B dz \int d\mu_\tau(x) p^f(\tau, x; s, z).$$

Therefore,

$$\int_{A \times C} d\mu(x, y) \int_B \frac{p(\tau, x; s, z) p(s, z; t, y)}{p(\tau, x; t, y)} dz$$
$$= \int_A d\mu_\tau(x) \int_B p^f(\tau, x; s, z) dz \int_C \overline{P}^f(s, z; t, y) \, d\mu_t(y). \tag{1.142}$$

It is not hard to see that (1.142) implies

$$d\mu(x, y) \frac{p(\tau, x; s, z) p(s, z; t, y)}{p(\tau, x; t, y)} dz = d\mu_\tau(x) d\mu_t(y) p^f(\tau, x; s, z) \overline{P}^f(s, z; t, y) dz.$$

Hence, there exists $z_0 \in E$ such that the following representation holds for μ:

$$d\mu(x, y) = p(\tau, x; t, y) \frac{p^f(\tau, x; s, z_0)}{p(\tau, x; s, z_0)} d\mu_\tau(x) \times \frac{\overline{P}^f(s, z_0; t, y)}{p(s, z_0; t, y)} d\mu_t(y).$$

This shows that the implication (1) $\implies$ (2) holds.

The proof of Theorem 1.14 is thus completed. $\square$

Remark 1.9 Assume that the conditions in Theorem 1.14 hold. Then (1.130) implies

$$\mathbb{E}_{(\tau,t),\mu}\left[F \mid \mathcal{F}_s^\tau\right] = \frac{\int_E p(s, X_s; t, y)\, d\nu_t(y)\, \mathbb{E}_{(s,t),(X_s,y)}[F]}{\int_E p(s, X_s; t, y)\, d\nu_t(y)}, \quad (1.143)$$

where the random variable F is measurable with respect to the σ-algebra $\mathcal{F}_t^s$. Indeed, the function f appearing in the proof of Theorem 1.14 can be represented as follows:

$$\begin{aligned} f(z_{n-1}) &= \frac{\int_{A_n \times E} p(s_{n-1}, z_{n-1}; t, y)\, q(s_{n-1}, z_{n-1}; s_n, z_n; t, y)\, dz_n d\nu_t(y)}{\int_E p(s_{n-1}, z_{n-1}; t, y) d\nu_t(y)} \\ &= \frac{\int_E p(s_{n-1}, z_{n-1}; t, y)\, d\nu_t(y)\, Q(s_{n-1}, z_{n-1}; s_n, A_n; t, y)\, dz_n}{\int_E p(s_{n-1}, z_{n-1}; t, y) d\nu_t(y)}. \end{aligned} \quad (1.144)$$

It is not hard to see that (1.130) and (1.144) imply

$$\begin{aligned} &\mathbb{P}_{(\tau,t),\mu}\left[X_{s_n} \in A_n \mid \mathcal{F}_{s_{n-1}}^\tau\right] \\ &= \frac{\int_E p(s_{n-1}, X_{s_{n-1}}; t, y)\, d\nu_t(y)\, Q(s_{n-1}, X_{s_{n-1}}; s_n, A_n; t, y)\, dz_n}{\int_E p(s_{n-1}, X_{s_{n-1}}; t, y) d\nu_t(y)} \\ &= \frac{\int_E p(s_{n-1}, X_{s_{n-1}}; t, y)\, d\nu_t(y)\, \mathbb{P}_{(s_{n-1},t),(X_{s_{n-1}},y)}[X_{s_n} \in A_n]}{\int_E p(s_{n-1}, X_{s_{n-1}}; t, y) d\nu_t(y)}. \end{aligned} \quad (1.145)$$

Therefore, (1.143) follows from (1.145) and from the Markov property of the process X_s.

A formula, similar to formula (1.143), holds in the case of a general probability measure μ on $\mathcal{E} \times \mathcal{E}$.

Lemma 1.23 *Suppose that the conditions in Theorem 1.14 hold. Then the following are true:*

(1) For all $\tau < r \le s < t$ and all $\mathcal{F}_t^s$-measurable random variables F,

$$\begin{aligned} &\mathbb{E}_{(\tau,t),\mu}\left[F \mid \mathcal{F}_s^r\right] \\ &= \frac{\displaystyle\int_{E \times E} p(\tau, x; r, X_r) p(s, X_s; t, y)\, \mathbb{E}_{(s,t),(X_s,y)}[F]\, \frac{d\mu(x,y)}{p(\tau, x; t, y)}}{\displaystyle\int_{E \times E} p(\tau, x; r, X_r) p(s, X_s; t, y)\, \frac{d\mu(x,y)}{p(\tau, x; t, y)}}. \end{aligned} \quad (1.146)$$

(2) For all $\tau < r \le s < t$ and all $\mathcal{F}_r^\tau$-measurable random variables F,

$$\mathbb{E}_{(\tau,t),\mu}\left[F \mid \mathcal{F}_s^\tau\right]$$
$$= \frac{\int_{E \times E} p(\tau, x; r, X_r) p(s, X_s; t, y) \mathbb{E}_{(\tau,r),(x,X_r)}[F] \dfrac{d\mu(x,y)}{p(\tau, x; t, y)}}{\int_{E \times E} p(\tau, x; r, X_r) p(s, X_s; t, y) \dfrac{d\mu(x,y)}{p(\tau, x; t, y)}}. \quad (1.147)$$

Note that the measurability properties of the random variable F in parts (1) and (2) of Lemma 1.23 are different.

Proof. Part (1). Let $\tau < s_1 < s_2 < \cdots < s_k < s_{k+1} \cdots < s_n < t$, and let $A_i \in \mathcal{E}$ for all $1 \le i \le n$. Then, using (1.124) and (1.127) and making cancellations, we get

$$\mathbb{P}_{(\tau,t),\mu}[X_{s_i} \in A_i, 1 \le i \le n]$$
$$= \int_{E \times E \times A_1 \times \cdots \times A_n} q(\tau, x; s_1, z_1; t, y) \ldots q(s_{n-1}, z_{n-1}; s_n, z_n; t, y)$$
$$\quad q(s_{n-1}, z_{n-1}; s_n, z_n; t, y) d\mu(x,y) dz_1 \ldots dz_{n-1} dz_n$$
$$= \int_{E \times E \times A_1 \times \cdots \times A_n} \frac{d\mu(x,y)}{p(\tau, x; t, y)} p(\tau, x; s_1, z_1) \ldots p(s_{n-1}, z_{n-1}; s_n, z_n)$$
$$\quad p(s_n, z_n; t, y) dz_1 \ldots dz_{n-1} dz_n. \quad (1.148)$$

For the sake of shortness, we will use the following notation:

$$B = E \times E \times A_{k+1} \times \cdots \times A_n \text{ and } d\nu(x,y) = \frac{d\mu(x,y)}{p(\tau, x; t, y)}.$$

Let us define the function f by

$$f(z_1, z_k)$$
$$= \frac{\int_B p(\tau, x; s_1, z_1) p(s_k, z_k; s_{k+1}, z_{k+1}) \cdots p(s_n, z_n; t, y) d\nu dz_{k+1} \cdots dz_n}{\int_{E \times E} p(\tau, x; s_1, z_1) p(s_k, z_k; t, y) d\nu}.$$

This function depends on s_1, s_i with $k \le i \le n$ and A_i with $k+1 \le i \le n$. It is easy to see from (1.148) that

$$\mathbb{P}_{(\tau,t),\mu}[X_{s_i} \in A_i, 1 \le i \le n] = \int_{E \times E \times A_1 \times \cdots \times A_k} p(\tau, x; s_1, z_1) \ldots$$
$$p(s_k, z_k; t, y) f(z_1, z_k) d\nu(x,y) dz_1 \ldots dz_k.$$

$$= \int_{E \times E \times A_1 \times \cdots \times A_k} q(\tau, x; s_1, z_1; t, y)$$
$$\cdots q(s_{k-1}, z_{k-1}; s_k, z_k; t, y) f(z_1, z_k) d\mu(x, y) dz_1 \ldots dz_k$$
$$= \mathbb{E}_{(\tau,t),\mu} \left[f(X_{s_1}, X_{s_k}), X_{s_i} \in A_i, 1 \leq i \leq k \right]. \tag{1.149}$$

It follows from (1.149) that

$$\mathbb{P}_{(\tau,t),\mu} \left[X_{s_{k+1}} \in A_{k+1}, \ldots, X_{s_n} \in A_n \mid \mathcal{F}_{s_k}^{s_1} \right] = f(X_{s_1}, X_{s_k}). \tag{1.150}$$

Moreover, we have

$$\int_B p(\tau, x; s_1, z_1) p(s_k, z_k; s_{k+1}, z_{k+1}) \cdots p(s_n, z_n; t, y) d\nu(x, y) dz_{k+1} \cdots dz_n$$
$$= \int_{E \times E} p(\tau, x; s_1, z_1) \frac{d\mu(x, y)}{p(\tau, x; t, y)}$$
$$\int_{A_{k+1} \times \cdots \times A_n} p(s_k, z_k; t, y) q(s_k, z_k; s_{k+1}, z_{k+1}; t, y)$$
$$\cdots q(s_{n-1}, z_{n-1}; s_n, z_n; t, y) dz_{k+1} \cdots dz_n$$
$$= \int_{E \times E} p(\tau, x; s_1, z_1) p(s_k, z_k; t, y) \frac{d\mu(x, y)}{p(\tau, x; t, y)}$$
$$\mathbb{P}_{(s_k,t),(z_k,y)} \left[X_{s_{k+1}} \in A_{k+1}, \ldots, X_{s_n} \in A_n \right]. \tag{1.151}$$

Now the definition of the function f, (1.150), and (1.151) in the case where $s_1 = r$ and $s_n = s$ give (1.146) for all random variables of the form

$$F = \prod_{i=k+1}^{n} \chi_{A_i}(X_{s_i}).$$

By the monotone class theorem, equality (1.146) holds for all bounded $\mathcal{F}_t^s$-measurable random variables F. It is not hard to see that this implies equality (1.146) for all $\mathcal{F}_t^s$-measurable random variables F. Equality (1.147) can be obtained from equality (1.146) by using time-reversal and noting that the reciprocality is preserved under time-reversal.

This completes the proof of Lemma 1.23. □

Let ν_0 and ν_T be σ-finite Borel measures on E, and let p be a strictly positive transition density (not necessarily normal). A pair (ν_0, ν_T) is called an entrance-exit law provided that

$$\int_{E \times E} p(0, x; T, y) d\nu_0(x) d\nu_T(y) = 1. \tag{1.152}$$

It follows from Jamison's theorem (Theorem 1.14) that there exists a one-to-one correspondence between entrance-exit laws and initial-final conditions μ on $\mathcal{E} \times \mathcal{E}$, for which the process X_s associated with the derived reciprocal density q is Markovian with respect to the measure $\mathcal{P}_{(0,T),\mu}$. This one-to-one correspondence is given by

$$(\nu_0, \nu_T) \iff \mu(A \times B) = \int_{A \times B} p(0, x; T, y) d\nu_0(x) d\nu_T(y) \quad (1.153)$$

for all $A \in \mathcal{E}$ and $B \in \mathcal{E}$.

Given an entrance-exit law (ν_0, ν_T), we define the functions h and $\widetilde{h}$ by

$$h(s, x) = \int_E p(s, x; T, y) \, d\nu_T(y)$$

for all s with $0 \le s < T$ and $x \in E$; and

$$\widetilde{h}(s, y) = \int_E p(0, x; s, y) \, d\nu_0(x)$$

for all s with $0 < s \le T$ and $y \in E$. These functions are strictly positive, but it is not excluded that $h(s,x)$ or $\widetilde{h}(s,x)$ may be infinite. However, for all s with $0 < s < T$, the function $x \mapsto h(s,x)$ is finite m-almost everywhere. This follows from the equalities

$$\int_E d\nu_0(z) \int_E dx\, p(0,z;s,x) h(s,x) = \int_{E \times E} p(0,z;T,y) d\nu_0(z) d\nu_T(y) = 1.$$

Here we used the strict positivity of p, the Chapman-Kolmogorov equation, and the definition of the entrance-exit law. If $s = 0$, then we can only claim that $h(0,x)$ is finite ν_0-almost everywhere. Similarly, for all $0 < s < T$, the function $x \mapsto \widetilde{h}(s,x)$ is finite m-almost everywhere, and the function $x \mapsto \widetilde{h}(T,x)$ is finite ν_T-almost everywhere.

Next, we will introduce the transition functions P^1 and $\widetilde{P}_1$. Unlike the transition functions in Section 1.3, the functions P^1 and $\widetilde{P}_1$ are defined almost everywhere. More exactly, we put

$$P^1(\tau, x; t, A) = \frac{1}{h(\tau, x)} \int_A p(\tau, x; t, y) h(t, y) \, dy \quad (1.154)$$

for $0 \le \tau < t < T$, $x \in E$, and $A \in \mathcal{E}$. Moreover, we set

$$P^1(\tau, x; T, A) = \frac{1}{h(\tau, x)} \int_A p(\tau, x; T, y) \, d\nu_T(y) \quad (1.155)$$

for $0 \leq \tau < T$ and $x \in E$. For every τ and t with $0 \leq \tau < t < T$ and every $A \in \mathcal{E}$, the function $x \mapsto P^1(\tau, x; t, A)$ is defined m-almost everywhere with the exceptional set independent of t and A. On the other hand, for all $0 \leq \tau < T$, the function $x \mapsto P^1(\tau, x; T, A)$ is defined ν_0-almost everywhere with the exceptional set independent of A. Put

$$\widetilde{P}_1(\tau, A; t, y) = \int_A \widetilde{h}(\tau, x) p(\tau, x; t, y) \, dx \frac{1}{\widetilde{h}(t, y)} \tag{1.156}$$

for $0 < \tau < t \leq T$ and $y \in E$, and

$$\widetilde{P}_1(0, A; t, y) = \int_A p(0, x; t, y) \, d\nu_0(x) \frac{1}{\widetilde{h}(t, y)} \tag{1.157}$$

for $0 < t \leq T$ and $y \in E$. Then for every τ and t with $0 \leq \tau < t < T$ and every $A \in \mathcal{E}$, the function $y \mapsto \widetilde{P}_1(\tau, A; t, y)$ is defined m-almost everywhere with the exceptional set independent of τ and A. On the other hand, for all $0 < t \leq T$, the function $y \mapsto P_1(0, A; t, y)$ is defined ν_T-almost everywhere with the exceptional set independent of A. It is easy to check that the function P^1 is a transition probability function and the function $\widetilde{P}_1$ is a backward transition probability function, if we exclude the exceptional sets described above.

Theorem 1.15 *Let p be a strictly positive transition density, and let (ν_0, ν_T) be an entrance-exit law with respect to p. Let μ be the probability measure defined in (1.153), and denote by q the derived reciprocal transition probability density associated with p. Let P^1 and $\widetilde{P}_1$ be given by (1.154)–(1.157). Then the process $X_t(\omega) = \omega(t)$, $\omega \in \Omega$, $0 \leq t \leq T$, where $\Omega = E^{[0,T]}$, has the following three realizations: a reciprocal process with the entrance-exit law μ and the reciprocal transition density q; a Markov process with the initial condition $h(0, x) \, d\nu_0(x)$ and the transition function P^1; and a backward Markov process with the final condition $\widetilde{h}(T, x) \, d\nu_T(x)$ and the backward transition function $\widetilde{P}_1$.*

Proof. Let $0 < s_1 < s_2 < \cdots < s_n < T$ and $A_i \in \mathcal{E}$ with $0 \leq i \leq n+1$. Then it is not difficult to see that the finite-dimensional distributions of all the processes described in the formulation of Theorem 1.15 are given by

$$\int_{A_0 \times A_{n+1}} d\mu(x,y) \int_{A_1 \times \cdots \times A_n} q(0,x;s_1,z_1;T,y) q(s_1,z_1;s_2,z_2;T,y)$$
$$\cdots q(s_{n-1},z_{n-1};s_n,z_n;T,y) dz_1 \ldots dz_n$$
$$= \int_{A_0 \times \cdots \times A_{n+1}} d\nu_0(x) p(0,x;s_1,z_1) p(s_1,z_1;s_2,z_2)$$
$$\cdots p(s_n,z_n;T,y) dz_1 \ldots dz_n d\nu_T(y).$$

This completes the proof of Theorem 1.15. □

Theorem 1.15 is taken from [Nagasawa (2000)] (see Theorem 3.3.1 in [Nagasawa (2000)]). A triplicate nature of the stochastic process X_t is clearly seen from Theorem 1.15. The three representations of the process X_t, that is, the reciprocal, the Markov, and the backward Markov are called the Schrödinger, the forward Kolmogorov, and the backward Kolmogorov representation of the process X_t, respectively. Many of the ideas discussed in this section go back to Schrödinger, Bernstein, and Kolmogorov. These ideas found applications in quantum mechanics (see the references in Section 1.13).

1.11 Path Properties of Reciprocal Processes

Let p be a strictly positive transition density, and denote by q the corresponding derived transition probability density given by (1.124). Fix $(\tau,x) \in [0,T] \times E$ and $(t,y) \in [0,T] \times E$, and suppose that the measure $\mu = \delta_x \times \delta_y$ is the initial-final distribution in formula (1.115). The resulting measure will be denoted by $\mathbb{P}_{(\tau,t),(x,y)}$. This measure satisfies

$$\mathbb{P}_{(\tau,t),(x,y)}[X_\tau \in A_0, X_{s_1} \in A_1, \ldots, X_{s_n} \in A_n, X_t \in A_{n+1}]$$
$$= \chi_{A_0}(x) \chi_{A_{n+1}}(y) \int_{A_1} Q(\tau,x;s_1,dz_1;t,y)$$
$$\int_{A_2} \cdots \int_{A_n} Q(s_{n-1},z_{n-1};s_n,dz_n;t,y)$$

(see (1.113)). Using the definition of the derived density q and making cancellations, we obtain

$$\mathbb{P}_{(\tau,t),(x,y)}\left[X_\tau \in A_0, X_{s_1} \in A_1, \ldots, X_{s_n} \in A_n, X_t \in A_{n+1}\right]$$
$$= \frac{\chi_{A_0}(x)\chi_{A_{n+1}}(y)}{p(\tau,x;t,y)} \int_{A_1 \times \cdots \times A_n} p(\tau, x; s_1, z_1) p(s_1, z_1; s_2, z_2)$$
$$\cdots p(s_n, z_n; t, y) dz_1 \cdots dz_n. \tag{1.158}$$

If the transition density p satisfies the normality condition, then an equivalent form of the previous equality is

$$\mathbb{P}_{(\tau,t),(x,y)}\left[X_\tau \in A_0, X_{s_1} \in A_1, \ldots, X_{s_n} \in A_n, X_t \in A_{n+1}\right]$$
$$= \frac{\chi_{A_0}(x)\chi_{A_{n+1}}(y)}{p(\tau,x;t,y)} \mathbb{E}_{\tau,x}\left[p\left(s_n, X_{s_n}; t, y\right), X_{s_1} \in A_1, \ldots, X_{s_n} \in A_n\right].$$

Moreover, using the monotone class theorem, we get

$$\mathbb{P}_{(\tau,t),(x,y)}[A] = \frac{\mathbb{E}_{\tau,x}\left[\chi_A p\left(s, X_s; t, y\right)\right]}{p\left(\tau, x; t, y\right)} \tag{1.159}$$

for all $A \in \mathcal{F}_s^\tau$ with $\tau \le s < t$. If, in addition, a strictly positive function $p(\tau, x; t, y)$ is simultaneously a forward and a backward transition probability density, then

$$\mathbb{P}_{(\tau,t),(x,y)}[B] = \frac{\widetilde{\mathbb{E}}^{t,y}\left[\chi_B p\left(\tau, x; s, X_s\right)\right]}{p\left(\tau, x; t, y\right)}, \tag{1.160}$$

for all $B \in \mathcal{F}_t^s$ with $\tau < s \le t$.

Example 1.1 Let $x_0 \in E$ and $y_0 \in E$ be given, and consider the following measures on E:

$$\nu_0 = \frac{\delta_{x_0}}{p\left(0, x_0; T, y_0\right)} \quad \text{and} \quad \nu_T = \delta_{y_0}. \tag{1.161}$$

Then it is easy to see that (ν_0, ν_T) is an entrance-exit law. Hence Theorem 1.15 can be applied. Since

$$h(s,x) = p(s,x;T,y_0) \quad \text{and} \quad \widetilde{h}(s,y) = \frac{p\left(0, x_0; s, y\right)}{p\left(0, x_0; T, y_0\right)},$$

we see that the process $X_s(\omega) = \omega(s)$, $\omega \in \Omega$, $0 \le s \le T$, in Theorem 1.15 has the following three properties. It is a reciprocal process with $\mu = \delta_{x_0} \times \delta_{y_0}$ as the entrance-exit law and q as the reciprocal density; a

Markov process with the initial condition δ_{x_0} and the transition function P^1_{T,y_0} given by

$$P^1_{T,y_0}(\tau, x; t, A) = \frac{1}{p(\tau, x; T, y_0)} \int_A p(\tau, x; t, y) p(t, y; T, y_0)\, dy$$

$$= \int_A q(\tau, x; t, y; T, y_0)\, dy \qquad (1.162)$$

for $0 \le \tau < t < T$, $x \in E$, and $A \in \mathcal{E}$, and by

$$P^1_{T,y_0}(\tau, x; T, A) = \chi_A(y_0) \qquad (1.163)$$

for $0 \le \tau < T$, $x \in E$, and $A \in \mathcal{E}$; and, finally, a backward Markov process with the final condition δ_{y_0} and the backward transition function $\widetilde{P}^{0,x_0}_1$ given by

$$\widetilde{P}^{0,x_0}_1(\tau, A; t, y) = \frac{1}{p(0, x_0; t, y)} \int_A p(0, x_0; \tau, x) p(\tau, x; t, y)\, dx$$

$$= \int_A q(0, x_0; \tau, x; t, y)\, dx \qquad (1.164)$$

for $0 < \tau < t \le T$ and $y \in E$, and by

$$\widetilde{P}^{0,x_0}_1(0, A; t, y) = \chi_A(x_0), \qquad (1.165)$$

for $0 < t \le T$, $y \in E$, and $A \in \mathcal{E}$.

As before, we denote by $\mathbb{P}_{(0,T),(x_0,y_0)}$ the measure on $\Omega = E^{[0,T]}$ associated with the entrance-exit law (1.161) and the reciprocal transition density q; by $\mathbb{P}^{T,y_0}_{0,x_0}$ the measure on Ω corresponding to the transition function P^1_{T,y_0} given by (1.162) and (1.163) and the initial condition δ_{x_0}; and by $\widetilde{\mathbb{P}}^{T,y_0}_{0,x_0}$ the measure on Ω associated with the backward transition function $\widetilde{P}^{0,x_0}_1$ defined by (1.164) and (1.165) and with the final condition δ_{y_0}. If p is a transition density, then we denote by $\mathbb{P}_{0,x_0}$ the measure on Ω associated with the density p and the initial distribution δ_{x_0}. If p is also a backward transition probability density, then we denote by $\widetilde{\mathbb{P}}^{T,y_0}$ the measure on Ω associated with the backward transition density p and the final distribution δ_{y_0}. By Theorem 1.15, the measures $\mathbb{P}_{(0,T),(x_0,y_0)}$, $\mathbb{P}^{T,y_0}_{0,x_0}$, and $\widetilde{\mathbb{P}}^{T,y_0}_{0,x_0}$ coincide on the σ-algebra $\mathcal{F}^0_T$.

It is not hard to prove that if p is a transition probability density, then the measure $\mathbb{P}^{T,y_0}_{0,x_0}$ is absolutely continuous with respect to the measure $\mathbb{P}_{0,x_0}$ on every σ-algebra $\mathcal{F}^0_r$ with $0 \le r < T$. Similarly, if, in addition, p is a backward transition probability density, then the measure $\widetilde{\mathbb{P}}^{T,y_0}_{0,x_0}$ is absolutely continuous with respect to the measure $\widetilde{\mathbb{P}}^{T,y_0}$ on every σ-algebra $\mathcal{F}_T^r$ with $0 < r \le T$.

Suppose that under certain restrictions on the transition density p the following conditions hold: (a) There exists a continuous $\mathbb{P}_{(0,T),(x_0,y_0)}$-modification of the forward Kolmogorov representation of the process X_t on the half-open interval $[0, T)$; (b) $\lim_{t \uparrow T} X_t = y_0$ $\mathbb{P}_{(0,T),(x_0,y_0)}$-almost everywhere. Then the same restrictions guarantee the existence of a continuous modification of the reciprocal process X_t on the closed interval $[0, T]$. Similarly, suppose that under certain restrictions on p, the following conditions hold: (a) There exists a $\mathbb{P}_{(0,T),(x_0,y_0)}$-modification of the forward Kolmogorov representation of the process X_t, which is right-continuous and has left limits on the half-open interval $[0, T)$; (b) $\lim_{t \uparrow T} X_t = y_0$ $\mathbb{P}_{(0,T),(x_0,y_0)}$-almost everywhere. Then the same restrictions imply the existence of a $\mathbb{P}_{(0,T),(x_0,y_0)}$-modification of the reciprocal process X_s that is right-continuous and has left limits on the closed interval $[0, T]$. We will use these ideas in the proof of the following assertion.

Theorem 1.16 *Let $p(\tau, x; t, y)$ be a strictly positive function that is simultaneously a forward and a backward transition probability density, and let $x_0 \in E$ and $y_0 \in E$ be given. Denote by q the derived reciprocal transition probability density associated with p, and consider the Schrödinger representation of the process $X_t(\omega) = \omega(t)$, $\omega \in \Omega$, $0 \le t \le T$, on the space $\Omega = E^{[0,T]}$ with respect to entrance-exit law (1.161) and the reciprocal transition density q. Suppose that for all $\epsilon > 0$,*

$$\lim_{t \downarrow 0} \int_{G_\epsilon(x_0)} p(0, x_0; t, y) \, dy = 0, \tag{1.166}$$

$$\lim_{t-\tau \downarrow 0; 0 < \tau < t < T} \sup_{x \in E} \int_{G_\epsilon(x)} p(\tau, x; t, z) \, dz = 0, \tag{1.167}$$

$$\lim_{t-\tau \downarrow 0; 0 < \tau < t < T} \sup_{y \in E} \int_{G_\epsilon(y)} p(\tau, z; t, y) \, dz = 0, \tag{1.168}$$

and

$$\lim_{t \uparrow T} \int_{G_\epsilon(y_0)} p(t, y; T, y_0) \, dy = 0. \tag{1.169}$$

Then there exists a $\mathbb{P}_{(0,T),(x_0,y_0)}$-modification $\widehat{X}_t$ of the process X_t that is right-continuous and has left limits on $[0, T]$.

Proof. Conditions (1.166) and (1.167) in the formulation of Theorem 1.16 guarantee the existence of a $\mathbb{P}_{0,x_0}$-modification Y_t of the process X_t that is right-continuous and has left limits on the half-open interval $[0,T)$ (see Remark 1.7). Since the measure $\mathbb{P}_{0,x_0}^{T,y_0}$ is absolutely continuous with respect to the measure $\mathbb{P}_{0,x_0}$ on every σ-algebra $\mathcal{F}_r^0$ where $0 \leq r < T$, there exists a $\mathbb{P}_{0,x_0}^{T,y_0}$-modification $\widetilde{Y}_t$ of the process X_t which is right-continuous and has left limits on the half-open interval $[0,T)$. Since

$$\mathbb{P}_{(0,T),(x_0,y_0)} = \mathbb{P}_{0,x_0}^{T,y_0},$$

the process $\widetilde{Y}_t$ is a $\mathbb{P}_{(0,T),(x_0,y_0)}$-modification of the process X_t on $[0,T)$.

On the other hand, since p is a backward transition probability density, conditions (1.168) and (1.169) imply the existence of a $\widetilde{\mathbb{P}}^{T,y_0}$-modification Z_t of the process X_t that is left-continuous and has right limits on the half-open interval $(0,T]$ (see Remark 1.7). Using the absolute continuity of the measure $\widetilde{\mathbb{P}}_{0,x_0}^{T,y_0}$ with respect to the measure $\widetilde{\mathbb{P}}^{T,y_0}$ on every σ-algebra $\mathcal{F}_T^r$ with $0 < r \leq T$ and the equality

$$\mathbb{P}_{(0,T),(x_0,y_0)} = \widetilde{\mathbb{P}}_{0,x_0}^{T,y_0},$$

we see that there exists a $\mathbb{P}_{(0,T),(x_0,y_0)}$-modification $\widetilde{Z}_t$ of the process X_t on the interval $(0,T]$ that is left-continuous and has right limits on the interval $(0,T]$. Using the fact that the processes $\widetilde{Y}_t$ and $\widetilde{Z}_t$ are $\mathbb{P}_{(0,T),(x_0,y_0)}$-modifications of the process X_t, we see that there exists a $\mathbb{P}_{(0,T),(x_0,y_0)}$-modification $\widetilde{X}_t$ of the process X_t having right and left limits on the closed interval $[0,T]$.

Finally, by redefining the process $\widetilde{X}_t$ as we did at the end of the proof of Lemma 1.16, we get a process $\widehat{X}_t$ satisfying the conditions in Theorem 1.16. □

The next assertion concerns continuous modifications of reciprocal processes. Its proof is similar to that of Theorem 1.16, and we leave it as an exercise for the reader.

Theorem 1.17 *Let $p(\tau,x;t,y)$ be a strictly positive function which is simultaneously a forward and a backward transition probability density, and let $x_0 \in E$ and $y_0 \in E$. Denote by q the derived reciprocal transition probability density associated with p, and consider the Schrödinger representation of the process $X_t(\omega) = \omega(t)$, $\omega \in \Omega$, $0 \leq t \leq T$, on the space $\Omega = E^{[0,T]}$ with respect to the entrance-exit law (1.161) and the reciprocal transition* density q. Suppose that the following conditions hold for every $\epsilon > 0$:

$$\lim_{t \downarrow 0} \frac{1}{t} \int_{G_\epsilon(x_0)} p(0, x_0; t, z) \, dz = 0, \tag{1.170}$$

$$\lim_{t-\tau \downarrow 0; 0 < \tau < t < T} \frac{1}{t-\tau} \sup_{y \in E} \int_{G_\epsilon(y)} p(\tau, y; t, z) \, dz = 0, \tag{1.171}$$

$$\lim_{t-\tau \downarrow 0; 0 < \tau < t < T} \frac{1}{t-\tau} \sup_{y \in E} \int_{G_\epsilon(y)} p(\tau, z; t, y) \, dz = 0, \tag{1.172}$$

and

$$\lim_{t \uparrow T} \frac{1}{T-t} \int_{G_\epsilon(y_0)} p(t, z; T, y_0) \, dz = 0. \tag{1.173}$$

Then there exists a continuous $\mathbb{P}_{(0,T),(x_0,y_0)}$-modification $\widehat{X}_t$ of the process X_t.

In [Jamison (1975)], a different condition is used to ensure the validity of the equality $\lim_{t \uparrow T} X_t = y_0$. In the next theorem, Jamison's condition is employed instead of the conditions in (1.168), (1.169), (1.172), and (1.173):

Theorem 1.18 Let $p(\tau, x; t, y)$ be a strictly positive transition probability density, and let $x_0 \in E$ and $y_0 \in E$. Denote by q the derived reciprocal transition probability density associated with p, and consider the Schrödinger representation of the process $X_t(\omega) = \omega(t)$, $\omega \in \Omega$, $0 \leq t \leq T$, on the space $\Omega = E^{[0,T]}$ with respect to the entrance-exit law (1.161) and the reciprocal transition density q. Suppose that conditions (1.166) and (1.167) hold and that

$$\lim_{t \uparrow T} \sup_{y \in G_\epsilon(y_0)} p(t, y; T, y_0) = 0 \tag{1.174}$$

for all $\epsilon > 0$. Then there exists a $\mathbb{P}_{(0,T),(x_0,y_0)}$-modification $\widehat{X}_t$ of the process X_t that is right-continuous and has left limits on $[0, T]$.

Theorem 1.19 Let $p(\tau, x; t, y)$ be a strictly positive transition probability density, and let $x_0 \in E$ and $y_0 \in E$. Denote by q the derived reciprocal transition probability density associated with p, and consider the Schrödinger representation of the process $X_t(\omega) = \omega(t)$, $\omega \in \Omega$, $0 \leq t \leq T$, on the space $\Omega = E^{[0,T]}$ with respect to the entrance-exit law (1.161) and the reciprocal transition density q. Suppose that conditions (1.170), (1.171), and (1.174) hold. Then there exists a continuous $\mathbb{P}_{(0,T),(x_0,y_0)}$-modification $\widehat{X}_t$ of the process X_t.

Remark 1.10 It is not assumed in Theorems 1.18 and 1.19 that p is a backward transition probability density.

Proof. Arguing as in the beginning of the proof of Theorem 1.16, we see that there exists a $\mathbb{P}_{(0,T),(x_0,y_0)}$-modification $\widetilde{Y}_t$ of the process X_t that is right-continuous and has left limits on the interval $[0,T)$. We will next show that Jamison's condition (1.174) implies the equality $\lim_{t \uparrow T} X_t = y_0$ $\mathbb{P}_{(0,T),(x_0,y_0)}$-almost everywhere. The following lemma has an independent interest. A special case of this lemma will be used in the proof of Theorem 1.18.

Lemma 1.24 *For every τ and t with $0 \leq \tau < t \leq T$, the process $\dfrac{p(\tau, X_\tau; t, X_t)}{p(s, X_s; t, X_t)}$, $\tau \leq s < t$, is a $\mathbb{P}_{(0,T),(x_0,y_0)}$-martingale with respect to the filtration $\mathcal{F}_s^\tau \vee \sigma(X_t)$, $\tau \leq s \leq t$.*

Proof. We will first show that

$$\mathbb{E}_{(0,T),(x_0,y_0)} \left[\frac{p(\tau, X_\tau; t, X_t)}{p(s, X_s; t, X_t)} \right] = 1. \tag{1.175}$$

Indeed, using the normality condition for p and the Chapman-Kolmogorov equation twice, we get

$$\mathbb{E}_{(0,T),(x_0,y_0)} \left[\frac{p(\tau, X_\tau; t, X_t)}{p(s, X_s; t, X_t)} \right]$$

$$= \int_{E^3} q(0, x_0; \tau, z_1; T, y_0) \, q(\tau, z_1; s, z_2; T, y_0) \, q(s, z_2; t, z_3; T, y_0)$$
$$\quad \frac{p(\tau, z_1; t, z_3)}{p(s, z_2; t, z_3)} dz_1 dz_2 dz_3$$

$$= \int_{E^3} \frac{p(\tau, z_1; t, z_3) \, p(0, x_0; \tau, z_1) \, p(\tau, z_1; s, z_2) \, p(t, z_3; T, y_0)}{p(0, x_0; T, y_0)} dz_1 dz_2 dz_3$$

$$= \frac{1}{p(0, x_0; T, y_0)} \int_{E \times E} dz_1 dz_3 \, p(\tau, z_1; t, z_3) \, p(0, x_0; \tau, z_1) \, p(t, z_3; T, y_0)$$

$$= \frac{1}{p(0, x_0; T, y_0)} \int_E p(t, z_3; T, y_0) \, dz_3 \int_E p(\tau, z_1; t, z_3) \, p(0, x_0; \tau, z_1) \, dz_1$$

$$= \frac{1}{p(0, x_0; T, y_0)} \int_E p(t, z_3; T, y_0) \, p(0, x_0; t, z_3) \, dz_3 = 1.$$

This gives (1.175).

We will next prove that for all $\tau \leq s_1 < s_2 < t$,

$$\mathbb{E}_{(0,T),(x_0,y_0)}\left[\frac{p(\tau, X_\tau; t, X_t)}{p(s_2, X_{s_2}; t, X_t)} \,\bigg|\, \mathcal{F}_{s_1}^\tau \vee \sigma(X_t)\right] = \frac{p(\tau, X_\tau; t, X_t)}{p(s_1, X_{s_1}; t, X_t)} \quad (1.176)$$

$\mathbb{P}_{(0,T),(x_0,y_0)}$-a.s. Indeed, using the normality condition for p, the measurability of the random variable $p(\tau, X_\tau; t, X_t)$ with respect to the σ-algebra $\mathcal{F}_{s_1}^\tau \vee \sigma(X_t)$, and conditions (1.119) and (1.120), we obtain

$$\mathbb{E}_{(0,T),(x_0,y_0)}\left[\frac{p(\tau, X_\tau; t, X_t)}{p(s_2, X_{s_2}; t, X_t)} \,\bigg|\, \mathcal{F}_{s_1}^\tau \vee \sigma(X_t)\right]$$

$$= p(\tau, X_\tau; t, X_t)\,\mathbb{E}_{(0,T),(x_0,y_0)}\left[\frac{1}{p(s_2, X_{s_2}; t, X_t)} \,\bigg|\, \mathcal{F}_{s_1}^\tau \vee \sigma(X_t)\right]$$

$$= p(\tau, X_\tau; t, X_t)\,\mathbb{E}_{(0,T),(x_0,y_0)}\left[\frac{1}{p(s_2, X_{s_2}; t, X_t)} \,\bigg|\, \mathcal{F}_{s_1}^0 \vee \mathcal{F}_T^t\right]$$

$$= p(\tau, X_\tau; t, X_t)\,\mathbb{E}_{(s_1,t),(X_{s_1},X_t)}\left[\frac{1}{p(s_2, X_{s_2}; t, X_t)}\right]$$

$$= p(\tau, X_\tau; t, X_t)\int_E \frac{q(s_1, X_{s_1}; s_2, z; t, X_t)}{p(s_2, z; t, X_t)}\,dz$$

$$= \frac{p(\tau, X_\tau; t, X_t)}{p(s_1, X_{s_1}; t, X_t)}\int_E p(s_1, X_{s_1}; s_2, z)\,dz$$

$$= \frac{p(\tau, X_\tau; t, X_t)}{p(s_1, X_{s_1}; t, X_t)}. \quad (1.177)$$

This establishes (1.176).

It follows from (1.175) and (1.176) that Lemma 1.24 holds. □

We are now ready to finish the proof of Theorem 1.18. Using Lemma 1.24 with $\tau = 0$ and $t = T$, we see that the process $\dfrac{1}{p(s, X_s; T, y_0)}$ is a nonnegative $\mathbb{P}_{(0,T),(x_0,y_0)}$-martingale. By Theorem 1.3 and equality (1.175), this martingale converges to a finite limit as $s \uparrow T$ $\mathbb{P}_{(0,T),(x_0,y_0)}$-almost surely. It follows that $p(s, X_s; T, y_0)$ converges to a strictly positive limit $\mathbb{P}_{(0,T),(x_0,y_0)}$-almost surely as $s \uparrow T$. By condition (1.174), we see that this can only happen if X_s converges to y_0 $\mathbb{P}_{(0,T),(x_0,y_0)}$-almost surely as $s \uparrow T$. Combining the previous assertion with the fact that there exists a $\mathbb{P}_{(0,T),(x_0,y_0)}$-modification $\widetilde{Y}_s$ of the process X_s that is right-continuous and has left limits on the interval $[0, T)$, we see that there exists a $\mathbb{P}_{(0,T),(x_0,y_0)}$-modification $\widehat{X}_s$ of the process X_s that is right-continuous and has left limits on the closed interval $[0, T]$.

This completes the proof of Theorem 1.18. The proof of Theorem 1.19 is similar. □

If the transition density p is not normal, then the measures $\mathbb{P}_{\tau,x_0}$ and $\widetilde{\mathbb{P}}^{T,y_0}$ employed in the proof of Theorem 1.16 do not exist. However, by changing the assumptions in Theorem 1.16 and taking into account the remarks before the formulation of Theorem 1.16, we see that the following assertion holds.

Theorem 1.20 *Let $p(\tau, x; t, y)$ be a strictly positive transition density, and let $x_0 \in E$ and $y_0 \in E$. Denote by q the derived reciprocal transition probability density associated with p, and consider the Schrödinger representation of the process $X_t(\omega) = \omega(t)$, $\omega \in \Omega$, $0 \le t \le T$, on the space $\Omega = E^{[0,T]}$ with respect to the entrance-exit law (1.161) and the reciprocal transition density q. Suppose that for all $\varepsilon > 0$,*

$$\lim_{t \downarrow 0} \int_{G_\varepsilon(x_0)} p(0, x_0; t, y) \, p(t, y; T, y_0) \, dy = 0, \tag{1.178}$$

$$\lim_{t-\tau \downarrow 0; 0 < \tau < t < T} \sup_{x \in E} \frac{1}{p(\tau, x; T, y_0)} \int_{G_\varepsilon(x)} p(\tau, x; t, z) \, p(t, z; T, y_0) \, dz = 0, \tag{1.179}$$

$$\lim_{t-\tau \downarrow 0; 0 < \tau < t < T} \sup_{y \in E} \frac{1}{p(0, x_0; t, y)} \int_{G_\varepsilon(y)} p(0, x_0; \tau, z) \, p(\tau, z; t, y) \, dz = 0, \tag{1.180}$$

and

$$\lim_{t \uparrow T} \int_{G_\varepsilon(y_0)} p(0, x_0; t, z) \, p(t, z; T, y_0) \, dy = 0. \tag{1.181}$$

Then there exists a $\mathbb{P}_{(0,T),(x_0,y_0)}$-modification $\widehat{X}_t$ of the process X_t that is right-continuous and has left limits on $[0,T]$.

A similar theorem holds for continuous modifications. It is based on Theorem 1.17. We leave this case as an exercise for the reader.

In the next definition, we introduce pinned measures.

Definition 1.20 Let $(\tau, x) \in [0, T] \times E$, $(t, y) \in [0, T] \times E$, and let p be a strictly positive transition density on E. Denote by X_s the process $X_s(\omega) = \omega(s)$ on the space $\Omega = E^{[0,T]}$. Then the measure $\mu_{\tau,x}^{t,y}$ on $(\Omega, \mathcal{F}_t^\tau)$

satisfying

$$\mu_{\tau,x}^{t,y}\left(X_\tau \in A_0, X_{s_1} \in A_1, \ldots, X_{s_n} \in A_n, X_t \in A_{n+1}\right)$$
$$= p\left(\tau, x; t, y\right) \mathbb{P}_{(\tau,t),(x,y)}\left[X_\tau \in A_0, X_{s_1} \in A_1, \ldots, X_{s_n} \in A_n, X_t \in A_{n+1}\right]$$
$$= \chi_{A_0}(x)\chi_{A_{n+1}}(y) \int_{A_1 \times \ldots \times A_n} p(\tau, x; s_1, z_1) p(s_1, z_1; s_2, z_2)$$
$$\cdots p(s_n, z_n; t, y) dz_1 \ldots dz_n \qquad (1.182)$$

where $\tau < s_1 < s_2 < \cdots < s_n < t$ and $A_i \in \mathcal{E}$ for $1 \leq i \leq n$, is called the pinned measure associated with p, (τ, x), and (t, y).

It follows from (1.158) and (1.182) that

$$\mu_{\tau,x}^{t,y} = p(\tau, x; t, y) \mathbb{P}_{(\tau,t),(x,y)}.$$

Remark 1.11 If the density p is normal, then in addition to (1.182) the following formula holds:

$$\mu_{\tau,x}^{t,y}\left(X_\tau \in A_0, X_{s_1} \in A_1, \ldots, X_{s_n} \in A_n, X_t \in A_{n+1}\right)$$
$$= \chi_{A_0}(x)\chi_{A_{n+1}}(y) \mathbb{E}_{\tau,x}\left[p\left(s_n, X_{s_n}; t, y\right), X_{s_1} \in A_1, \ldots, X_{s_n} \in A_n\right]. \qquad (1.183)$$

The pinned measure $\mu_{\tau,x}^{t,y}$ is defined on the measurable space $(E^{[0,T]}, \mathcal{F}_t^\tau)$. However, if X_s is a Markov process on a smaller path space, e.g., on the space of continuous paths or on the space of left- or right-continuous paths, then certain difficulties may arise at the endpoints τ and t. To avoid these difficulties, the measure $\mu_{\tau,x}^{t,y}$ is usually restricted to the σ-algebra

$$\mathcal{F}_{t-}^\tau = \sigma\left(X_s : \tau \leq s < t\right)$$

in the case of right-continuous processes X_s having left limits, while in the case of left-continuous processes having right limits, the measure $\mu_{\tau,x}^{t,y}$ is restricted to the σ-algebra

$$\mathcal{F}_t^{\tau+} = \sigma\left(X_s : \tau < s \leq t\right).$$

Lemma 1.25 Let $0 \leq \tau \leq u < t \leq T$, $x \in E$, and $y \in E$. Suppose that a random variable F is measurable with respect to the σ-algebra $\mathcal{F}_u^\tau$. Then

$$\int_\Omega F d\mu_{t,y}^{\tau,x} = \mathbb{E}_{\tau,x}\left[Fp\left(u, X_u; t, y\right)\right].$$

The proof of Lemma 1.25 is not difficult, and we leave it as an exercise for the reader.

1.12 Examples of Transition Functions and Markov Processes

In this section we discuss several well-known examples of stochastic processes (see Section 1.13 where more references can be found). We also introduce pinned processes or "bridges".

1.12.1 *Brownian motion and Brownian bridge*

The state space E of Brownian motion is d-dimensional Euclidean space $\mathbb{R}^d$ equipped with the Borel σ-algebra $\mathcal{B}_{\mathbb{R}^d}$. The d-dimensional Lebesgue measure plays the role of the reference measure m on $\mathbb{R}^d$.

Recall that the d-dimensional Gaussian transition probability density p_d is defined by

$$p_d(\tau, x; t, y) = \frac{1}{[2\pi(t-\tau)]^{d/2}} \exp\left\{-\frac{|x-y|^2}{2(t-\tau)}\right\}$$

(see Section 1.11). Since p_d depends only on the differences $s = t - \tau$ and $z = x - y$, we will use the notation $g_d(s, z) = p_d(\tau, x; t, y)$. The function p_d is a transition probability density associated with a time- and space-homogeneous transition probability function. The density p_d is normal, since

$$\frac{1}{(2\pi)^{d/2}} \int_{\mathbb{R}^d} e^{-\frac{|x|^2}{2}} dx = 1.$$

The fact that g_d satisfies the Chapman-Kolmogorov equation can be obtained from the following well-known formula for the Fourier transform of the Gaussian density:

$$\widehat{g_d}(t, \xi) = \frac{1}{(2\pi t)^{d/2}} \int_{\mathbb{R}^d} e^{-\frac{|x|^2}{2t}} e^{-i\xi \cdot x} dx = e^{-\frac{t|\xi|^2}{2}}.$$

For any $(\tau, x) \in [0, \infty) \times \mathbb{R}^d$, there exists a measure $\mathbb{P}_{\tau,x}$ on the path space

$[\mathbb{R}^d]^{[0,\infty)}$ with the finite-dimensional distributions given by

$$\mathbb{P}_{\tau,x}[X_\tau \in A_0, X_{t_1} \in A_1, \ldots, X_{t_n} \in A_n]$$
$$= \chi_{A_0}(x) \int_{A_1 \times \cdots \times A_n} g_d(t_1, x_1 - x) g_d(t_2 - t_1, x_2 - x_1) \cdots$$
$$g_d(t_n - t_{n-1}, x_n - x_{n-1}) dx_1 \cdots dx_n, \tag{1.184}$$

where $\tau < t_1 < t_2 < \cdots < t_n$, $x \in \mathbb{R}^d$, and $A_i \in \mathcal{B}_{\mathbb{R}^d}$ for $0 \le i \le n$. The Gaussian transition function P is defined by

$$P(\tau, x; t, A) = \int_A g_d(t - \tau, x - y) dy.$$

The function P satisfies condition (1.87) in Corollary 1.2. Indeed,

$$\frac{P(s, y; t, G_\epsilon(y))}{t - s} = \frac{1}{t - s} \int_{x:|x-y|>\epsilon} \frac{1}{(2\pi(t-s))^{d/2}} \exp\left\{-\frac{|x-y|^2}{2(t-s)}\right\} dx$$
$$= \frac{1}{t - s} \int_{z:|z|>\epsilon} \frac{1}{(2\pi(t-s))^{d/2}} \exp\left\{-\frac{|z|^2}{2(t-s)}\right\} dz$$
$$\le \frac{1}{t-s} \frac{1}{(2\pi)^{d/2}} \int_{u:|u|>\epsilon(t-s)^{-1/2}} \exp\left\{-\frac{|u|^2}{2}\right\} du.$$
$$\le \frac{1}{\epsilon^2} \frac{1}{(2\pi)^{d/2}} \int_{u:|u|>\epsilon(t-s)^{-1/2}} |u|^2 \exp\left\{-\frac{|u|^2}{2}\right\} du, \tag{1.185}$$

and it follows from (1.185) that

$$\lim_{t-s \to 0+} \sup_{y \in \mathbb{R}^d} \frac{P(s, y; t, G_\epsilon(y))}{t - s} = 0$$

for all $\epsilon > 0$.

By Corollary 1.2, there exists a continuous stochastic process $(B_t, \mathcal{F}_t^\tau, \mathbb{P}_{\tau,x})$ with state space $\mathbb{R}^d$ such that its finite-dimensional distributions are given by (1.184). The space Ω of all continuous paths from $[0, \infty)$ into $\mathbb{R}^d$ can be taken as the sample space in this case. The process B_t is called Brownian motion in $\mathbb{R}^d$. It follows from (1.184) that

$$\mathbb{P}_{\tau,x}[B_\tau \in A_0, B_{t_1} \in A_1, \ldots, B_{t_n} \in A_n]$$
$$= \mathbb{P}_{0,0}[B_0 \in A_0 - x, B_{t_1-\tau} \in A_1 - x, \ldots, B_{t_n-\tau} \in A_n - x].$$

The measure $\mathbb{P}_{0,x}$ is denoted by $\mathbb{P}_x$. Brownian motion starting at $x = 0$ at moment $\tau = 0$ is called a standard Brownian motion, and the measure space $\left(\mathbb{C}\left(\mathbb{R}^+;\mathbb{R}^d\right), \mathcal{C}, \mathbb{P}_{0,0}\right)$, where $\mathcal{C}$ is the cylinder σ-algebra of $\mathbb{C}\left(\mathbb{R}^+;\mathbb{R}^d\right)$, is called the Wiener space.

We denote the components of Brownian motion B_t by B_t^i, $1 \le i \le d$. For a given pair $(\tau, x) \in [0, \infty) \times \mathbb{R}^d$, Brownian motion has the following properties:

(1) Every component B_t^i of B_t is a one-dimensional Brownian motion with transition density g_1. The components of B_t are $\mathbb{P}_{\tau,x}$-independent.
(2) $\mathbb{P}_{\tau,x}[B_\tau = x] = 1$.
(3) For $\tau \le t_1 < t_2 < \cdots < t_n$, the increments B_{t_1}, $B_{t_2} - B_{t_1}$, $\cdots$, $B_{t_n} - B_{t_{n-1}}$ of Brownian motion are $\mathbb{P}_{\tau,x}$-independent.
(4) The increments of B_t are stationary; that is, for all $\tau \le s \le t$ and $h \ge 0$, $B_t - B_s$ has the same $\mathbb{P}_{\tau,x}$-distribution as $B_{t+h} - B_{s+h}$.
(5) Brownian motion is a Gaussian process with mean $\mathbb{E}_{\tau,x} B_t = x$, $\tau \le t$, $x \in \mathbb{R}^d$, and the correlation matrix given by

$$\mathbb{E}_{\tau,x}\left(B_s^i - x^i\right)\left(B_t^j - x^j\right) = (s - \tau) \wedge (t - \tau),$$

where $\tau \le s$, $\tau \le t$, $1 \le i, j \le d$, and $x = \left(x^1, \ldots, x^d\right) \in \mathbb{R}^d$ (see [Revuz and Yor (1991)] for the properties of Gaussian processes).

If B_t is a d-dimensional Brownian motion starting at x at time τ, then $-B_t$ is a d-dimensional Brownian motion starting at $-x$ at time τ. For a standard Brownian motion B_t, the finite-dimensional distributions of the processes B_t and $tB_{1/t}$, $t > 0$, coincide. They also coincide with the finite-dimensional distributions of the process $B_{t+h} - B_h$ for every $h > 0$. Moreover, for any $a > 0$, the finite-dimensional distributions of the process B_t are the same as those of the process $a^{-1/2} B_{at}$ (scale-invariance). If Q is an orthogonal $d \times d$ matrix, then the process QB_t is a d-dimensional standard Brownian motion (orthogonal invariance). If Brownian motion B_t starts at x at time τ, then $B_t + y$ is a Brownian motion starting at $x + y$ at time τ (translation-invariance).

A d-dimensional Brownian motion B_t is a $\mathbb{P}_{\tau,x}$-martingale with respect to the filtration $\mathcal{F}_t^\tau$. There are several other interesting martingales related to Brownian motion, for instance, if B_t is Brownian motion starting at x at time τ, then the process $|B_t|^2 - dt$, $t \ge \tau$, is a $\mathbb{P}_{\tau,x}$-martingale with respect to the filtration $\mathcal{F}_t^\tau$, $\tau \le t$. If $d = 1$, then for every $\sigma > 0$, the process $\exp\left\{\sigma B_t - \frac{1}{2}\sigma^2 t\right\}$ is a $\mathbb{P}_{\tau,x}$-martingale with respect to the filtration $\mathcal{F}_t^\tau$, $\tau \le t$. Let us also recall that the process $p_d(s, B_s; t, y)$ is a $\mathbb{P}_{\tau,x}$-martingale with respect to the filtration $\mathcal{F}_s^\tau$, $\tau \le s < t$.

Next, we will discuss the reciprocal process and the pinned measure associated with Brownian motion. Fix $T > 0$. Since the Gaussian density is strictly positive, we can define the derived transition probability density q_d using formula (1.124). An important link between the reciprocal process with transition density q_d and the original Brownian motion is provided by the following formula:

$$q_d(\tau, x; s, z; t, y)$$
$$= \frac{1}{\left(2\pi \frac{(s-\tau)(t-s)}{t-\tau}\right)^{d/2}} \exp\left\{-\frac{\left|z - \frac{(t-s)x + (s-\tau)y}{t-\tau}\right|^2}{2\frac{(s-\tau)(t-s)}{t-\tau}}\right\}. \quad (1.186)$$

This formula can be established as follows:

$$q_d(\tau, x; s, z; t, y)$$
$$= \frac{g_d(s-\tau, z-x) g_d(t-s, z-y)}{g_d(t-\tau, x-y)}$$
$$= \left[\frac{t-\tau}{2\pi(s-\tau)(t-s)}\right]^{d/2} \exp\left\{\frac{|x-y|^2}{2(t-\tau)} - \frac{|x-z|^2}{2(s-\tau)} - \frac{|z-y|^2}{2(t-s)}\right\}$$
$$= \frac{1}{\left(2\pi \frac{(s-\tau)(t-s)}{t-\tau}\right)^{d/2}} \exp\left\{-\frac{\left|z - \frac{(t-s)x + (s-\tau)y}{t-\tau}\right|^2}{2\frac{(s-\tau)(t-s)}{t-\tau}}\right\}$$

where $0 \leq \tau < s < t \leq T$ and $x, y, z \in \mathbb{R}^d$. The equality in (1.186) allows one to prove that the Chapman-Kolmogorov equation holds for the Gaussian density g_d without using the Fourier transform. Indeed, it is not hard to see from (1.186) that

$$\int_{\mathbb{R}^d} g_d(s-\tau, z-x) g_d(t-s, z-y) \, dz$$
$$= g_d(t-\tau, x-y) \int_{\mathbb{R}^d} q_d(\tau, x; s, z; t, y) \, dz$$
$$= g_d(t-\tau, x-y).$$

Let $(\tau, x) \in [0, T] \times \mathbb{R}^d$ and $(t, y) \in [0, T] \times \mathbb{R}^d$. Since the density g_d satisfies the conditions in Theorem 1.17, there exists a continuous reciprocal stochastic process $B^{(t,y)}_{(\tau,x),s}$, $\tau \leq s \leq t$, such that its reciprocal transition probability density is equal to the derived density q_d obtained from the density g_d as in formula (1.124). The process $B^{(t,y)}_{(\tau,x),s}$ is called the Brownian bridge between (τ, x) and (t, y), or Brownian motion pinned at x at time τ and at y at time t.

Lemma 1.26 *The finite-dimensional distributions of the process*

$$X_s = B^{(t,y)}_{(\tau,x),s}, \quad \tau \leq s \leq t,$$

with respect to the measure $\mathbb{P}_{(\tau,t),(x,y)}$ coincide with the finite-dimensional distributions of the following processes:
(a) The process

$$X_s^1 = \sqrt{t-\tau} B^{(1,0)}_{(0,0),\frac{s-\tau}{t-\tau}} + \left(1 - \frac{s-\tau}{t-\tau}\right) x + \frac{s-\tau}{t-\tau} y, \quad \tau < s < t, \quad (1.187)$$

with respect to the measure $\mathbb{P}_{(0,1),(0,0)}$.
(b) The process

$$X_s^2 = \frac{t-s}{t-\tau} B_{\frac{(s-\tau)(t-\tau)}{t-s}} + \left(1 - \frac{s-\tau}{t-\tau}\right) x + \frac{s-\tau}{t-\tau} y, \quad \tau < s < t, \quad (1.188)$$

with respect to the measure $\mathbb{P}_{0,0}$.
(c) The process

$$X_s^3 = \sqrt{t-\tau}\left(B_{\frac{s-\tau}{t-\tau}} - \frac{s-\tau}{t-\tau} B_1\right) + \left(1 - \frac{s-\tau}{t-\tau}\right) x + \frac{s-\tau}{t-\tau} y, \quad \tau < s < t,$$
$$(1.189)$$

with respect to the measure $\mathbb{P}_{0,0}$.

Proof. The finite-dimensional distributions of the process X_s, $\tau \leq s \leq t$, are given by

$$\mathbb{P}_{(\tau,t),(x,y)}\left[X_{s_i} \in A_i : 1 \leq i \leq n\right]$$
$$= \int_{A_1 \times \cdots \times A_n} q_d(\tau, x; s_1, z_1; t, y) \prod_{i=1}^{n-1} q_d(s_i, z_i; s_{i+1}, z_{i+1}; t, y)\, dz_1 \ldots dz_n$$
$$(1.190)$$

where $\tau < s_1 < s_2 < \cdots < s_n < t$ and $A_i \in \mathcal{B}_{\mathbb{R}^d}$ for $1 \leq i \leq n$. Put

$$C_i = \frac{t-\tau}{t-s_i}\left(A_i - \frac{t-s_i}{t-\tau} x - \frac{s_i-\tau}{t-\tau} y\right),$$

$$w_i = \frac{t-\tau}{t-s_i}\left(z_i - \frac{t-s_i}{t-\tau} x - \frac{s_i-\tau}{t-\tau} y\right) = \frac{t-\tau}{t-s_i} z_i - x - \frac{s_i-\tau}{t-s_i} y,$$

and

$$\delta_i = \frac{(s_i - \tau)(t - \tau)}{t - s_i}$$

for $1 \leq i \leq n$. Then the finite-dimensional distributions of the process X_s^2 are given by

$$\mathbb{P}_{0,0}\left[B_{\frac{(s_i-\tau)(t-\tau)}{t-s_i}} \in \frac{t-\tau}{t-s_i}\left(A_i - \frac{t-s_i}{t-\tau}x - \frac{s_i-\tau}{t-\tau}y\right) : 1 \leq i \leq n\right]$$

$$= \mathbb{P}_{0,0}\left[B_{\frac{(s_i-\tau)(t-\tau)}{t-s_i}} \in C_i : 1 \leq i \leq n\right]$$

$$= \int_{\prod_{i=1}^n C_i} g_d(\delta_1, w_1) \prod_{i=1}^{n-1} g_d(\delta_{i+1} - \delta_i, w_{i+1} - w_i)\, dw_1 \ldots dw_n$$

$$= \prod_{i=1}^n \left[\frac{t-\tau}{t-s_i}\right]^d \int_{\prod_{i=1}^n A_i} g_d(\delta_1, w_1) \prod_{i=1}^{n-1} g_d(\delta_{i+1} - \delta_i, w_{i+1} - w_i)\, dz_1 \ldots dz_n$$

$$= \prod_{i=1}^n \left[\frac{t-\tau}{t-s_i}\right]^d \int_{\prod_{i=1}^n A_i} dz_1 \ldots dz_n$$

$$g_d\left(\frac{(s_1-\tau)(t-\tau)}{t-s_1}, \frac{t-\tau}{t-s_1}z_1 - x - \frac{s_1-\tau}{t-s_1}y\right)$$

$$\prod_{i=1}^{n-1} g_d\left(\frac{(t-\tau)^2(s_{i+1}-s_i)}{(t-s_{i+1})(t-s_i)}, \frac{(t-\tau)z_{i+1}}{t-s_{i+1}} - \frac{(t-\tau)z_i}{t-s_i} - \frac{(t-\tau)(s_{i+1}-s_i)y}{(t-s_{i+1})(t-s_i)}\right) \quad (1.191)$$

where $\tau < s_1 < s_2 < \cdots < s_n < t$ and $A_i \in \mathcal{B}_{\mathbb{R}^d}$ for $1 \leq i \leq n$.

Put $\rho_i = \frac{t-\tau}{t-s_i}$ where $1 \leq i \leq n$. Then, it is not hard to see that formula (1.186) implies the equalities

$$q_d(\tau, x; s_1, z_1; t, y) = \rho_1^d g_d\left(\rho_1(s_1 - \tau), \rho_1 z_1 - x - \frac{s_1 - \tau}{t - s_1}y\right) \quad (1.192)$$

and

$$q_d(s_i, z_i; s_{i+1}, z_{i+1}; t, y)$$
$$= \rho_{i+1}^d g_d\left(\rho_{i+1}\rho_i(s_{i+1} - s_i), \rho_{i+1}z_{i+1} - \rho_i z_i - \rho_{i+1}\rho_i \frac{(s_{i+1} - s_i)}{t - \tau}y\right) \quad (1.193)$$

for $1 \leq i \leq n-1$. It follows from formulae (1.190)–(1.193) that the processes X_s and X_s^2 are stochastically equivalent.

By the previous results, the process $B_{(0,0),\lambda}^{1,0}$ is stochastically equivalent to the process $(1-\lambda) B_{\frac{\lambda}{1-\lambda}}$. Hence, the process $B_{(0,0),\frac{s-\tau}{t-\tau}}^{1,0}$ is stochastically equivalent to the process $\frac{t-s}{t-\tau} B_{\frac{s-\tau}{t-s}}$. Now the scaling properties of Brownian motion imply that the process $\sqrt{t-\tau} B_{(0,0),\frac{s-\tau}{t-\tau}}^{1,0}$ is stochastically equivalent to the process $\frac{t-s}{t-\tau} B_{\frac{(s-\tau)(t-\tau)}{t-s}}$. It follows that the process X_s^1 is stochastically equivalent to the process X_s^2.

Finally, we will prove the stochastic equivalence of the processes X_s^2 and X_s^3. In the proof, we will use the fact that the process $(1-\lambda) B_{\frac{\lambda}{1-\lambda}}$ is stochastically equivalent to the process $B_\lambda - \lambda B_1$ where $0 < \lambda < 1$. Indeed, it follows from the properties of Brownian motion that the processes $B_\lambda - \lambda B_1$, $\lambda B_{\frac{1}{\lambda}} - \lambda B_1$, $\lambda B_{\frac{1}{\lambda}-1}$, and $(1-\lambda) B_{\frac{\lambda}{1-\lambda}}$ are all stochastically equivalent. Moreover, the process $B_{\frac{s-\tau}{t-\tau}} - \frac{s-\tau}{t-\tau} B_1$ is stochastically equivalent to the process $\frac{t-s}{t-\tau} B_{\frac{s-\tau}{t-s}}$. This follows from the property of Brownian motion formulated above. Here we take $\lambda = \frac{s-\tau}{t-\tau}$. Multiplying by $\sqrt{t-\tau}$ and using the scaling property of Brownian motion, we see that the processes X_s^2 and X_s^3 are stochastically equivalent.

This completes the proof of Lemma 1.26. □

The stochastic equivalence of the processes X_s and X_s^2 in Lemma 1.26 provides an alternative proof of the existence of continuous versions of Brownian bridges.

1.12.2 *Cauchy process and Cauchy bridge*

The transition density p_d^c of a d-dimensional Cauchy process is defined by

$$p_d^c(\tau, x; t, y) = c_d(t-\tau, x-y), \quad 0 \leq \tau < t, \quad x \in \mathbb{R}^d, \quad y \in \mathbb{R}^d,$$

where

$$c_d(s,x) = \Gamma\left(\frac{d+1}{2}\right) \frac{s}{[\pi(s^2+|x|^2)]^{\frac{1}{2}(d+1)}}, \quad (s,x) \in [0,\infty) \times \mathbb{R}^d.$$

It is known that the following formula holds for the Fourier transform of c_d:

$$\widehat{c_d}(t,\xi) = e^{-t|\xi|}, \ t > 0, \xi \in \mathbb{R}^d. \tag{1.194}$$

It is not hard to see using (1.194) that the density p_d^c is normal and satisfies the Chapman-Kolmogorov equation.

A Markov process associated with the transition density p_d^c is called a Cauchy process. Since the Cauchy density satisfies the conditions in Corollary 1.1, there exists a realization X_t of a Cauchy process that is right-continuous and has left limits. Cauchy processes have the following scaling property. For every $a > 0$, the process $a^{-1}X_{at}$ is $\mathbb{P}_{0,0}$-stochastically equivalent to the process X_t.

The next lemma provides an example of a martingale related to a Cauchy process.

Lemma 1.27 *Let X_t be a Cauchy process. Then for every $\xi \in \mathbb{R}^d$, $x \in \mathbb{R}^d$, and $\tau \geq 0$, the process $t \mapsto \exp\{i\xi \cdot X_t + t|\xi|\}$ is a complex $\mathbb{P}_{\tau,x}$-martingale on the interval $[\tau, \infty)$.*

Proof. By the Markov property and formula (1.194), we see that for $0 \leq s < t$,

$$\mathbb{E}_{\tau,x}\left[\exp\{i\xi \cdot X_t + t|\xi|\} \mid \mathcal{F}_s\right] = e^{t|\xi|}\mathbb{E}_{s,X_s}\left[e^{i\xi \cdot X_t}\right]$$
$$= e^{t|\xi|}\int e^{i\xi \cdot y} c_d(t-s, X_s - y)\, dy = e^{t|\xi|}e^{i\xi \cdot X_s}e^{-(t-s)|\xi|}$$
$$= \exp\{i\xi \cdot X_s + s|\xi|\}.$$

This completes the proof of Lemma 1.27. □

Our next goal is to define a pinned Cauchy process. Since the density p_d^c is strictly positive, formula (1.124) can be used for a Cauchy process. According to this formula, the derived transition probability density q_d^c is given by

$$q_d^c(\tau, x; s, z; t, y) = \Gamma\left(\frac{d+1}{2}\right)\frac{(s-\tau)(t-s)}{t-\tau}$$
$$\left[\frac{(t-\tau)^2 + |x-y|^2}{\pi\left((s-\tau)^2 + |z-x|^2\right)\left((t-s)^2 + |y-z|^2\right)}\right]^{\frac{1}{2}(d+1)} \tag{1.195}$$

where $0 \leq \tau < s < t \leq T$ and $x, y, z \in \mathbb{R}^d$. Fix $(\tau, x) \in [0, T] \times \mathbb{R}^d$ and $(t, y) \in [0, T] \times \mathbb{R}^d$ with $\tau < t$. Since the Cauchy density c_d satisfies the conditions in Theorem 1.16, there exists a reciprocal process $X^{(t,y)}_{(\tau,x),s}$, $\tau \leq s \leq t$, associated with the transition density q_d^c that is right-continuous and has left limits. This process is called the Cauchy bridge between (τ, x) and (t, y), or Cauchy process pinned at x at time τ and at y at time t.

1.12.3 Forward Kolmogorov representation of Brownian bridges

Brownian motion and Cauchy process are homogeneous Markov processes. Next we give an example of a non-homogeneous transition density and non-homogeneous Markov process. Such examples can be obtained from the forward Kolmogorow representation of Brownian bridges. Similar examples can be constructed using the forward Kolmogorow representation of Cauchy bridges and other pinned processes.

Let $x_0 \in \mathbb{R}^d$ and $y_0 \in \mathbb{R}^d$. In Subsection 1.12.1, we defined the Brownian bridge $B^{(T,y_0)}_{(0,x_0),s}$, $s \in [0, T]$, between $(0, x_0)$ and (T, y_0). In this case, the reciprocal transition density is given by the formula

$$q_d(\tau, x; s, z; t, y)$$
$$= \frac{1}{\left(2\pi \frac{(s-\tau)(t-s)}{t-\tau}\right)^{d/2}} \exp\left\{-\frac{\left|z - \frac{(t-s)x + (s-\tau)y}{t-\tau}\right|^2}{2\frac{(s-\tau)(t-s)}{t-\tau}}\right\}, \quad (1.196)$$

where $0 \leq \tau < s < t \leq T$. The family of measures corresponding to this case is the family $\{\mathbb{P}_{(0,T),(x_0,y_0)}\}$. It follows from Example 1.1 and from the continuity criteria for reciprocal processes that the Brownian bridge $B^{(T,y_0)}_{(0,x_0),s}$ is a non-homogeneous continuous Markov process on any interval $[0, T']$, $0 < T' < T$, with respect to the measure $\mathbb{P}_{(0,T),(x_0,y_0)}$. This Markov process is associated with the non-homogeneous transition function P^1_{T,y_0} given by

$$P^1_{T,y_0}(\tau, x; t, A) = \int_A q_d(\tau, x; t, y; T, y_0)\, dy$$

where $0 \leq \tau < t < T$ and q_d is defined in (1.196).

1.13 Notes and Comments

(a) The following list is a sample of books devoted to Markov processes [Dynkin (1960); Dynkin (1965); Dynkin (1973); Dynkin (1982); Dynkin (2000); Dynkin and Yushkevich (1969); Bhattacharya and Waymire (1990); Blumenthal and Getoor (1968); Chung (1982); Chung and Zhao (1995); Doob (2001); Ethier and Kurtz (1986); Gihman and Skorohod (1974); Gihman and Skorohod (1975); Gihman and Skorohod (1979); Jacob (2001); Jacob (2002); Jacob (2005); Meyer (1966); Revuz and Yor (1991); Sharpe (1988); Stroock (2005)]. Our presentation of the path properties of Markov processes in Sections 1.8 and 1.9 is similar to that in [Gihman and Skorohod (1975)].

(b) Kolmogorov's papers [Kolmogorov (1931); Kolmogorov (1933)] are important early contributions to the theory of non-homogeneous Markov processes.

(c) Progressively measurable stochastic processes were first introduced and studied in [Chung and Doob (1965)]. Theorem 1.4 concerning the existence of progressively measurable modifications of measurable processes was established in [Chung and Doob (1965)] (see also [Meyer (1966); Doob (2001)]).

(d) Reciprocal processes were studied in [Jamison (1970); Jamison (1974); Jamison (1975)]. The concept of a reciprocal process and a reciprocal transition function goes back to Schrödinger (see [Schrödinger (1931)]) and Bernstein (see [Bernstein (1932)]). Reciprocal processes are used in stochastic quantum mechanics to model some aspects of the behavior of quantum mechanical systems. The readers who would like to learn more about reciprocal processes and their use in quantum mechanics may consult the following books: [Nelson (1967); Nelson (1988); Aebi (1996); Nagasawa (1993); Nagasawa (2000); Chung and Zambrini (2003)], and the following articles: [Cruzeiro and Zambrini (1994); Cruzeiro, Wu, and Zambrini (2000); Privault and Zambrini (2004); Privault and Zambrini (2005); Roelly and Thieullen (2002); Roelly and Thieullen (2005); Thieullen (1993); Thieullen (1998); Thieullen and Zambrini (1997); Thieullen (2002); Truman and Davies (1988); van Casteren (2000)]. An important paper on time reversal of Markov processes is [Chung and Walsh (1969)]. For more information on time symmetries of Markov processes see [Chung and Walsh (2005)].

(e) Brownian motion is probably the most popular example of a Markov process. For more information on Brownian motion see [Chung (1982);

Chung and Zhao (1995); Chung and Walsh (2005); Durrett (1984); Johnson and Lapidus (2000); Kahane (1997); Kahane J.-P. (1998); Karatzas and Shreve (1991); Revuz and Yor (1991)].

(f) Cauchy processes belong to the class of Levy processes. This means that they are right-continuous, have left limits, and their increments are independent and stationary. More precisely, a Cauchy process is a symmetric stable process of index 1 (see [Bertoin (1996); Sato (2000); Barndorff et al (2001); Schoutens (2003); Applebaum (2004)] for more information on Levy processes).

Chapter 2

Propagators: General Theory

2.1 Propagators and Backward Propagators on Banach Spaces

Propagators are two-parameter families of bounded linear operators on a Banach space satisfying the flow condition (forward propagators) or the backward flow condition (backward propagators). Let B be a Banach space, and denote by $L(B, B)$ the space of all bounded linear operators on B. The symbol I will stand for the identity operator on B.

Definition 2.1 A two-parameter family
$$\{W(t, \tau) \in L(B, B) : 0 \leq \tau \leq t \leq T\}$$
is called a propagator on B provided that the following conditions hold:

(1) $W(t, \tau) = W(t, \lambda)W(\lambda, \tau)$ for $0 \leq \tau \leq \lambda \leq t \leq T$.
(2) $W(\tau, \tau) = I$ for $0 \leq \tau \leq T$.

Conditions (1) and (2) in Definition 2.1 are called the flow conditions.

Definition 2.2 A two-parameter family of operators
$$\{Q(\tau, t) \in L(B, B) : 0 \leq \tau \leq t \leq T\}$$
is called a backward propagator on B provided that the following conditions hold:

(1) $Q(\tau, t) = Q(\tau, \lambda)Q(\lambda, t)$ for $0 \leq \tau \leq \lambda \leq t \leq T$.
(2) $Q(t, t) = I$ for $0 \leq t \leq T$.

Conditions (1) and (2) in Definition 2.2 are called the backward flow conditions.

There are simple relations between propagators and backward propagators. If $T > 0$ is given, and Q is a backward propagator on a Banach space B, then the family of operators defined by

$$W(t,\tau) = Q(T-t, T-\tau), \quad 0 \leq \tau \leq t \leq T, \tag{2.1}$$

is a propagator. Moreover, if Q is a backward propagator on a Banach space B, and a family of operators is defined by $W(t,\tau) = Q^*(\tau,t)$ where $Q^*(\tau,t)$ is the adjoint of $Q(\tau,t)$, then W is a propagator on the space B^*. Here B^* stands for the dual space of B. Similarly, if W is a propagator on B, and $Q(\tau,t) = W^*(t,\tau)$, then Q is a backward propagator on B^*.

A propagator W is called strongly continuous if for every $x \in B$, the B-valued function $(t,\tau) \to W(t,\tau)x$, $0 \leq \tau \leq t \leq T$, is continuous. A propagator W is called uniformly bounded if

$$\|W(t,\tau)\|_{B \to B} \leq M \quad \text{for all } 0 \leq \tau \leq t \leq T.$$

If τ and t are such that $0 \leq \tau \leq t < \infty$, and for every compact subset K of the set $\{(\tau,t) : 0 \leq \tau \leq t < \infty\}$, the estimate

$$\|W(t,\tau)\|_{B \to B} \leq M_K$$

holds for all $(t,\tau) \in K$, then W is called a locally uniformly bounded propagator. A propagator W is called separately strongly continuous if for every fixed t and $x \in B$, the function $\tau \to W(t,\tau)x$ is continuous on $[0,t]$, and for every fixed τ and $x \in B$, the function $t \to W(t,\tau)x$ is continuous on $[\tau,T]$ (if $T = \infty$, then we consider the interval $[t,\infty)$ instead of the interval $[t,T]$). Similar definitions apply in the case of backward propagators.

The next theorem states that the joint continuity and the separate continuity are equivalent if forward or backward propagators are locally uniformly bounded.

Theorem 2.1 *Let W be a propagator on a Banach space B. Then the following are equivalent for W:*

(i) The strong continuity.
(ii) The strong separate continuity and the uniform local boundedness.

The same assertion holds for a backward propagator Q on B.

Proof. We will prove Theorem 2.1 for a backward propagator Q. The case of propagators is similar.

By the uniform boundedness principle, condition (i) implies condition (ii). Next, let Q be a strongly separately continuous and locally uniformly bounded backward propagator. Let $0 \leq \tau \leq t \leq T$, and suppose t' and τ' are close to t and τ, respectively. We will first assume that $t > \tau$. Then for τ' close to τ, we have $t > \tau'$. Using the local uniform boundedness condition and assuming that $t' \geq t$, we see that for every $x \in B$,

$$\begin{aligned} I &= \|Q(\tau',t')x - Q(\tau,t)x\|_B \\ &\leq \|Q(\tau',t')x - Q(\tau',t)x\|_B + \|Q(\tau',t)x - Q(\tau,t)x\|_B \\ &\leq \|Q(\tau',t)(Q(t,t')x - x)\|_B + \|Q(\tau',t)x - Q(\tau,t)x\|_B \\ &\leq M \|Q(t,t')x - x\|_B + \|Q(\tau',t)x - Q(\tau,t)x\|_B . \end{aligned}$$

It follows from the separate continuity condition that

$$\lim_{t' \to t, \tau' \to \tau} I = 0. \tag{2.2}$$

If $t' < t$, then

$$\begin{aligned} I &\leq \|Q(\tau',t')x - Q(\tau',t)x\|_B + \|Q(\tau',t)x - Q(\tau,t)x\|_B \\ &\leq \|Q(\tau',t')(x - Q(t',t)x)\|_B + \|Q(\tau',t)x - Q(\tau,t)x\|_B \\ &\leq M \|Q(t',t)x - x\|_B + \|Q(\tau',t)x - Q(\tau,t)x\|_B , \end{aligned}$$

and we see that formula (2.2) also holds for $t' < t$.

Finally, let $\tau = t < \tau' \leq t'$. Then the separate continuity condition implies that for every $\epsilon > 0$ there exists $\lambda > 0$ such that $\lambda > \tau$ and

$$\|Q(\tau,\lambda)x - x\|_B \leq \epsilon. \tag{2.3}$$

If follows from the local uniform boundedness condition and from (2.3) that

$$\begin{aligned} I &= \|Q(\tau',t')x - x\|_B \\ &\leq \|Q(\tau',t')x - Q(\tau',\lambda)x\|_B + \|Q(\tau',\lambda)x - Q(\tau,\lambda)x\|_B + \|Q(\tau,\lambda)x - x\|_B \\ &\leq \|Q(\tau',t')(x - Q(t',\lambda)x)\|_B + \|Q(\tau',\lambda)x - Q(\tau,\lambda)x\|_B + \epsilon \\ &\leq M \|Q(t',\lambda)x - x\|_B + \|Q(\tau',\lambda)x - Q(\tau,\lambda)x\|_B + \epsilon \\ &\leq M \|Q(t',\lambda)x - Q(\tau,\lambda)x\|_B + M \|Q(\tau,\lambda)x - x\|_B \\ &\quad + \|Q(\tau',\lambda)x - Q(\tau,\lambda)x\|_B + \epsilon \\ &\leq M \|Q(t',\lambda)x - Q(\tau,\lambda)x\|_B + \|Q(\tau',\lambda)x - Q(\tau,\lambda)x\|_B \\ &\quad + (M+1)\epsilon. \end{aligned} \tag{2.4}$$

In (2.4), M depends on t. Next we get from (2.4) and from the separate continuity condition that there exists $\delta > 0$ such that for $\tau \leq \tau' \leq t' < \tau + \delta$, we have $I \leq (2M + 2)\epsilon$. Therefore, (2.2) holds for $\tau = t < \tau' \leq t'$.

This completes the proof of Theorem 2.1. □

A simple example of a propagator is as follows. Let S_t be a semigroup of linear operators on a Banach space B, and consider the following two-parameter family of operators: $S_{t-\tau}$, $0 \le \tau \le t$. Put

$$U(t,\tau) = Y(\tau,t) = S_{t-\tau}.$$

Then U is a propagator on B, and Y is a backward propagator on B.

2.2 Free Propagators and Free Backward Propagators

Recall that we denoted by E a locally compact Hausdorff topological space satisfying the second axiom of countability. Let $\rho : E \times E \to [0,\infty)$ be a metric generating the topology of E, and let $\mathcal{E}$ be the σ-algebra of Borel subsets of E. The symbol m will stand for the reference measure. The measure m is a nonnegative Borel measure on $(E,\mathcal{E})$, and we always assume that $0 < m(A) < \infty$ for any compact subset A of E with nonempty interior.

We will next define various function spaces on the space E. The symbol BC will stand for the space of all bounded continuous functions on E equipped with the norm

$$\|f\|_\infty = \sup_{x \in E} |f(x)|, \quad f \in BC. \tag{2.5}$$

We denote by C_0 the subspace of the space BC consisting of all bounded continuous functions on E vanishing at infinity. More precisely, a function f belongs to the space C_0 if for every $\epsilon > 0$ there exists a compact set $K_{\epsilon,f}$ in E such that $|f(x)| \le \epsilon$ for all $x \in E \backslash K_{\epsilon,f}$. It is not hard to prove that C_0 is a closed subspace of the space BC. The symbol BUC will stand for the space of all bounded uniformly continuous functions on E. More precisely, a function f belongs to the space BUC provided that for every $\epsilon > 0$ there exists $\delta > 0$ such that for all $x \in E$ and $y \in E$ with $\rho(x,y) < \delta$, the inequality $|f(x) - f(y)| < \epsilon$ holds. The space BUC is a closed subspace of BC. For $1 \le r < \infty$, we denote by $L^r_\mathcal{E}$ the space of all Borel functions on E such that

$$\|f\|_r = \left\{ \int_E |f(x)|^r \, dx \right\}^{1/r} < \infty. \tag{2.6}$$

The symbol $L^\infty_\mathcal{E}$ will stand for the space of all bounded Borel functions on

E equipped with the norm

$$\|f\|_\infty = \sup_{x \in E} |f(x)|.$$

We denote by L^r, $1 \leq r < \infty$, the Lebesgue space on E with respect to the measure m. The norm on the space L^r is defined by (2.6). We denote by $\bar{\mathcal{E}}^m$ the σ-algebra obtained by completing the Borel σ-algebra $\mathcal{E}$ with respect to the reference measure m. As usual, it is assumed that the elements of Lebesgue spaces are classes of equivalence mod 0 of $\bar{\mathcal{E}}^m$-measurable functions on E with respect to the measure m. For $r = \infty$, the norm of a function $f \in L^\infty$ is given by

$$\|f\|_\infty = \operatorname{ess\,sup}_{x \in E} |f(x)|$$

where

$$\operatorname{ess\,sup}_{x \in E} |f(x)| = \inf \{a : |f(x)| \leq a \quad \text{for } m\text{-almost all } x \in E\}.$$

In the remaining part of the present section, we discuss backward propagators generated by transition functions. Let $P(\tau, x; t, A)$ be a transition subprobability function, and define a family of operators on $L_{\mathcal{E}}^\infty$ by

$$Y(\tau, t)f(x) = \begin{cases} \int_E f(y) P(\tau, x; t, dy), & \text{if } 0 \leq \tau < t \leq T \\ f(x), & \text{if } \tau = t \text{ and } 0 \leq \tau \leq T \end{cases} \quad (2.7)$$

for all $x \in E$ and $f \in L_{\mathcal{E}}^\infty$. It follows from the subnormality condition that Y is a family of contraction operators on $L_{\mathcal{E}}^\infty$. Moreover, the Chapman-Kolmogorov equation shows that Y is a backward propagator on $L_{\mathcal{E}}^\infty$. The family Y defined by (2.7) will be called the free backward propagator associated with the transition function P.

Suppose that the transition subprobability function P possesses a density p; that is, there exists a nonnegative function $p(r, x; s, y)$ such that

$$P(r, x; s, A) = \int_A p(r, x; s, y) dy$$

for all $A \in \mathcal{E}$. In this case, the free backward propagator Y is defined on the space L^∞ by

$$Y(\tau, t)f(x) = \begin{cases} \int_E f(y) p(\tau, x; t, y) dy, & \text{if } 0 \leq \tau < t \leq T \\ f(x), & \text{if } \tau = t \text{ and } 0 \leq \tau \leq T \end{cases} \quad (2.8)$$

for all $x \in E$ and $f \in L^\infty$. The operator $Y(\tau, t)$ maps the space L^∞ into the space $L_{\mathcal{E}}^\infty$.

If $\widetilde{P}$ is a backward transition subprobability function, then we define the free propagator associated with $\widetilde{P}$ by

$$U(t,\tau)f(y) = \begin{cases} \int_E f(x)\widetilde{P}(\tau, dx; t, y), & \text{if } 0 \leq \tau < t \leq T \\ f(y), & \text{if } \tau = t \text{ and } 0 \leq \tau \leq T \end{cases} \quad (2.9)$$

for all $x \in E$ and $f \in L_{\mathcal{E}}^\infty$. If $\widetilde{p}$ is a backward subprobability transition density, then the free propagator has the following form:

$$U(t,\tau)f(y) = \begin{cases} \int_E f(x)\widetilde{p}(\tau, x; t, y) dx, & \text{if } 0 \leq \tau < t \leq T \\ f(y), & \text{if } \tau = t \text{ and } 0 \leq \tau \leq T \end{cases} \quad (2.10)$$

for all $x \in E$ and $f \in L^\infty$. Let us recall that in the case of a finite time-interval $[0, T]$, there exists a one-to-one correspondence

$$P(\tau, x; t, A) = \widetilde{P}(T-t, A; T-\tau, x), \quad 0 \leq \tau < t \leq T, y \in E, A \in \mathcal{E},$$

between forward and backward transition functions. Therefore, if $\widetilde{P}$ is a given backward transition subprobability function, then the free propagator associated with $\widetilde{P}$ can be defined by

$$U(t,\tau) = Y(T-t, T-\tau), \quad (2.11)$$

where Y is the backward free propagator associated with P.

2.3 Generators of Propagators and Kolmogorov's Forward and Backward Equations

In this section we continue the study of propagators on Banach spaces. Our goal is to show that propagators are related to non-homogeneous evolution equations. We will first discuss general propagators and then consider free propagators generated by transition functions. The non-homogeneous equations which are studied in the present section are called Kolmogorov's forward and backward equations.

Let B be a Banach space and let φ be a B-valued function on the interval $(0, T)$. If the function φ is differentiable from the right, then the symbol $\dfrac{\partial^+ \varphi}{\partial \tau}(\tau)$ will stand for its derivative from the right, that is,

$$\frac{\partial^+ \varphi}{\partial \tau}(\tau) = \lim_{h \downarrow 0} \frac{\varphi(\tau+h) - \varphi(\tau)}{h}.$$

Similarly, if the function φ is differentiable from the left, then the symbol $\dfrac{\partial^- \varphi}{\partial \tau}(\tau)$ will stand for its derivative from the left,

$$\frac{\partial^- \varphi}{\partial \tau}(\tau) = \lim_{h \downarrow 0} \frac{\varphi(\tau - h) - \varphi(\tau)}{-h}.$$

Suppose that $Q = Q(\tau, t)$, $0 \leq \tau \leq t \leq T$ is a backward propagator on a Banach space B. For every τ with $0 < \tau \leq T$, consider a linear operator on B defined as follows:

$$A_-(\tau)x = \lim_{h \downarrow 0} \frac{Q(\tau - h, \tau)x - x}{h}. \tag{2.12}$$

The domain of this operator is the set $D(A_-(\tau))$ of all $x \in B$ for which the limit in (2.12) exists. The operators $A_-(\tau)$, $0 < \tau \leq T$, are called the left generators of the backward propagator Q. For every $t \in (0, T]$, denote by $D_-(t)$ the set of all $x \in B$ such that the function $\tau \mapsto Q(\tau, t)$ is differentiable from the left on $(0, t)$, and by $F(t)$ the set of all $x \in B$ for which

$$\lim_{h \downarrow 0} Q(t - h, t)x = x. \tag{2.13}$$

The next result explains in what sense the left generators are related to the backward propagator Q.

Theorem 2.2 *Let Q be a backward propagator on B, and let $t \in (0, T]$. Then for every $x \in D_-(t) \cap F(t)$, the function $u(\tau) = Q(\tau, t)x$ is a solution to the following final value problem on $(0, t)$:*

$$\begin{cases} \dfrac{\partial^- u}{\partial \tau}(\tau) = -A_-(\tau) u(\tau), \\ \lim\limits_{\tau \uparrow t} u(\tau) = x. \end{cases} \tag{2.14}$$

Proof. Let $x \in D_-(t) \cap F(t)$. Then, using the properties of backward propagators, we obtain

$$\begin{aligned}
\frac{\partial^- u}{\partial \tau}(\tau) &= \lim_{h \downarrow 0} \frac{Q(\tau - h, t)x - Q(\tau, t)x}{-h} \\
&= \lim_{h \downarrow 0} \frac{Q(\tau - h, \tau)Q(\tau, t)x - Q(\tau, t)x}{-h} \\
&= \lim_{h \downarrow 0} \frac{(Q(\tau - h, \tau) - I) Q(\tau, t)x}{-h} \\
&= -A_-(\tau) Q(\tau, t)x = -A_-(\tau) u(\tau).
\end{aligned}$$

In addition, the equality $\lim_{\tau\uparrow t} u(\tau) = x$ follows from the definition of the set $F(t)$.

This completes the proof of Theorem 2.2. $\square$

Our next goal is to introduce the family of right generators of the backward propagator Q. For every τ with $0 \le \tau < T$, consider a linear operator on the space B given by

$$A_+(\tau)x = \lim_{h\downarrow 0} \frac{Q(\tau,\tau+h)x - x}{h}. \qquad (2.15)$$

The domain $D(A_+(\tau))$ of this operator is the set of points $x \in B$ for which the limit in (2.15) exists. The operators $A_+(\tau)$, $0 < \tau \le T$, are called the right generators of the backward propagator Q. For every $t \in (0, T]$, denote by $D_+(t)$ the set of all $x \in B$ such that the function $\tau \mapsto Q(\tau, t)$ is differentiable from the right on $(0, t)$.

Theorem 2.3 *Let $Q(\tau, t)$, $0 \le \tau \le t \le T$, be a strongly continuous backward propagator on B, and fix t with $0 < t \le T$. Then for every $x \in D_+(t)$, the function $u(\tau) = Q(\tau, t)x$ is a solution to the following final value problem on $(0, t)$:*

$$\begin{cases} \dfrac{\partial^+ u}{\partial \tau}(\tau) = -A_+(\tau)u(\tau), \\ \lim_{\tau\uparrow t} u(\tau) = x. \end{cases} \qquad (2.16)$$

Proof. Let $x \in D_+(t)$. Then, using the strong continuity of the backward propagator Q, the Banach-Steinhaus theorem, and the definition of the set $D_+(t)$, we obtain

$$\begin{aligned}
\frac{\partial^+ u}{\partial \tau} &= \lim_{h\downarrow 0} Q(\tau, \tau+h) \frac{\partial^+ u}{\partial \tau} \\
&= \lim_{h\downarrow 0} Q(\tau, \tau+h) \left[\frac{\partial^+ u}{\partial \tau} - \frac{Q(\tau+h, t)x - Q(\tau, t)x}{h} \right] \\
&\quad + \lim_{h\downarrow 0} Q(\tau, \tau+h) \frac{Q(\tau+h, t)x - Q(\tau, t)x}{h} \\
&= \lim_{h\downarrow 0} Q(\tau, \tau+h) \frac{Q(\tau+h, t)x - Q(\tau, t)x}{h} \\
&= \lim_{h\downarrow 0} \frac{Q(\tau, t)x - Q(\tau, \tau+h)Q(\tau, t)x}{h} \\
&= -\lim_{h\downarrow 0} \frac{Q(\tau, \tau+h) - I}{h} Q(\tau, t)x. \qquad (2.17)
\end{aligned}$$

It follows from (2.15) and (2.17) that $Q(\tau,t)x \in D(A_+(\tau))$ and the equation in (2.16) is satisfied. In addition, the equality $\lim_{\tau \uparrow t} u(\tau) = x$ follows from the strong continuity of Q.

This completes the proof of Theorem 2.3. $\square$

Now we turn our attention to propagators on B. The generators in this case are defined exactly as in the case of backward propagators. Suppose that W is a propagator on a Banach space B. For every t with $0 \leq t < T$, consider a linear operator on the space B given by

$$\widetilde{A}_+(t)x = \lim_{h \downarrow 0} \frac{W(t+h,t)x - x}{h}. \tag{2.18}$$

The domain $D\left(\widetilde{A}_+(t)\right)$ of this operator is the set of points $x \in B$ for which the limit in (2.18) exists. The operators $\widetilde{A}_+(t)$, $0 \leq t < T$, are called the right generators of the propagator W. For every $\tau \in [0,T)$, denote by $\widetilde{D}_+(\tau)$ the set of all $x \in B$ such that the function $t \mapsto W(t,\tau)x$ is differentiable from the right on (τ, T), and by $\widetilde{F}(\tau)$ the set of all $x \in B$ for which

$$\lim_{h \downarrow 0} W(\tau + h, \tau)x = x. \tag{2.19}$$

Theorem 2.4 *Let W be a propagator on B, and fix τ with $0 \leq \tau < T$. Then for every $x \in \widetilde{D}_+(\tau) \cap \widetilde{F}(\tau)$, the function $\widetilde{u}(t) = W(t,\tau)x$ is a solution to the following initial value problem on (τ, T):*

$$\begin{cases} \dfrac{\partial^+ \widetilde{u}}{\partial t}(t) = \widetilde{A}_+(t)\widetilde{u}(t), \\ \lim_{t \downarrow \tau} \widetilde{u}(t) = x. \end{cases} \tag{2.20}$$

Proof. Let $x \in \widetilde{D}_+(\tau) \cap \widetilde{F}(\tau)$. Then, using the properties of propagators and the definition of the set $\widetilde{D}_+(\tau)$, we obtain

$$\begin{aligned}
\frac{\partial^+}{\partial t}\widetilde{u}(t) &= \lim_{h \downarrow 0} \frac{W(t+h,t)x - W(t,\tau)x}{h} \\
&= \lim_{h \downarrow 0} \frac{W(t+h,t)W(t,\tau)x - W(t,\tau)x}{h} \\
&= \lim_{h \downarrow 0} \frac{(W(t+h,t) - I)W(t,\tau)x}{h} \\
&= \widetilde{A}_+(t)W(t,\tau)x = \widetilde{A}_+(t)\widetilde{u}(t).
\end{aligned}$$

In addition, the equality $\lim_{t\downarrow\tau}\widetilde{u}(t) = x$ follows from the definition of the set $\widetilde{F}(\tau)$.

This completes the proof of Theorem 2.4. $\square$

Let W be a propagator on B. For every t with $0 < t \leq T$, consider a linear operator on the space B given by

$$\widetilde{A}_-(t)x = \lim_{h\downarrow 0} \frac{W(t, t-h)x - x}{h}. \qquad (2.21)$$

The domain $D\left(\widetilde{A}_-(t)\right)$ of this operator is the set of points $x \in B$ for which the limit in (2.21) exists. The operators $\widetilde{A}_-(\tau)$, $0 < t \leq T$, are called the left generators of the propagator W. For every $\tau \in [0, T)$, denote by $\widetilde{D}_-(\tau)$ the set of all $x \in B$ such that the function $t \mapsto W(t, \tau)$ is differentiable from the left on (τ, T).

Theorem 2.5 *Let W be a strongly continuous propagator on B, and fix τ with $0 < \tau < T$. Then for every $x \in \widetilde{D}_-(\tau)$, the function $\widetilde{u}(t) = W(t, \tau)x$ is a solution to the following initial value problem on (τ, T):*

$$\begin{cases} \dfrac{\partial^- \widetilde{u}}{\partial t}(t) = \widetilde{A}_-(t)\widetilde{u}(t), \\ \lim_{t\downarrow\tau} u(t) = x. \end{cases} \qquad (2.22)$$

Proof. Let $x \in \widetilde{D}_-(t)$. Then, using the strong continuity of the propagator W, the Banach-Steinhaus theorem, and the definition of the set $\widetilde{D}_-(t)$, we obtain

$$\begin{aligned}
\frac{\partial^- \widetilde{u}}{\partial t}(t) &= \lim_{h\downarrow 0} W(t, t-h) \frac{\partial^- \widetilde{u}}{\partial t} \\
&= \lim_{h\downarrow 0} W(t, t-h) \left[\frac{\partial^- \widetilde{u}}{\partial t} - \frac{W(t-h, \tau)x - W(t, \tau)x}{-h}\right] \\
&\quad + \lim_{h\downarrow 0} W(t, t-h) \frac{W(t-h, \tau)x - W(t, \tau)x}{-h} \\
&= \lim_{h\downarrow 0} W(t, t-h) \frac{W(t-h, \tau)x - W(t, \tau)x}{-h} \\
&= \lim_{h\downarrow 0} \frac{W(t, \tau)x - W(t, t-h)W(t, \tau)x}{-h} \\
&= \lim_{h\downarrow 0} \frac{W(t, t-h) - I}{h} W(t, \tau)x. \qquad (2.23)
\end{aligned}$$

It is not hard to see that (2.21) and (2.23) imply $W(t,\tau)x \in \widetilde{D}\left(\widetilde{A}_-(t)\right)$. Moreover, the equation in (2.22) is satisfied. In addition, the equality $\lim_{t\downarrow\tau}\widetilde{u}(t) = x$ follows from the strong continuity of W.

This completes the proof of Theorem 2.5. □

Our next goal is to discuss what happens if we differentiate Q and W with respect to "wrong" time variables. Let Q be a backward propagator on a Banach space B. For every τ with $0 \le \tau < T$, put

$$D_+^*(\tau) = \bigcap_{t:\tau<t<T} D(A_+(t)),$$

where $D(A_+(t))$ is the domain of the operator $A_+(t)$ defined by (2.15).

Theorem 2.6 *Let Q be a backward propagator on B, and fix τ with $0 < \tau < T$. Then for every $x \in D_+^*(\tau)$ and t with $\tau < t < T$,*

$$\frac{\partial^+ Q(\tau,t)x}{\partial t} = Q(\tau,t)A_+(t)x.$$

Proof. Let $x \in D_+^*(\tau)$. Then, using the properties of backward propagators and the definition of $D_+^*(\tau)$, we obtain

$$\begin{aligned}\frac{\partial^+ Q(\tau,t)x}{\partial t} &= \lim_{h\downarrow 0}\frac{Q(\tau,t+h)x - Q(\tau,t)x}{h}\\ &= \lim_{h\downarrow 0}\frac{Q(\tau,t)Q(t,t+h)x - Q(\tau,t)x}{h}\\ &= Q(\tau,t)\lim_{h\downarrow 0}\frac{Q(t,t+h)x - x}{h}\\ &= Q(\tau,t)A_+(t)x.\end{aligned}$$

This completes the proof of Theorem 2.6. □

For a propagator W on a Banach space B and $t \in (0,T]$, put

$$\widetilde{D}_-^*(t) = \bigcap_{\tau:0<\tau<t} D\left(\widetilde{A}_-(\tau)\right),$$

where $D\left(\widetilde{A}_-(\tau)\right)$ is the domain of the operator $\widetilde{A}_-(\tau)$ defined by (2.21).

Theorem 2.7 *Let W be a propagator on B, and fix t with $0 < t \le T$. Then for every $x \in \widetilde{D}_-^*(t)$ and τ with $0 < \tau < t$,*

$$\frac{\partial^- W(t,\tau)x}{\partial \tau} = -W(t,\tau)\widetilde{A}_-(\tau)x.$$

Proof. Let $x \in \widetilde{D}_-^*(t)$. Then, using the properties of propagators and the definition of $\widetilde{D}_-^*(t)$, we obtain

$$\begin{aligned}\frac{\partial^- W(t,\tau)x}{\partial \tau} &= -\lim_{h\downarrow 0}\frac{W(t,\tau-h)x - W(t,\tau)x}{h}\\ &= -\lim_{h\downarrow 0}\frac{W(t,\tau)W(\tau,\tau-h)x - W(t,\tau)x}{h}\\ &= -W(t,\tau)\lim_{h\downarrow 0}\frac{W(\tau,\tau-h)x - x}{h}\\ &= -W(t,\tau)\widetilde{A}_-(\tau)x.\end{aligned}$$

This completes the proof of Theorem 2.7. $\square$

Next we will discuss Kolmogorov's forward and backward equations for transition probability functions. Given such a function P, let us consider the Banach space $B = L_\mathcal{E}^\infty$ and the free backward propagator Y defined by formula (2.7). Instead of the strong generators $A_-(\tau)$ (see formula (2.12)) and $A_+(\tau)$ (see formula (2.15)), we will consider generators in a weaker sense. The new generators have larger domains than those of the generators $A_-(\tau)$ and $A_+(\tau)$. We will use the topology $\sigma(L_\mathcal{E}^\infty, M)$ on the space $L_\mathcal{E}^\infty$, where the symbol M stands for the space of all finite signed Borel measures on E. For every $\tau \in [0,T)$, denote by $D^w\left(A_-^M(\tau)\right)$ the set of all $f \in L_\mathcal{E}^\infty$ for which the $\sigma(L_\mathcal{E}^\infty, M)$-limit

$$A_-^M(\tau)f = \sigma(L_\mathcal{E}^\infty, M)\text{-}\lim_{h\downarrow 0}\frac{Y(\tau-h,\tau)f - f}{h}$$

exists. This means that there is a function $A_-^M(\tau)f \in L_\mathcal{E}^\infty$ such that

$$\int_E A_-^M(\tau)f\,d\nu = \lim_{h\downarrow 0}\int_E \frac{Y(\tau-h,\tau)f - f}{h}d\nu \qquad (2.24)$$

for all $\nu \in M$. The operators $A_-^M(\tau)$, $0 < \tau < T$, are called the left $\sigma(L_\mathcal{E}^\infty, M)$-generators of the backward propagator Y. The next simple lemma provides an equivalent condition for the convergence of a sequence of functions in the topology $\sigma(L_\mathcal{E}^\infty, M)$.

Lemma 2.1 *A sequence of functions $h_k \in L_\mathcal{E}^\infty$ converges to a function $h \in L_\mathcal{E}^\infty$ in the topology $\sigma(L_\mathcal{E}^\infty, M)$ if and only if $\lim_{k\to\infty} h_k(x)$ exists for all $x \in E$, and $\sup_{k,x}|h_k(x)| < \infty$.*

Proof. Suppose that $h_k \to h$ in the topology $\sigma(L_{\mathcal{E}}^{\infty}, M)$. This means that

$$\lim_{k \to \infty} \int_E h_k d\nu = \int_E h d\nu \qquad (2.25)$$

for all $\nu \in M$. Put $\nu = \delta_x$. Then the sequence h_k converges to h pointwise on E. Moreover, the uniform boundedness of the sequence $\{h_k\}$ follows from the Banach-Steinhaus theorem.

Now assume that the sequence h_k converges pointwise on E to h, and moreover, it is uniformly bounded. Then $h \in L_{\mathcal{E}}^{\infty}$, and by the dominated convergence theorem, we see that $h_k \to h$ in the topology $\sigma(L_{\mathcal{E}}^{\infty}, M)$. $\square$

For every τ with $0 \leq \tau < T$ and every $f \in D^w\left(A_{-}^{M}(\tau)\right)$, we have

$$\lim_{h \downarrow 0} \frac{Y(\tau - h, \tau) f(x) - f(x)}{h} = A_{-}^{M}(\tau) f(x) \qquad (2.26)$$

and

$$\sup_{(h,x) \in (0,\tau) \times E} \frac{|Y(\tau - h, \tau) f(x) - f(x)|}{h} < \infty. \qquad (2.27)$$

This follows from Lemma 2.1. By (2.26), (2.27), and the dominated convergence theorem, for all $f \in D^w\left(A_{-}^{M}(\tau)\right)$, the pointwise derivative from the left $\dfrac{\partial^{-} Y(\tau, t) f}{\partial \tau}(x)$ of the function $\tau \mapsto Y(\tau, t) f(x)$ coincides with its $\sigma(L_{\mathcal{E}}^{\infty}, M)$-derivative from the left. This means that for all $\nu \in M$,

$$\lim_{h \downarrow 0} \int_E \frac{Y(\tau - h, t) f(x) - Y(\tau, t) f(x)}{h} d\nu(x)$$
$$= \int_E \lim_{h \downarrow 0} \frac{Y(\tau - h, t) f(x) - Y(\tau, t) f(x)}{h} d\nu(x). \qquad (2.28)$$

Let $t \in (0, T]$, and denote by $F_{-}^{M}(t)$ the set of all functions $f \in L_{\mathcal{E}}^{\infty}$ such that

$$\lim_{h \downarrow 0} Y(t - h, t) f(x) = f(x)$$

for all $x \in E$. The uniform boundedness of the family of functions $Y(t - h, t) f$ in the space $L_{\mathcal{E}}^{\infty}$ follows from the definition of the free backward propagator Y. By Lemma 2.1, $Y(t - h, t) f \to f$ in the topology $\sigma(L_{\mathcal{E}}^{\infty}, M)$ as $h \downarrow 0$. For any t with $0 < t \leq T$, denote by $D_{-}^{M}(t)$ the set of all functions $f \in L_{\mathcal{E}}^{\infty}$ such that the function $\tau \mapsto Y(\tau, t) f$ is $\sigma(L_{\mathcal{E}}^{\infty}, M)$-differentiable from the left on $(0, t)$.

Theorem 2.8 Let $0 < t \leq T$ and $f \in D^M_-(t) \cap F^M_-(t)$. Then the function $u(\tau, x) = Y(\tau, t) f(x)$ is a solution to the following final value problem on $(0, t)$:

$$\begin{cases} \dfrac{\partial^- u}{\partial \tau}(\tau, x) = -A^M_-(\tau) u(\tau)(x) \\ \lim_{\tau \uparrow t} u(\tau, x) = f(x) \end{cases} \quad (2.29)$$

for all $x \in E$.

Remark 2.1 In final value problem (2.29), the derivative $\dfrac{\partial^- u}{\partial \tau}(\tau, x)$ and the equality $\lim_{\tau \uparrow t} u(\tau, x) = f(x)$ are understood in the sense of pointwise convergence, or in the sense of convergence in the space $\sigma\left(L^\infty_{\mathcal{E}}, M\right)$.

Proof. Let $f \in D^M_-(t) \cap F^M_-(t)$. Then, using the properties of backward propagators and the definition of the set $D^M_-(t)$, we see that for every $\nu \in M$,

$$\begin{aligned}
\lim_{h \downarrow 0} \int_E \frac{u(\tau - h) - u(\tau)}{-h} d\nu &= -\lim_{h \downarrow 0} \int_E \frac{Y(\tau - h, t) f - Y(\tau, t) f}{h} d\nu \\
&= -\lim_{h \downarrow 0} \int_E \frac{(Y(\tau - h, \tau) - I) Y(\tau, t) f}{h} d\nu \\
&= -\int_E A^M_-(\tau) Y(\tau, t) f d\nu \\
&= -\int_E A^M_-(\tau) u(\tau) d\nu. \quad (2.30)
\end{aligned}$$

Now it is easy to see that Theorem 2.8 follows from (2.26), (2.28), (2.30), and the definition of the set $F^M_-(t)$.

This completes the proof of Theorem 2.8. $\square$

The equation in (2.29) is called Kolmogorov's backward equation. Kolmogorov's equation is a linearization of the Chapman-Kolmogorov equation. One can find examples of transition probability functions by solving Kolmogorov's backward equation. This can be done, for instance, if the set $D^M_-(t)$ is large enough, and if final value problem (2.29) is uniquely solvable. More precisely, suppose that for every open set $O \subset E$, there exists a sequence $f_n \in D^M_-(t)$ such that $\sup_{n,x} |f_n(x)| < \infty$ and $\lim_{n \to \infty} f_n(x) = \chi_O(x)$ for all $x \in E$. By Lemma 2.1, the sequence f_n converges to χ_O in the topology $\sigma\left(L^\infty_{\mathcal{E}}, M\right)$. Therefore, the sequence $Y(\tau, t) f_n(x)$ converges as $n \to \infty$ for all $x \in E$ and $0 \leq \tau < t$. Here we use the dominated convergence theorem. Put

$$Y(\tau,t)\chi_O(x) = \lim_{n\to\infty} Y(\tau,t) f_n(x).$$

Then it is not hard to see that $Y(\tau,t)\chi_O$ does not depend on the approximating sequence f_n. It is also clear that

$$P(\tau,x;t,O) = Y(\tau,t)\chi_O(x).$$

Fix τ, t, and x such that $0 \leq \tau < t \leq T$. By the definition of a transition probability function, $A \mapsto P(\tau,x;t,A)$ is a Borel measure. Define a family $\mathcal{D}$ of Borel sets by $\mathcal{D} = \{B \in \mathcal{E} : P(\tau,x;t,B) = Y(\tau,t)\chi_B(x)\}$. Then $\mathcal{D}$ is a d-class and contains all open subsets of E. By the monotone class theorem, $\mathcal{D} = \mathcal{E}$. This allows us to recover the transition function P from Y.

Next we will introduce the family of right $\sigma(L_\mathcal{E}^\infty, M)$-generators of the backward propagator Y. As in formula (2.15), for every τ with $0 \leq \tau < T$, we consider the set $D^w\left(A_+^M(\tau)\right)$ of all functions $f \in L_\mathcal{E}^\infty$ for which the limit

$$A_+^M(\tau)f = \sigma(L_\mathcal{E}^\infty, M)\text{-}\lim_{h\downarrow 0} \frac{Y(\tau,\tau+h)f - f}{h} \qquad (2.31)$$

exists. The operators $A_+^M(\tau)$ are called the right $\sigma(L_\mathcal{E}^\infty, M)$-generators of the backward propagator Y. For any t with $0 < t \leq T$, denote by $D_+^M(t)$ the set of all functions $f \in L_\mathcal{E}^\infty$ such that the function $\tau \to Y(\tau,t)f$ is $\sigma(L_\mathcal{E}^\infty, M)$-differentiable from the right on $(0,t)$.

Theorem 2.9 *Let Y be an $\sigma(L_\mathcal{E}^\infty, M)$-continuous backward propagator on the space $L_\mathcal{E}^\infty$, and fix t with $0 < t \leq T$. Then for every $f \in D_+^M(t)$, the function $u(\tau) = Y(\tau,t)f$ is a solution to the following final value problem on $(0,t)$:*

$$\begin{cases} \dfrac{\partial^+ u}{\partial \tau}(\tau, x) = -A_+^M(\tau)u(\tau)(x), \\ \lim_{\tau\uparrow t} u(\tau, x) = f(x) \end{cases} \qquad (2.32)$$

for all $x \in E$.

Remark 2.2 The derivative $\dfrac{\partial^+ u}{\partial \tau}(\tau, x)$ and the equality

$$\lim_{\tau\uparrow t} u(\tau)(x) = f(x)$$

in final value problem (2.32) are understood in the sense of convergence in the space $(L_{\mathcal{E}}^\infty, \sigma(L_{\mathcal{E}}^\infty, M))$.

Proof. Let $f \in D_+^M(t)$, and suppose that h_n is a sequence of positive numbers such that $h_n \downarrow 0$. Let $\nu \in M$ be a nonnegative Borel measure on E. Then, using the fact that Y is a $\sigma(L_{\mathcal{E}}^\infty, M)$-continuous backward propagator on the space $L_{\mathcal{E}}^\infty$ and reasoning as in the proof of (2.17), we get

$$\int_E \frac{\partial^+ u}{\partial \tau} d\nu = \lim_{n \to \infty} \int_E Y(\tau, \tau + h_n) \frac{\partial^+ u}{\partial \tau} d\nu$$

$$= \lim_{n \to \infty} \int_E Y(\tau, \tau + h_n) \left[\frac{\partial^+ u}{\partial \tau} - \frac{Y(\tau + h_n, t)f - Y(\tau, t)f}{h_n} \right] d\nu$$

$$+ \lim_{n \to \infty} \int_E Y(\tau, \tau + h_n) \frac{Y(\tau + h_n, t)f - Y(\tau, t)f}{h_n} d\nu. \quad (2.33)$$

Our next goal is to prove that

$$\lim_{n \to \infty} \int_E Y(\tau, \tau + h_n) \left[\frac{\partial^+ u}{\partial \tau} - \frac{Y(\tau + h_n, t)f - Y(\tau, t)f}{h_n} \right] d\nu = 0. \quad (2.34)$$

It follows from the $\sigma(L_{\mathcal{E}}^\infty, M)$-continuity of Y that for every $n \in \mathbb{N}$, the set function

$$\nu_n(B) = \int_E Y(\tau, \tau + h_n) \chi_B(x) d\nu(x), \quad B \in \mathcal{E},$$

is a Borel measure. It is not hard to see that the set function μ defined by

$$\mu(B) = \sum_{n=1}^\infty \frac{1}{2^n} \nu_n(B), \quad B \in \mathcal{E},$$

is a finite Borel measure on E. Define a sequence of functions on E by

$$g_n = \frac{\partial^+ u}{\partial \tau} - \frac{Y(\tau + h_n, t)f - Y(\tau, t)f}{h_n}.$$

Then,

$$\int_E Y(\tau, \tau + h_n) \left[\frac{\partial^+ u}{\partial \tau} - \frac{Y(\tau + h_n, t)f - Y(\tau, t)f}{h_n} \right] d\nu = \int_E g_n(x) d\nu_n(x). \quad (2.35)$$

Since $f \in D_+^M(t)$ and Lemma 2.1 holds,

$$\lim_{n \to \infty} g_n(x) = 0 \quad (2.36)$$

for all $x \in E$, and

$$\sup_n |g_n|_\infty < \infty. \tag{2.37}$$

It follows from (2.36) and Egorov's theorem that for every $\varepsilon > 0$ there exists $B \in \mathcal{E}$ such that

$$\mu(E\backslash B) \leq \varepsilon \text{ and } |g_n(x)| \leq \varepsilon,\ x \in B,\ n \in \mathbb{N}. \tag{2.38}$$

Moreover, since every measure ν_n is absolutely continuous with respect to the measure μ, and $\lim_{n \to \infty} \nu_n(B) = \nu(B)$ for all $B \in \mathcal{E}$, the Vitali-Hahn-Saks theorem (see Section 5.5) implies that

$$\lim_{\mu(B)\downarrow 0} \sup_n \nu_n(B) = 0. \tag{2.39}$$

Now it is not hard to see that equality (2.34) follows from (2.35), (2.37), and (2.39).

Next, using (2.34) and the properties of backward propagators, we see that

$$\begin{aligned}
\int_E \frac{\partial^+ u}{\partial \tau} d\nu &= \lim_{n \to \infty} \int_E Y(\tau, \tau + h_n) \frac{Y(\tau + h_n, t)f - Y(\tau, t)f}{h_n} d\nu \\
&= \lim_{n \to \infty} \int_E \frac{Y(\tau, t)f - Y(\tau, \tau + h_n)Y(\tau, t)f}{h_n} d\nu \\
&= -\lim_{n \to \infty} \int_E \frac{Y(\tau, \tau + h_n) - I}{h_n} Y(\tau, t) f d\nu.
\end{aligned} \tag{2.40}$$

It follows from (2.40) that $Y(\tau, t)f \in D^w(A_+^M(\tau))$ and that the equation in (2.32) holds in the space $(L_{\mathcal{E}}^\infty, \sigma(L_{\mathcal{E}}^\infty, M))$. In addition, the equality $\lim_{\tau \uparrow t} \int_E u(\tau) d\nu = \int_E f d\nu$ follows from the continuity of Y on the space $(L_{\mathcal{E}}^\infty, \sigma(L_{\mathcal{E}}^\infty, M))$.

This completes the proof of Theorem 2.3. $\square$

Next we will discuss Kolmogorov's forward equation. Let Y be the free backward propagator on the space $L_{\mathcal{E}}^\infty$ corresponding to the transition function P. Then the family of operators given by $W(t, \tau) = Y(\tau, t)^*$ is a propagator on the space $(L_{\mathcal{E}}^\infty)^*$. Since the space M of finite signed Borel measures on E is a subspace of the space $(L_{\mathcal{E}}^\infty)^*$, the operator $W(t, \tau)$ is defined on the space M. Moreover, it maps the space M into itself, since

$$W(t, \tau)\mu(B) = \int_E d\mu(x) \int_B P(\tau, x; t, dy) = \int_E P(\tau, x; t, B) d\mu(x),$$

and the last expression is a Borel measure on E.

Let $0 \le t < T$, and denote by $D^w\left(\widetilde{A}_+^{\mathcal{E}}(t)\right)$ the set of all $\mu \in M$ such that the limit
$$\widetilde{A}_+^{\mathcal{E}}(t)\mu(B) = \lim_{h\downarrow 0} \frac{W(t+h,t)\mu(B) - \mu(B)}{h} \tag{2.41}$$
exists for all $B \in \mathcal{E}$. Then, by the corollary to the Vitali–Hahn–Saks theorem formulated in Subsection 5.5 of the Appendix (see Corollary 5.2), $\widetilde{A}_+^{\mathcal{E}}(t)\mu$ is a Borel measure on E for all $\mu \in D^w\left(\widetilde{A}_+^{\mathcal{E}}(t)\right)$. We will denote by $\widetilde{F}^{\mathcal{E}}(\tau)$ where $0 \le \tau < T$ the set consisting of all $\mu \in M$ such that
$$\lim_{h\downarrow 0} W(\tau+h,\tau)\mu(B) = \mu(B)$$
for all $B \in \mathcal{E}$. For every $\tau \in [0,T)$, denote by $\widetilde{D}_+^{\mathcal{E}}(\tau)$ the set of measures $\mu \in M$ such that the function $t \mapsto W(t,\tau)\mu$ is setwise differentiable from the right on (τ, T).

The next result concerns Kolmogorov's forward equation.

Theorem 2.10 *Let Y be the free backward $L_{\mathcal{E}}^\infty$-propagator corresponding to a transition probability function P, and let W be the propagator on the space M given by $W(t,\tau) = Y(\tau,t)^*$, $0 \le \tau \le t \le T$. Suppose that $0 \le \tau < T$ and $\mu \in \widetilde{D}_+^{\mathcal{E}}(\tau) \cap \widetilde{F}^{\mathcal{E}}(\tau)$. Then the M-valued function ν defined for all $t \in (\tau, T)$ by the formula $\nu(t)(B) = W(t,\tau)\mu(B)$, $B \in \mathcal{E}$, is a solution to the following initial value problem:*
$$\begin{cases} \dfrac{\partial^+ \nu}{\partial t}(t)(B) = \widetilde{A}_+^{\mathcal{E}}(t)\nu(t)(B), \\ \lim_{t\downarrow\tau} \nu(t)(B) = \mu(B) \end{cases} \tag{2.42}$$
for all $B \in \mathcal{E}$.

Proof. We have already established that the family W is a propagator on the space M. If $\mu \in \widetilde{D}_+^{\mathcal{E}}(\tau)$, then using the properties of propagators and the definition of the set $\widetilde{D}_+^{\mathcal{E}}(\tau)$, we get
$$\begin{aligned}\frac{\partial^+ \nu}{\partial t}(t)(B) &= \lim_{h\downarrow 0} \frac{W(t+h,t)\mu(B) - W(t,\tau)\mu(B)}{h} \\ &= \lim_{h\downarrow 0} \frac{W(t+h,t)W(t,\tau)\mu(B) - W(t,\tau)\mu(B)}{h} \\ &= \lim_{h\downarrow 0} \frac{(W(t+h,t) - I)W(t,\tau)\mu(B)}{h}\end{aligned}$$

$$= \widetilde{A}_t^{\mathcal{E}} W(t,\tau)\mu(B) = \widetilde{A}_t^{\mathcal{E}} \nu(t)(B)$$

for all $B \in \mathcal{E}$. In addition, the equality $\lim_{t \downarrow \tau} \nu(t)(B) = \mu(B)$, $B \in \mathcal{E}$, follows from the definition of the set $\widetilde{F}^{\mathcal{E}}(\tau)$.

This completes the proof of Theorem 2.10. □

The equation in Theorem 2.10 is called Kolmogorov's forward equation, or the Fokker-Planck equation. If the Dirac measure δ_x belongs to the set $\widetilde{D}_+^{\mathcal{E}}(\tau)$ for every $x \in E$, and if problem (2.42) is uniquely solvable for all $\mu \in \widetilde{D}_+^{\mathcal{E}}(\tau)$, then we can recover the transition function P from the formula

$$P(\tau, x; t, A) = W(t,\tau)\delta_x(A).$$

Let Y be the free backward propagator on the space $L_{\mathcal{E}}^{\infty}$ associated with a transition probability function P. For every τ with $0 \leq \tau < T$, put

$$D_+^{M,*}(\tau) = \bigcap_{t:\tau<t<T} D^w\left(A_+^M(t)\right),$$

where the operator $A_+^M(t)$ is defined by (2.31). Then the following theorem holds:

Theorem 2.11 *Let Y be the free backward propagator on the space $L_{\mathcal{E}}^{\infty}$ associated with a transition probability function P. Then for every τ with $0 < \tau < T$ and $f \in D_+^{M,*}(\tau)$,*

$$\frac{\partial^+ Y(\tau,t)f}{\partial t} = Y(\tau,t)A_+^M(t)f \qquad (2.43)$$

for all $t \in (\tau, T)$.

Proof. Let $f \in D_+^{M,*}(\tau)$. Then for every $\mu \in M$,

$$\lim_{h \downarrow 0} \int_E \frac{Y(\tau, t+h)f - Y(\tau,t)f}{h} d\mu = \lim_{h \downarrow 0} \int_E \frac{Y(\tau,t)(Y(t, t+h)f - f)}{h} d\mu. \qquad (2.44)$$

By Lemma 2.1, the $\sigma(L_{\mathcal{E}}^{\infty}, M)$-convergence of a sequence $h_k \in L_{\mathcal{E}}^{\infty}$ to a function $h \in L_{\mathcal{E}}^{\infty}$ is equivalent to the pointwise convergence and the uniform boundedness of the sequence h_k in the space $L_{\mathcal{E}}^{\infty}$. It is not hard to prove that

$$\sigma(L_{\mathcal{E}}^{\infty}, M) - \lim_{h \downarrow 0} \frac{Y(\tau,t)(Y(t, t+h)f - f)}{h} = Y(\tau,t)A_+^M(t)f. \qquad (2.45)$$

Next, passing to the limit under the integral sign on the left-hand side of equality (2.44) and using (2.45), we get

$$\int_E \lim_{h\downarrow 0} \frac{Y(\tau,t+h)f - Y(\tau,t)f}{h} d\mu = \int_E Y(\tau,t) A_+^M(t) f d\mu \qquad (2.46)$$

for all $\mu \in M$. It is clear that (2.46) implies (2.43).
This completes the proof of Theorem 2.11. $\square$

Finally, let Y be the free backward propagator on the space $L_\mathcal{E}^\infty$ associated with a transition probability function P, and let $W(t,\tau) = Y(\tau,t)^*$. Then W is a propagator on the space M (see the discussion before the formulation of Theorem 2.10). For every τ with $0 < \tau < T$, consider the operator

$$\widetilde{A}_-^\mathcal{E}(\tau)\mu(B) = \lim_{h\downarrow 0} \frac{W(\tau,\tau-h)\mu(B) - \mu(B)}{h}, \quad B \in \mathcal{E}, \qquad (2.47)$$

and denote by $D\left(\widetilde{A}_-^\mathcal{E}(\tau)\right)$ the subspace of the space M consisting of all measures μ for which the limit in (2.47) exists for all $B \in \mathcal{E}$. By the corollary to the Vitali-Hahn-Saks theorem formulated in Subsection 5.5 of the Appendix (see Corollary 5.2), for $\mu \in D\left(\widetilde{A}_-^\mathcal{E}(\tau)\right)$, $\widetilde{A}_-^\mathcal{E}(\tau)\mu$ is a Borel measure on E. Put

$$\widetilde{D}_-^{\mathcal{E},*}(t) = \bigcap_{\tau: 0 < \tau < t} D\left(\widetilde{A}_-^\mathcal{E}(\tau)\right). \qquad (2.48)$$

Then the following theorem holds:

Theorem 2.12 *Let Y be the free backward $L_\mathcal{E}^\infty$-propagator associated with a transition probability function P, and let W be the propagator on the space M given by $W(t,\tau) = Y(\tau,t)^*$ where $0 \leq \tau \leq t \leq T$. Then for any t with $0 < t \leq T$ and $\mu \in \widetilde{D}_-^{\mathcal{E},*}(t)$,*

$$\frac{\partial^- W(t,\tau)\mu}{\partial \tau} = -W(t,\tau)\widetilde{A}_-^\mathcal{E}(\tau)\mu. \qquad (2.49)$$

Proof. Let $\mu \in M$. Then we have

$$W(t,\tau)\mu(B) = \int_E P(\tau,x;t,B) d\mu(x) \qquad (2.50)$$

for all $B \in \mathcal{E}$. It follows from (2.50) that if a sequence $\mu_k \in M$ converges to $\mu \in M$ setwise, then the sequence $W(t,\tau)\mu_k$ converges to $W(t,\tau)\mu$ setwise.

For $\mu \in \widetilde{D}_-^{\mathcal{E},*}(t)$, we have

$$\frac{W(t,\tau-h)\mu - W(t,\tau)\mu}{-h}(B) = -\frac{W(t,\tau)(W(\tau,\tau-h)\mu - \mu)}{h}(B). \quad (2.51)$$

Now using the previous remark and passing to the limit as $h \downarrow 0$ in (2.51), we get (2.49).

This completes the proof of Theorem 2.12. □

2.4 Howland Semigroups

Propagators are two-parameter generalizations of semigroups. Conversely, if a propagator is given, then, under certain restrictions, one can define a semigroup by introducing an extra time variable. For instance, let Q be a backward propagator on a Banach space B, and let $T > 0$ be a positive number. Denote by $L_B^\infty([0,T], B)$ the space of all bounded B-valued strongly measurable functions on $[0,T]$ equipped with the norm

$$\|f\|_\infty = \sup_{t \in [0,T]} \|f(t)\|_B.$$

Definition 2.3 Let Q be a backward propagator on a Banach space B, and let $T > 0$. The Howland semigroup $S_Q(t)$ on the space $L_B^\infty([0,T], B)$ is defined by

$$S_Q(t)f(\tau) = Q(\tau, (\tau+t) \wedge T)f((\tau+t) \wedge T), \quad t \in [0,T], \quad (2.52)$$

where $f \in L_B^\infty([0,T], B)$.

Lemma 2.2 *The family of operators $S_Q(t)$ is a semigroup on the space $L_B^\infty([0,T], B)$.*

Remark 2.3 Note that the semigroup $S_Q(t)$ is defined on a finite interval $[0,T]$. If $T = \infty$, then the Howland semigroup is defined by

$$S_Q(t)f(\tau) = Q(\tau, \tau+t)f(\tau+t), \quad 0 \le t < \infty, \quad 0 \le \tau < \infty.$$

Proof. We will first prove the following two assertions:

(1) If $f \in L_B^\infty([0,T], B)$, then $S_Q(t)(f) \in L_B^\infty([0,T], B)$.
(2) For all $s > 0$ and $t > 0$, $S_Q((s+t) \wedge T) = S_Q(s) \circ S_Q(t)$, and moreover $S_Q(0) = I$, where I stands for the identity operator on the space $L_B^\infty([0,T], B)$.

Let $f \in L_B^\infty([0,T], B)$. Then there exists a sequence of simple functions $s_n : [0,T] \to B$ such that for all $u \in [0,T]$, $\|f(u) - s_n(u)\|_B \to 0$ as $n \to \infty$. Fix $t \in [0,T]$, and define a sequence of simple functions by

$$\tilde{s}_n(\tau) = Q(\tau, (\tau + t) \wedge T) s_n((\tau + t) \wedge T)$$

where $\tau \in [0,T]$. Then, for all $\tau \in [0,T]$,

$$\|S_Q(t)f(\tau) - \tilde{s}_n(\tau)\|_B$$
$$= \|Q(\tau, (\tau+t) \wedge T)f((\tau+t) \wedge T) - Q(\tau, (\tau+t) \wedge T)s_n((\tau+t) \wedge T)\|_B$$
$$\leq \|Q(\tau, (\tau+t) \wedge T)\|_{B \to B} \|f((\tau+t) \wedge T) - s_n((\tau+t) \wedge T)\|_B \to 0$$

as $n \to \infty$. Therefore, the B-valued function $S_Q(t)f$ is strongly measurable. Moreover,

$$\|S_Q(t)f(\tau)\|_B \leq \|Q(\tau, (\tau+t) \wedge T)\|_{B \to B} \|f((\tau+t) \wedge T)\|_B,$$

and hence,

$$\sup_{\tau \in [0,T]} \|S_Q(t)f(\tau)\|_B < \infty.$$

It follows that $S_Q(t)f \in L_B^\infty([0,T], B)$. This establishes condition (1) above. We will next prove condition (2). It is clear that

$$S_Q(0)f(\tau) = Q(\tau, \tau)f(\tau) = f(\tau).$$

Moreover, using the properties of backward propagators, we see that for all $s \in [0,T]$ and $t \in [0,T]$,

$$S_Q(s) \circ S_Q(t)f(\tau) = Q(\tau, (\tau+s) \wedge T)(S_Q(t)f)((\tau+s) \wedge T)$$
$$= Q(\tau, (\tau+s) \wedge T) Q((\tau+s) \wedge T, ((\tau+s) \wedge T + t) \wedge T)$$
$$\quad f(((\tau+s) \wedge T + t) \wedge T)$$
$$= Q(\tau, (\tau+s) \wedge T) Q((\tau+s) \wedge T, (\tau+s+t) \wedge T) f((\tau+s+t) \wedge T)$$
$$= Q(\tau, (\tau+s+t) \wedge T) f((\tau+s+t) \wedge T) = S_Q((s+t) \wedge T)f(\tau).$$

This establishes condition (2).

The proof of Lemma 2.2 is thus completed. □

An element x of the space B can be identified with the constant function $f_x(\tau) = x$, $\tau \in [0,T]$. By taking this identification into account, we get the following formula connecting the backward propagator Q with the Howland semigroup S_Q:

$$S_Q(t)f_x(\tau) = Q(\tau, (\tau + t) \wedge T)x \qquad (2.53)$$

for all $t \in [0, T]$, $\tau \in [0, T]$, and $x \in B$.

Next, we will discuss Howland semigroups associated with propagators.

Definition 2.4 Let W be a propagator on a Banach space B, and let $T > 0$. The Howland semigroup S_W on the $L^\infty_{\mathcal{B}}([0,T], B)$ is defined for all $\tau > 0$ by

$$S_W(\tau)f(t) = W(t, (t - \tau) \vee 0)f((t - \tau) \vee 0) \qquad (2.54)$$

where $f \in L^\infty_{\mathcal{B}}([0,T], B)$.

Arguing as in the proof of Lemma 2.2, we can show that $S_W(\tau)$ is a semigroup on the space $L^\infty_{\mathcal{B}}([0,T], B)$. The propagator W is related to the semigroup S_W as follows:

$$S_W(\tau)f_x(t) = W(t, (t - \tau) \vee 0)x \qquad (2.55)$$

for all $t \in [0, T]$, $\tau \in [0, T]$, and $x \in B$.

Howland semigroups associated with free propagators and free backward propagators admit a probabilistic description. Let P be a transition probability function, and let X_t be a corresponding Markov process with state space $(E, \mathcal{E})$. Recall that in Section 1.5 we discussed space-time processes associated with the process X_t (see Section 1.5 for the definition of the transition function $\widehat{P}$, the family of measures $\widehat{\mathbb{P}}_{(\tau,x)}$, and the sample space $\widehat{\Omega}$ of the space-time process $\widehat{X}_t$). Let Y be the free backward propagator on the space $L^\infty_{\mathcal{E}}$ associated with the transition function P. As we already know, the backward propagator Y can be expressed in probabilistic terms as follows:

$$Y(\tau, t)f(x) = \mathbb{E}_{\tau,x} f(X_t) \qquad (2.56)$$

for all $0 \le \tau \le t \le T$, $x \in E$, and $f \in L^\infty_{\mathcal{E}}$. In addition, for the Howland semigroup $S_Y(t)$ associated with the backward propagator Y, the following probabilistic characterization is valid:

$$S_Y(t)F(\tau, x) = \mathbb{E}_{\tau,x} F\left((\tau + t) \wedge T, X_{(\tau+t)\wedge T}\right) = \widehat{\mathbb{E}}_{(\tau,x)} F\left(\widehat{X}_t\right) \qquad (2.57)$$

for all $\tau \in [0, T]$, $t \in [0, T]$, $x \in E$, and $F \in L^\infty_{\mathcal{B}}([0,T], L^\infty_{\mathcal{E}})$. This can be derived from (2.52) and (2.56).

Now let $\widetilde{P}$ be a backward transition probability function, and let $\widetilde{X}_\tau$ be a corresponding backward Markov process. In Section 1.5, we constructed a time-homogeneous transition probability function $\widehat{P}$ and the family of measures $\widehat{\mathbb{P}}^{(t,x)}$. The function $\widehat{P}$ was used as the transition function of the space-time process $\widehat{X}_\tau$. Let U be the free propagator on the space $L_{\mathcal{E}}^\infty$ corresponding to the backward transition probability function $\widetilde{P}$. Then the following formula is valid:

$$U(t,\tau)f(x) = \widetilde{\mathbb{E}}^{t,x} f\left(\widetilde{X}_\tau\right) \tag{2.58}$$

for all $0 \leq \tau \leq t \leq T$, $x \in E$, and $f \in L_{\mathcal{E}}^\infty$. For the Howland semigroup $S_U(\tau)$ associated with the propagator U, we have

$$S_U(\tau)F(t,x) = \widetilde{\mathbb{E}}^{t,x} F\left((t-\tau) \vee 0, \widetilde{X}_{(t-\tau)\vee 0}\right) = \widehat{\mathbb{E}}^{(t,x)} F\left(\widehat{X}_\tau\right) \tag{2.59}$$

for all $\tau \in [0,T]$, $t \in [0,T]$, $x \in E$, and $F \in L_{\mathcal{B}}^\infty([0,T], L_{\mathcal{E}}^\infty)$. This follows from (2.54) and (2.58).

2.5 Feller-Dynkin Propagators and the Continuity Properties of Markov Processes

Let us recall that in Section 2.2 we defined the following spaces of continuous functions on the space E: the space BC of all bounded continuous functions on E; the space BUC of all bounded uniformly continuous functions on E, and the space C_0 of all functions from BC which vanish at infinity.

Definition 2.5 A backward BC-propagator is called a backward Feller propagator. A backward C_0-propagator is called a backward Feller-Dynkin propagator. If a backward $L_{\mathcal{E}}^\infty$-propagator Q is such that

$$Q(\tau,t) \in L(L_{\mathcal{E}}^\infty, BC)$$

for all $0 \leq \tau < t \leq T$, then it is said that Q possesses the strong Feller property. If a backward $L_{\mathcal{E}}^\infty$-propagator Q is such that

$$Q(\tau,t) \in L(L_{\mathcal{E}}^\infty, BUC)$$

for all $0 \leq \tau < t \leq T$, then it is said that Q satisfies the strong BUC-condition.

Remark 2.4 If Q is a backward L^∞-propagator, then one can replace the space $L_{\mathcal{E}}^\infty$ by the space L^∞ in the definition of the strong Feller property and the strong BUC-condition.

It is not hard to see how to define forward Feller and Feller-Dynkin propagators, and also propagators possessing the strong Feller property, or satisfying the strong BUC-condition.

It this section, we begin the study of Markov processes associated with free backward Feller-Dynkin propagators. Let P be a transition probability function, and let Y be the corresponding free backward propagator. Our first goal is to prove that if Y is a strongly continuous backward Feller-Dynkin propagator, then the class of all Markov processes associated with P contains a process X_t such that its sample functions are right-continuous and have left limits. In the following sections, we will show that the process X_t possesses additional properties.

Given a transition probability function P, we start with the standard realization $\widetilde{X}_t$ of a corresponding Markov process on the probability space $\left(\widetilde{\Omega}, \widetilde{\mathcal{F}}_t^\tau, \widetilde{\mathbb{P}}_{\tau,x}\right)$ where $\widetilde{\Omega} = E^{[0,T]}$. It is defined by $\widetilde{X}_t(\omega) = \omega(t)$ where $\omega \in \widetilde{\Omega}$. By Ω, will be denoted the space of all E-valued functions defined on the interval $[0,T]$, which are right-continuous and have left limits in E. Then we have $\Omega \subset \widetilde{\Omega}$. Put $X_t(\omega) = \omega(t)$, $\omega \in \Omega$, $t \in [0,T]$, and let $\mathcal{F}_t^\tau$, $0 \leq \tau \leq t \leq T$, be the σ-algebra generated by X_s with $\tau \leq s \leq t$. For every $\tau \in [0,T)$ and $x \in E$, denote by $\mathbb{P}_{\tau,x}$ the probability measure on $\mathcal{F}^\tau$ determined by

$$\mathbb{P}_{\tau,x}[X_{t_1} \in B_1, \ldots, X_{t_n} \in B_n] = \widetilde{\mathbb{P}}_{\tau,x}\left[\widetilde{X}_{t_1} \in B_1, \ldots, \widetilde{X}_{t_n} \in B_n\right]. \quad (2.60)$$

Here $\tau \leq t_1 < \cdots < t_n \leq T$ and $B_j \in \mathcal{E}$, $1 \leq j \leq n$. Let us denote by $\mathcal{F}_{t+}^\tau$, $t \in [\tau, T)$, the σ-algebra defined by

$$\mathcal{F}_{t+}^\tau = \bigcap_{t < s \leq T} \mathcal{F}_s^\tau. \quad (2.61)$$

Theorem 2.13 *Let P be a transition probability function such that the corresponding free backward propagator Y is a strongly continuous backward Feller-Dynkin propagator. Then the processes $\left(X_t, \mathcal{F}_{t+}^\tau, \mathbb{P}_{\tau,x}\right)$ and $\left(\widetilde{X}_t, \widetilde{\mathcal{F}}_t^\tau, \widetilde{\mathbb{P}}_{\tau,x}\right)$ are stochastically equivalent.*

Remark 2.5 It follows from Theorem 2.13 that the process $\left(X_t, \mathcal{F}_{t+}^\tau, \mathbb{P}_{\tau,x}\right)$ is a Markov process with P as its transition function. Moreover, all the sample paths of X_t are right-continuous and have left limits.

It also follows from the proof of Theorem 2.13 that for every $\omega \in \Omega$, the set $\{X_t(\omega) : t \in [0, T]\}$ is relatively compact in E.

Proof. For every function $f \in C_0$ and every ϵ with $0 \le \epsilon < T$, put

$$N_\epsilon(f)(t, z) = \int_{t+\epsilon}^T Y(t, s) f(z) ds$$

where $0 \le t \le T - \epsilon$ and $z \in E$.

Lemma 2.3 *Fix τ with $0 \le \tau < T$ and $x \in E$. Then for every nonnegative function $f \in C_0$ and every ϵ with $0 \le \epsilon < T - \tau$, the process $t \mapsto N_\epsilon(f)\left(t, \widetilde{X}_t\right)$, $t \in [\tau, T - \epsilon]$, is a supermartingale with respect to the filtration $\widetilde{\mathcal{F}}_t^\tau$ and the measure $\widetilde{\mathbb{P}}_{\tau,x}$.*

Proof. It is clear that the process $t \mapsto N_\epsilon(f)\left(t, \widetilde{X}_t\right)$ is adapted to the filtration $\widetilde{\mathcal{F}}_t^\tau$. Moreover, for $\tau \le t_1 < t_2 \le T - \epsilon$, the Markov property implies that

$$\mathbb{E}_{\tau,x}\left[N_\epsilon(f)\left(t_2, \widetilde{X}_{t_2}\right) \mid \mathcal{F}_{t_1}^\tau\right] = \mathbb{E}_{t_1, \widetilde{X}_{t_1}}\left[N_\epsilon(f)\left(t_2, \widetilde{X}_{t_2}\right)\right]. \quad (2.62)$$

Since Y is a backward propagator,

$$\mathbb{E}_{t_1,z}\left[N_\epsilon(f)\left(t_2, \widetilde{X}_{t_2}\right)\right] = \int_{t_2+\epsilon}^T \mathbb{E}_{t_1,z}\left[Y(t_2, s) f\left(\widetilde{X}_{t_2}\right)\right] ds$$

$$= \int_{t_2+\epsilon}^T Y(t_1, t_2) Y(t_2, s) f(z) ds = \int_{t_2+\epsilon}^T Y(t_1, s) f(z) ds$$

$$\le \int_{t_1+\epsilon}^T Y(t_1, s) f(z) ds = N_\epsilon(f)(t_1, z) \quad (2.63)$$

for all $z \in E$. Finally, (2.63) with $z = \widetilde{X}_{t_1}$ and (2.62) show that Lemma 2.3 holds. □

Let us continue the proof of Theorem 2.13. It follows from Lemma 2.3 and Theorem 1.10 that for every $f \in C_0$ and $\tau \in [0, T)$, there exists a set $\Lambda_\tau \in \widetilde{\mathcal{F}}$ such that $\widetilde{\mathbb{P}}_{\tau,x}(\Lambda_\tau) = 0$, and for every integer $n \ge 1$, the process

$$Z_t^n(f)(\omega) = n \int_t^{t+\frac{1}{n}} Y(t, s) f\left(\widetilde{X}_t(\omega)\right) ds, \quad \tau \le t \le T - \frac{1}{n}, \quad (2.64)$$

has left and right limits,

$$\lim_{s \uparrow t, s \in \mathbb{Q}} Z_s^n(f)(\omega) \quad \text{and} \quad \lim_{s \downarrow t, s \in \mathbb{Q}} Z_s^n(f)(\omega),$$

for all $\omega \in \widetilde{\Omega}\backslash\Lambda_\tau$ and $t \in [\tau, T - \frac{1}{n}]$. The set Λ_τ does not depend on n and t. Moreover, Λ_τ can be chosen independently of $f \in C_0$. Indeed, let f_k be an everywhere dense countable subset of C_0. Then it is not hard to see that there exists an exceptional set $\Lambda_\tau \in \widetilde{\mathcal{F}}$ with $\widetilde{\mathbb{P}}_{\tau,x}(\Lambda) = 0$ such that the limits in (2.64) with f replaced by f_k exist for all $\omega \in \widetilde{\Omega}\backslash\Lambda_\tau$, $t \in [\tau, T - \frac{1}{n}]$, and $k \geq 1$. It is also easy to prove that if a sequence $g_k \in C_0$ converges uniformly to a function g, then we have

$$n \int_t^{t+\frac{1}{n}} Y(t,s) g_k(z) \, ds \to n \int_t^{t+\frac{1}{n}} Y(t,s) g(z) \, ds$$

uniformly in z as $k \to \infty$. It follows that the exceptional set Λ_τ can be chosen independently of $f \in C_0$, $n \geq 1$, and t with $\tau \leq t \leq T - \frac{1}{n}$.

By the assumption that the backward propagator Y is strongly continuous on C_0, we get

$$\lim_{n \to \infty} \sup_{z \in E, 0 \leq t \leq T - \frac{1}{n}} \left| n \int_t^{t+\frac{1}{n}} Y(t,s) f(z) \, ds - f(z) \right| = 0$$

for every $f \in C_0$. Hence, there exists a set $\Lambda_\tau \in \mathcal{F}$ for which $\mathbb{P}_{\tau,x}(\Lambda_\tau) = 0$, and for every $f \in C_0$ the limits

$$\lim_{s \uparrow t, s \in \mathbb{Q}} f\left(\widetilde{X}_s(\omega)\right) \quad \text{and} \quad \lim_{s \downarrow t, s \in \mathbb{Q}} f\left(\widetilde{X}_s(\omega)\right) \tag{2.65}$$

exist for all $\omega \in \widetilde{\Omega}\backslash\Lambda_\tau$ and all t with $\tau \leq t \leq T$. The exceptional set Λ_τ does not depend on $f \in C_0$ and $t \in [\tau, T]$.

The following well-known fact concerning nonnegative supermartingales will be used in the proof of Theorem 2.13.

Lemma 2.4 *Let* $(\Omega, \mathcal{F}, \mathcal{F}_t, \mathbb{P})$ *be a filtered probability space, and let* Z_t *be a nonnegative* $\mathcal{F}_t$*-supermartingale. Denote by* $\mathbb{Q}$ *a countable dense subset of the interval* $[0, T]$. *Then*

$$\mathbb{P}\left[Z_t > 0, \inf_{s \leq t; s \in Q} Z_s = 0\right] = 0$$

for all $t \in [0, T]$.

Proof. Fix t with $0 \leq t \leq T$, and let $Q_j = \left\{s_1^j < \cdots < s_{m_j}^j\right\}$ be an increasing sequence of finite subsets of the set $\mathbb{Q} \cap [0, t]$ such that

$$\mathbb{Q} \cap [0, t] = \bigcup_j Q_j.$$

For all $j \geq 1$ and $n \geq 1$, put
$$\tau_{j,n} = \min\left\{ s_k^j : 1 \leq k \leq m_j, Z_{s_k^j} \leq \frac{1}{n} \right\}.$$

In addition, if
$$\omega \in \Omega \setminus \left[\min_{1 \leq k \leq m_j} Z_{s_k^j} \leq \frac{1}{n} \right],$$
then we put $\tau_{j,n}(\omega) = T$. Next, using the properties of supermartingales and the $\mathcal{F}_{s_k^j}$-measurability of the event $\left\{ \tau_{j,n} = s_k^j \right\}$, we get

$$\mathbb{E}\left[Z_t, \min_{1 \leq k \leq m_j} Z_{s_k^j} \leq \frac{1}{n} \right] = \mathbb{E}\left[Z_t, \tau_{j,n} < T, Z_{\tau_{j,n}} \leq \frac{1}{n} \right]$$
$$= \sum_{k=1}^{m_j} \mathbb{E}\left[Z_t, \tau_{j,n} = s_k^j, Z_{\tau_{j,n}} \leq \frac{1}{n} \right]$$
$$= \sum_{k=1}^{m_j} \mathbb{E}\left[\mathbb{E}\left[Z_t \mid \mathcal{F}_{s_k^j} \right], \tau_{j,n} = s_k^j, Z_{s_k^j} \leq \frac{1}{n} \right]$$
$$\leq \sum_{k=1}^{m_j} \mathbb{E}\left[Z_{s_k^j}, \tau_{j,n} = s_k^j, Z_{s_k^j} \leq \frac{1}{n} \right] \leq \frac{1}{n}. \quad (2.66)$$

Passing to the limit in (2.66) as $j \to \infty$ and then as $n \to \infty$, we get
$$\mathbb{E}\left[Z_t, \inf_{s \leq t, s \in Q} Z_s = 0 \right] = 0. \quad (2.67)$$

Now it is clear that Lemma 2.4 follows from (2.67). $\square$

Let us return to the proof of Theorem 2.13. Fix a countable dense subset $\mathbb{Q}$ of the interval $[0, T]$, and for every $\tau \in [0, T)$, consider the event Γ_τ consisting of all $\omega \in \tilde{\Omega}$ for which there exists $r \in \mathbb{Q} \cap [\tau, T)$ such that the set $\tilde{X}_s(\omega)$, $s \in \mathbb{Q} \cap [\tau, r]$, is not contained in any compact subset of E. Then the following lemma holds:

Lemma 2.5 *Let f be a strictly positive function from C_0. Then for every τ with $0 \leq \tau < T$,*

$$\Gamma_\tau = \bigcup_{r \in \mathbb{Q} \cap [\tau, T)} \left\{ \omega : \int_r^T Y(r, u) f\left(\tilde{X}_r(\omega) \right) du > 0, \right.$$
$$\left. \inf_{s \in \mathbb{Q} \cap [\tau, r]} \int_s^T Y(s, u) f\left(\tilde{X}_s(\omega) \right) du = 0 \right\}. \quad (2.68)$$

Proof. Our first goal is to prove that for every $r \in [\tau, T)$ and $z \in E$,

$$\int_r^T Y(r, u) f(z) du > 0. \tag{2.69}$$

Indeed, suppose that $\int_r^T Y(r, u) f(z) du = 0$. Then the equality

$$\int_r^T Y(r, u) f(z) du = \int_r^T du \int_E f(y) P(r, z; u, dy),$$

and the strict positivity of the function f imply that for almost all $u \in [r, T]$ with respect to the Lebesgue measure, we have $P(r, z; u, E) = 0$. This contradicts the definition of transition probability functions. Therefore, inequality (2.69) holds. Next, fix a sequence $s_n \in [\tau, T]$ such that $\lim_{n \to \infty} s_n = s$ where $\tau \le s < T$. We will show below that for a sequence $z_n \in E$, the following two conditions are equivalent:

(1) There exists a compact subset $C \subset E$ such that $z_n \in C$ for all $n \ge 1$.
(2) The inequality

$$\inf_{n \ge 1} \int_{s_n}^T Y(s_n, u) f(z_n) du > 0 \tag{2.70}$$

holds.

Indeed, if condition (1) does not hold, then there exists a subsequence z_{n_k} of z_n such that $\lim_{k \to \infty} g(z_{n_k}) = 0$ for all $g \in C_0$. It follows from the strong continuity of Y on C_0 and from the condition $\lim_{n \to \infty} s_n = s$ that

$$\int_s^T Y(s, u) f(z_{n_k}) du - \int_{s_{n_k}}^T Y(s_{n_k}, u) f(z_{n_k}) du \to 0 \tag{2.71}$$

as $k \to \infty$. Next, the strong continuity of Y on C_0 gives

$$\lim_{k \to \infty} \int_s^T Y(s, u) f(z_{n_k}) du = 0. \tag{2.72}$$

By (2.71) and (2.72), we see that inequality (2.70) does not hold. Hence, the implication (2) $\Longrightarrow$ (1) is valid.

On the other hand, if condition (1) above holds, and condition (2) does not hold, then there exists a sequence n_k such that $\lim_{k \to \infty} z_{n_k} = z$ where

$z \in C$; moreover,

$$\lim_{k \to \infty} \int_{s_{n_k}}^{T} Y(s_{n_k}, u) f(z_{n_k}) \, du = 0. \qquad (2.73)$$

Since Y is a strongly continuous backward propagator on C_0, and (2.69) holds, we have

$$\lim_{k \to \infty} \int_{s_{n_k}}^{T} Y(s_{n_k}, u) f(z_{n_k}) \, du = \int_{s}^{T} Y(s, u) f(z) du > 0.$$

This contradicts (2.73). Therefore, the implication (1) $\Longrightarrow$ (2) holds.
The proof of Lemma 2.5 is thus completed. $\square$

It follows from Lemma 2.3 with $\epsilon = 0$, Lemma 2.4, and Lemma 2.5 that for all $\tau \in [0, T)$ and all $x \in E$, the equality $\widetilde{\mathbb{P}}_{\tau,x}[\Gamma_\tau] = 0$ holds. Therefore, $\widetilde{\mathbb{P}}_{\tau,x}\left[\widetilde{\Omega}\backslash\Gamma_\tau\right] = 1$. Moreover, for all $\omega \in \widetilde{\Omega}\backslash\Gamma_\tau$ and $r \in \mathbb{Q} \cap [\tau, T)$, there exists a compact set $C \subset E$ such that $\widetilde{X}_s(\omega) \in C$ for all $s \in \mathbb{Q} \cap [\tau, r]$. The set C depends on r and ω.

Given $\tau \in [0, T)$, put

$$\widetilde{\Omega}_\tau = \widetilde{\Omega} \backslash (\Lambda_\tau \cup \Gamma_\tau),$$

where Λ_τ is the complement of the event consisting of all $\omega \in \widetilde{\Omega}$ for which the limits in (2.65) exist for all $f \in C_0$ and all t with $\tau \leq t \leq T$. Then, for every $\omega \in \widetilde{\Omega}_\tau$, the limits

$$\lim_{s \uparrow t, s \in \mathbb{Q}} \widetilde{X}_s(\omega) \quad \text{and} \quad \lim_{s \downarrow t, s \in \mathbb{Q}} \widetilde{X}_s(\omega) \qquad (2.74)$$

exist for all t with $\tau \leq t < T$. Indeed, suppose that $\omega \in \widetilde{\Omega}_\tau$ and $t \in [\tau, T)$. Let $r \in \mathbb{Q} \cap (t, T)$, and assume that $s \downarrow t$, $s \in \mathbb{Q}$, and $s < r$. Then there exists a compact subset $C_r(\omega)$ of the space E such that $\widetilde{X}_s(\omega) \in C_r(\omega)$ for all $s \in \mathbb{Q} \cap (t, r)$. It follows from the fact that the limits in (2.65) exist and from Urysohn's Lemma that the limit $\lim_{s \downarrow t, s \in \mathbb{Q}} \widetilde{X}_s(\omega)$ exists in E. The case of the first limit in (2.74) is similar.

Let us consider the subset $\widehat{\Omega}$ of the set $\widetilde{\Omega}$, consisting of all $\omega \in E^{[0,T]}$ such that the following conditions hold: the limits $\lim_{s \downarrow t, s \in \mathbb{Q}} \omega(s)$ and $\lim_{s \downarrow t, s \in \mathbb{Q}} \omega(s)$ exist for all $t \in [0, T)$, and for every $r \in \mathbb{Q} \cap [0, T)$ there exists a compact subset $C_r(\omega)$ of E for which $\omega(s) \in C_r(\omega)$ where $s \in \mathbb{Q} \cap [0, r]$. It is not difficult to show that for all τ and x,

$$\widetilde{\mathbb{P}}^*_{\tau,x}\left[\widehat{\Omega}\right] = 1, \qquad (2.75)$$

where $\widetilde{\mathbb{P}}^*_{\tau,x}$ denotes the outer measure generated by the measure $\widetilde{\mathbb{P}}_{\tau,x}$. Indeed, for every $\tau \in [0,T)$ and $x \in E$, the set $\widehat{\Omega}$ contains the set $\widehat{\Omega}_{\tau,x}$, consisting of those $\omega \in \widetilde{\Omega}_\tau$ for which $\omega(s) = x$ for $0 \le s \le \tau$. Since $\mathbb{P}_{\tau,x}\left[\widetilde{\Omega}_\tau\right] = 1$, we have $\widetilde{\mathbb{P}}^*_{\tau,x}\left[\widehat{\Omega}_{\tau,x}\right] = 1$. Now it is clear that (2.75) holds. Therefore, one can restrict the probability space structure from $\widetilde{\Omega}$ to $\widehat{\Omega}$. The resulting probability space is denoted by $\left(\widehat{\Omega}, \widehat{\mathcal{F}}, \widehat{\mathbb{P}}_{\tau,x}\right)$. Let us define a stochastic process on $\widehat{\Omega}$ by

$$\widehat{X}_t = \begin{cases} \lim_{s \downarrow t, s \in \mathbb{Q}} \widetilde{X}_s, & \text{if } 0 \le t < T, \\ \widetilde{X}_T, & \text{if } t = T. \end{cases} \qquad (2.76)$$

Then, it is not difficult to see that the process $\widehat{X}_t$ is right-continuous and has left limits on the interval $[0,T)$. It is also clear that the process $\widehat{X}_t$, $t \in [\tau, T)$, is $\widehat{\mathcal{F}}^\tau_{t+}$-adapted. Here

$$\widehat{\mathcal{F}}^\tau_{t+} = \bigcap_{t < s \le T} \widehat{\mathcal{F}}^\tau_s.$$

Lemma 2.6 *The process $\left(\widehat{X}_t, \widehat{\mathcal{F}}^\tau_{t+}, \widehat{\mathbb{P}}_{\tau,x}\right)$ is a Markov process with transition function P.*

Proof. Let $\tau \le s < t < T$, and choose $s_n \in \mathbb{Q} \cap (s,t)$ and $t_n \in \mathbb{Q} \cap (t, T)$ with $s_n \downarrow s$ and $t_n \downarrow t$ as $n \to \infty$. Let $A \in \widehat{\mathcal{F}}^\tau_{s+}$. Then $A \in \widehat{\mathcal{F}}^\tau_{s_n}$ for all $n \ge 1$, and since $\widetilde{X}_u$ is a Markov process,

$$\widetilde{\mathbb{E}}_{\tau,x}\left[f\left(\widetilde{X}_{t_n}\right)\chi_A\right] = \widetilde{\mathbb{E}}_{\tau,x}\left[\widetilde{\mathbb{E}}_{s_n,\widetilde{X}_{s_n}}\left[f\left(\widetilde{X}_{t_n}\right)\right]\chi_A\right]$$
$$= \widetilde{\mathbb{E}}_{\tau,x}\left[Y(s_n,t_n)f\left(\widetilde{X}_{s_n}\right)\chi_A\right] \qquad (2.77)$$

for every $f \in C_0$. It follows from the strong continuity of Y on C_0 and from (2.77) that

$$\widehat{\mathbb{E}}_{\tau,x}\left[f\left(\widehat{X}_t\right)\chi_A\right] = \widehat{\mathbb{E}}_{\tau,x}\left[Y(s,t)f\left(\widehat{X}_s\right)\chi_A\right]$$

for all $A \in \widehat{\mathcal{F}}_{s+}^\tau$. Here we used the fact that $\widehat{\mathbb{E}}_{\tau,x}$ is the restriction of $\widetilde{\mathbb{E}}_{\tau,x}$ to $\widehat{\Omega}$. Therefore,

$$\widetilde{\mathbb{E}}_{\tau,x}\left[f\left(\widetilde{X}_t\right) \mid \widehat{\mathcal{F}}_{s+}^\tau\right] = Y(s,t) f\left(\widetilde{X}_s\right) \tag{2.78}$$

for all $f \in C_0$. In the case where $t = T$, (2.78) is also true. Indeed, we may assume $t_n = T$ for all $n \geq t$, and use the equality $\widehat{X}_T = \widetilde{X}_T$.

For all $x \in E$ and $0 \leq \tau < t < T$, we have

$$\widehat{\mathbb{E}}_{\tau,x}\left[f\left(\widehat{X}_t\right)\right] = \widehat{\mathbb{E}}_{\tau,x}\left[\lim_{s\downarrow t, s\in\mathbb{Q}} f\left(\widetilde{X}_s\right)\right] = \lim_{s\downarrow t, s\in\mathbb{Q}} \widehat{\mathbb{E}}_{\tau,x}\left[f\left(\widetilde{X}_s\right)\right]$$
$$= \lim_{s\downarrow t, s\in\mathbb{Q}} \widetilde{\mathbb{E}}_{\tau,x}\left[f\left(\widetilde{X}_s\right)\right] = \lim_{s\downarrow t, s\in\mathbb{Q}} Y(\tau,s) f(x)$$
$$= Y(\tau,t) f(x) = \widetilde{\mathbb{E}}_{\tau,x}\left[f\left(\widetilde{X}_t\right)\right]. \tag{2.79}$$

Here we used the strong continuity of Y. In the case where $t = T$, (2.79) also holds since $\widehat{X}_T = \widetilde{X}_T$. Next, using Urysohn's Lemma, the monotone class theorem, and approximating bounded Borel functions by simple functions, we see that (2.78) and (2.79) hold for all bounded Borel functions f on E. Consequently, the process $\widehat{X}_t$ is an $\widehat{\mathbb{P}}_{\tau,x}$-Markov process with respect to the filtration $\widehat{\mathcal{F}}_{t+}^\tau$, $\tau \leq t < T$. Moreover, the processes $\left(\widehat{X}_t, \widehat{\mathcal{F}}_{t+}^\tau, \widehat{\mathbb{P}}_{\tau,x}\right)$ and $\left(\widetilde{X}_t, \widetilde{\mathcal{F}}_t^\tau, \widetilde{\mathbb{P}}_{\tau,x}\right)$ are stochastically equivalent.

This completes the proof of Lemma 2.6. □

Lemma 2.7 *The process $\left(\widehat{X}_t, \widehat{\mathcal{F}}_{t+}^\tau, \widehat{\mathbb{P}}_{\tau,x}\right)$ is a modification of the process $\left(\widetilde{X}_t, \widetilde{\mathcal{F}}_t^\tau, \widetilde{\mathbb{P}}_{\tau,x}\right)$; that is, for every $0 \leq \tau \leq t \leq T$ and $x \in E$, the equality $\widehat{\mathbb{P}}_{\tau,x}\left[\widetilde{X}_t = \widehat{X}_t\right] = 1$ holds.*

Proof. Let f and g be functions from the space C_0, and fix $x \in E$, $t \in [0,t]$, and τ with $0 \leq \tau \leq t$. Let s_n be a sequence in $\mathbb{Q} \cap (t,T)$ such that $s_n \downarrow t$. Then the Markov property of the process $\widetilde{X}_u$, $0 \leq u \leq T$, with respect to the measure $\widetilde{\mathbb{P}}_{\tau,x}$ implies

$$\widetilde{\mathbb{E}}_{\tau,x}\left[f\left(\widetilde{X}_{s_n}\right) g\left(\widetilde{X}_t\right)\right] = \widetilde{\mathbb{E}}_{\tau,x}\left[\widetilde{\mathbb{E}}_{\tau,x}\left[f\left(\widetilde{X}_{s_n}\right) \mid \widetilde{\mathcal{F}}_t^\tau\right] g\left(\widetilde{X}_t\right)\right]$$
$$= \widetilde{\mathbb{E}}_{\tau,x}\left[\widetilde{\mathbb{E}}_{t,\widetilde{X}_t}\left[f\left(\widetilde{X}_{s_n}\right)\right] g\left(\widetilde{X}_t\right)\right]$$
$$= \widetilde{\mathbb{E}}_{\tau,x}\left[\left(Y(t,s_n) f\left(\widetilde{X}_t\right)\right) g\left(\widetilde{X}_t\right)\right]. \tag{2.80}$$

Passing to the limit in (2.80) as $n \to \infty$ and taking into account that Y is strongly continuous on C_0, we obtain

$$\widehat{\mathbb{E}}_{\tau,x}\left[f\left(\widehat{X}_t\right)g\left(\widetilde{X}_t\right)\right] = \lim_{n\to\infty}\widetilde{\mathbb{E}}_{\tau,x}\left[f\left(\widetilde{X}_{s_n}\right)g\left(\widetilde{X}_t\right)\right]$$
$$= \lim_{n\to\infty}\widetilde{\mathbb{E}}_{\tau,x}\left[\left(Y(t,s_n)f\left(\widetilde{X}_t\right)\right)g\left(\widetilde{X}_t\right)\right]$$
$$= \widetilde{\mathbb{E}}_{\tau,x}\left[\lim_{n\to\infty}\left(Y(t,s_n)f\left(\widetilde{X}_t\right)\right)g\left(\widetilde{X}_t\right)\right]$$
$$= \widetilde{\mathbb{E}}_{\tau,x}\left[f\left(\widetilde{X}_t\right)g\left(\widetilde{X}_t\right)\right] = \widehat{\mathbb{E}}_{\tau,x}\left[f\left(\widetilde{X}_t\right)g\left(\widetilde{X}_t\right)\right]. \quad (2.81)$$

Next, formula (2.81) with $g = f$ and the stochastic equivalence of the processes $\widetilde{X}_t$ and $\widehat{X}_t$ (see Lemma 2.6) give

$$\widehat{\mathbb{E}}_{\tau,x}\left[\left|f\left(\widehat{X}_t\right) - f\left(\widetilde{X}_t\right)\right|^2\right]$$
$$= \widehat{\mathbb{E}}_{\tau,x}\left[f\left(\widehat{X}_t\right)^2\right] - 2\widehat{\mathbb{E}}_{\tau,x}\left[f\left(\widehat{X}_t\right)f\left(\widetilde{X}_t\right)\right] + \widehat{\mathbb{E}}_{\tau,x}\left[f\left(\widetilde{X}_t\right)^2\right]$$
$$= \widetilde{\mathbb{E}}_{\tau,x}\left[f\left(\widetilde{X}_t\right)^2\right] - 2\widetilde{\mathbb{E}}_{\tau,x}\left[f\left(\widetilde{X}_t\right)f\left(\widetilde{X}_t\right)\right] + \widetilde{\mathbb{E}}_{\tau,x}\left[f\left(\widetilde{X}_t\right)^2\right] = 0 \quad (2.82)$$

for all $f \in C_0$, $0 \le \tau \le t \le T$, and $x \in E$.

Let $f_k \in C_0$ with $\|f_k\|_\infty \le 1$, $k \ge 1$, be a sequence such that the closure of its linear span coincides with C_0. For all $x, y \in E$, put

$$\widetilde{\rho}(x,y) = \sum_{k=1}^{\infty} 2^{-k}\left|f_k(x) - f_k(y)\right|.$$

Then $\widetilde{\rho}$ is a metric on $E \times E$ generating the same topology as the metric ρ. It follows from (2.82) that

$$\widehat{\mathbb{E}}_{\tau,x}\left[\widetilde{\rho}\left(\widehat{X}_t, \widetilde{X}_t\right)\right] = 0 \quad (2.83)$$

for all $0 \le \tau \le t \le T$ and $x \in E$.

This completes the proof of Lemma 2.7. □

Now we are ready to finish the proof of Theorem 2.13. By equality (2.60) and Lemma 2.7,

$$\mathbb{P}_{\tau,x}\left[X_{t_1} \in B_1, \ldots, X_{t_n} \in B_n\right] = \widehat{\mathbb{P}}_{\tau,x}\left[\widehat{X}_{t_1} \in B_1, \ldots, \widehat{X}_{t_n} \in B_n\right] \quad (2.84)$$

for all $\tau \le t_1 < \cdots < t_n \le T$ and $B_j \in \mathcal{E}$ with $1 \le j \le n$. It is not hard to see that Theorem 2.13 follows from (2.84). □

2.6 Stopping Times and the Strong Markov Property

In this section we discuss the Markov property with respect to random times. In the next definition, an important family of random times is introduced.

Definition 2.6 Let $(\Omega, \mathcal{G}, \mathcal{G}_t, \mathbb{P})$, $\tau \leq t \leq T$, be a filtered probability space. A function $S : \Omega \to [\tau, T]$ is called a $\mathcal{G}_t$-stopping time if for every $t \in [\tau, T]$, the event $\{S \leq t\}$ belongs to the σ-algebra $\mathcal{G}_t$. If $T = \infty$, then it is assumed that S takes values in the extended real half-line $\bar{\mathbb{R}}_+$.

We will often write "a stopping time" instead of "a $\mathcal{G}_t$-stopping time" if the filtration is fixed. It is not hard to see that S is a stopping time if and only if the process $t \mapsto \chi_{\{S \leq t\}}$ is adapted to the filtration $\mathcal{G}_t$. If S_1 and S_2 are stopping times, then $S_1 \vee S_2$ and $S_1 \wedge S_2$ are stopping times. If the filtration $\mathcal{G}_t$ is right-continuous, then $(S_1 + S_2) \wedge T$ is a stopping time (see [Yeh (1995)], Section 3 in Chapter 1). For a stochastic process X_t and a stopping time S, the random variable X_S is defined by

$$X_S(\omega) = X_{S(\omega)}(\omega), \quad \omega \in \Omega.$$

Definition 2.7 Let $(\Omega, \mathcal{G}, \mathcal{G}_t, \mathbb{P})$, $\tau \leq t \leq T$, be a filtered probability space, and let S be a $\mathcal{G}_t$-stopping time. The σ-algebra $\mathcal{G}_S$ is defined as follows: An event $A \in \mathcal{G}$ belongs to $\mathcal{G}_S$ if and only if $A \cap \{S \leq t\} \in \mathcal{G}_t$ for every $t \in [\tau, T]$.

It is not hard to see that $\mathcal{G}_S$ is a σ-algebra. The σ-algebra $\mathcal{G}_S$ is usually interpreted as the information contained in the process X_t before or at time S. If S_1 and S_2 are stopping times such that $S_1 \leq S_2$, then

$$\mathcal{G}_{S_1} \subset \mathcal{G}_{S_2}. \tag{2.85}$$

Indeed, if $A \in \mathcal{G}_{S_1}$, then $A \cap \{S_2 \leq t\} = A \cap \{S_1 \leq t\} \cap \{S_2 \leq t\}$, and since the events $A \cap \{S_1 \leq t\}$ and $\{S_2 \leq t\}$ belong to the σ-algebra $\mathcal{G}_t$, the event $A \cap \{S_2 \leq t\}$ also belongs to $\mathcal{G}_t$.

Let S_1 and S_2 be stopping times with $S_1 \leq S_2$, and denote by $\mathcal{S}_{S_2}^{S_1}$ the family of all stopping times $\widetilde{S}$ such that $S_1 \leq \widetilde{S} \leq S_2$. In the next definition, we introduce σ-algebras containing the information about the process between the stopping times S_1 and S_2.

Definition 2.8 Let $(X_t, \mathcal{G}, \mathcal{G}_t, \mathbb{P})$ be a stochastic process with state space E, and let S_1 and S_2 be stopping times such that $S_1 \leq S_2$. The σ-algebras

$\widetilde{\mathcal{G}}_{S_2}^{S_1}$ and $\widehat{\mathcal{G}}_{S_2}^{S_1}$ are defined as follows:

$$\widetilde{\mathcal{G}}_{S_2}^{S_1} = \sigma\left(X_{S'} : S' \in \mathcal{S}_{S_2}^{S_1}\right)$$

and

$$\widehat{\mathcal{G}}_{S_2}^{S_1} = \sigma\left(S', X_{S'} : S' \in \mathcal{S}_{S_2}^{S_1}\right).$$

Put $\widetilde{\mathcal{G}}_S = \widetilde{\mathcal{G}}_S^0$ and $\widehat{\mathcal{G}}_S = \widehat{\mathcal{G}}_S^0$. It is not hard to see that for all stopping times S_1 and S_2 with $S_1 \le S_2$ the following inclusion holds: $\widetilde{\mathcal{G}}_{S_2}^{S_1} \subset \widehat{\mathcal{G}}_{S_2}^{S_1}$. On the other hand, it is not clear whether $\widetilde{\mathcal{G}}_S \subset \mathcal{G}_S$. However, if the process X_t is progressively measurable, then

$$\widetilde{\mathcal{G}}_S \subset \widehat{\mathcal{G}}_S \subset \mathcal{G}_S. \tag{2.86}$$

The inclusions in (2.86) follow from the fact that S is $\mathcal{G}_S$-measurable and from the following lemma:

Lemma 2.8 *Let $(X_t, \mathcal{G}_t^\tau)$, $0 \le \tau \le t \le T$, be a $\mathcal{G}_t^\tau$-progressively measurable stochastic process with state space $(E, \mathcal{E})$, where $\mathcal{G}_t^\tau$ is a two-parameter filtration. Fix $\tau \in [0, T]$, and let S be a $\mathcal{G}_t^\tau$-stopping time. Then the random variable X_S is $\mathcal{G}_S^\tau$-measurable.*

Proof. Without loss of generality we may assume that $S < T$. This follows from the possibility to extend the process X_t beyond the point $t = T$ by assuming that $X_t = X_T$ and $\mathcal{G}_t^\tau = \mathcal{G}_T^\tau$ for $t \ge T$. It is clear that the resulting process is progressively measurable on $[0, \infty)$.

Consider the stopped process $X_{S \wedge t}$, $t \in [\tau, T]$. For every $u \in [\tau, T]$ and $B \in \mathcal{E}$, we have

$$\{X_{S \wedge t} \in B\} \cap \{S \wedge t \le u\} = \{X_{S \wedge (t \wedge u)} \in B\} \cap \{S \wedge t \le u\}. \tag{2.87}$$

Since the composition of measurable functions is measurable, the random variable $X_{S \wedge (t \wedge u)}$ is $\mathcal{G}_{t \wedge u}^\tau/\mathcal{E}$-measurable. It follows that the event $\{X_{S \wedge (t \wedge u)} \in B\}$ belongs to the σ-algebra $\mathcal{G}_{t \wedge u}^\tau$, and therefore it also belongs to the σ-algebra $\mathcal{G}_u^\tau$. In addition, the event $\{S \wedge t \le u\}$ belongs to $\mathcal{G}_u^\tau$. This follows from the definition of stopping times. Therefore, the event on the left-hand side of equality (2.87) belongs to $\mathcal{G}_u^\tau$. Next, using the definition of the σ-algebra $\mathcal{G}_{S \wedge t}^\tau$ and equality (2.87), we see that $X_{S \wedge t}$ is $\mathcal{G}_{S \wedge t}^\tau$-measurable for all $t \in [\tau, T]$. Since $\mathcal{G}_{S \wedge t}^\tau \subset \mathcal{G}_S^\tau$, $X_{S \wedge t}$ is $\mathcal{G}_S^\tau$-measurable. Finally, we observe that since $S < T$, $X_{S \wedge t}(\omega) \to X_S(\omega)$ as $t \uparrow T$ for all $\omega \in \Omega$. This implies the $\mathcal{G}_S^\tau$-measurability of X_S.

The proof of Lemma 2.8 is thus completed. $\square$

Our next goal is to define and study σ-algebras related to the future behavior of a stochastic process X_t after a stopping time S, and also σ-algebras containing the present information about X_t. It seems promising to choose the σ-algebras $\widetilde{\mathcal{G}}_T^S$ and $\widehat{\mathcal{G}}_T^S$ to represent the future of the process after the stopping time S. However, this is not a good choice, if we want to use these σ-algebras in the formulation of the strong Markov property (see Definition 2.14 below), since the events belonging to the σ-algebras $\widetilde{\mathcal{G}}_T^S$ and $\widehat{\mathcal{G}}_T^S$ may depend on the past history of the process before the stopping time S.

It is easy to see that $\widetilde{\mathcal{G}}_T^S \subset \widehat{\mathcal{G}}_T^S$. On the other hand, it is not clear whether these σ-algebras coincide, or if the σ-algebras in (2.86) coincide. Note that the stopping time S is not necessarily measurable with respect to the σ-algebra $\sigma(X_S)$. As a result, we can choose the σ-algebra $\sigma(X_S)$ or the σ-algebra $\sigma(S, X_S)$ as a storage of information contained in the process X_t at the stopping time S.

Next, we turn our attention to σ-algebras representing the future of a stochastic process X_t after the stopping time S. One possibility to define such a σ-algebra is to imitate the definition of the σ-algebra $\mathcal{G}_S$ representing the past before S. Let us first note that if a stochastic process X_t is progressively measurable, then

$$\mathcal{G}_S = \bigcap_{t:0\leq t\leq T} \left\{ A \in \mathcal{G}_T^0 : A \cap \{S \leq t\} \in \sigma(S, X_S) \vee \mathcal{G}_t^0 \right\}. \tag{2.88}$$

Indeed, equality (2.88) follows from the equality

$$\left\{ A \in \mathcal{G}_T^0 : A \cap \{S \leq t\} \in \sigma(S, X_S) \vee \mathcal{G}_t^0 \right\}$$
$$= \left\{ A \in \mathcal{G}_T^0 : A \cap \{S \leq t\} \in \sigma(S \wedge t, X_{S\wedge t}) \vee \mathcal{G}_t^0 \right\}, \ 0 \leq t \leq T,$$

and from the fact that for progressively measurable processes,

$$\sigma(S \wedge t, X_{S\wedge t}) \subset \mathcal{G}_{S\wedge t}^0 \subset \mathcal{G}_t^0$$

(see Lemma 2.8). For a stopping time S, define the family of events $\mathcal{G}^S$ by the following:

$$A \in \mathcal{G}^S \iff A \cap \{S \leq t\} \in \sigma(S, X_S) \vee \mathcal{G}_T^t \tag{2.89}$$

for all t with $0 \leq t \leq T$. It is not hard to see that the family $\mathcal{G}^S$ in (2.89) is a σ-algebra. Note that an equality similar to the equality in (2.88) does not hold for this σ-algebra. Indeed, the event $\{S \leq t\}$ does not necessarily belong to the σ-algebra $\mathcal{G}_T^t$. The next equality shows that the σ-algebra $\mathcal{G}^S$

is not large enough to store the information about the process X_t between S and T:

$$\mathcal{G}^S = \sigma(S, X_S, X_T). \tag{2.90}$$

The equality in (2.90) can be obtained as follows. By (2.89) with $t = T$, we see that if $A \in \mathcal{G}^S$, then

$$A \in \sigma(S, X_S, X_T).$$

Moreover, if $A \in \sigma(S, X_S, X_T)$, then

$$A \in \sigma(S, X_S) \vee \mathcal{G}_T^t$$

for all $t \in [0, T]$. On the other hand, we have $\{S \le t\} \in \sigma(S, X_S)$ for all $t \in [0, T]$. It follows that

$$A \cap \{S \le t\} \in \sigma(S, X_S) \vee \mathcal{G}_T^t$$

for all $t \in [0, T]$, and hence $A \in \mathcal{G}^S$. This completes the proof of equality (2.90).

We will next look for larger σ-algebras than the σ-algebra $\mathcal{G}^S$ to represent the information about the future of a stochastic process after the stopping time S.

Definition 2.9 Let S be a stopping time. The class $\mathcal{N}(S)$ consists of all stopping times S' with $0 \le S \le S' \le T$ such that for every t,

$$\{S' > t \ge S\} \in \sigma(S, X_S) \vee \mathcal{G}_T^t. \tag{2.91}$$

Condition (2.91) is equivalent to the following condition: for all t_1 and t_2 with $0 \le t_1 \le t_2 \le T$,

$$\{S' > t_2 \ge t_1 \ge S\} \in \sigma(S, X_S) \vee \mathcal{G}_T^{t_2}. \tag{2.92}$$

Indeed, this equivalence can be easily established by taking into account the equality

$$\{S' > t_2 \ge t_1 \ge S\} = \{t_1 \ge S\} \cap \{S' > t_2 \ge S\}.$$

Example 2.1 Let S be a stopping time, and let a random variable S' be defined by $S' = \phi(S, X_S)$ where ϕ is a Borel function on $[0, T] \times E$. Assume that S' is a stopping time such that $S \le S'$. Then we have $S' \in \mathcal{N}(S)$. This follows from the fact that the event $\{S \le t < S'\}$ belongs to the σ-algebra $\sigma(S, X_S)$. As an example, we can take $S' = (\alpha + S) \wedge T$ where $\alpha > 0$. Another example is given by $S' = S \vee \rho$ where $0 \le \rho \le T$.

Definition 2.10 A stopping time S is called a terminal stopping time if for every pair of fixed times (t_1, t_2) with $\tau \leq t_1 < t_2 \leq T$, the event $\{t_1 < S \leq t_2\}$ belongs to the σ-algebra $\mathcal{G}_{t_2}^{t_1}$.

Remark 2.6 Not all stopping times are terminal. For instance, the stopping times $(\alpha + S) \wedge T$, where $\alpha > 0$ and S is a given stopping time, are not necessarily terminal stopping times.

Example 2.2 Let S and S' be stopping times with $\tau \leq S \leq S' \leq T$. Assume that S' is a terminal stopping time in the sense of Definition 2.10. Then $S' \in \mathcal{N}(S)$. Indeed,

$$\{S \leq t < S'\} = \{S \leq t\} \cap \{S' > t\} \in \sigma(S, X_S) \vee \mathcal{G}_T^t.$$

Lemma 2.9 Let S_1, S_2, and S_3 be stopping times such that $S_1 \leq S_2$ and $S_1 \leq S_3$. If $S_2 \in \mathcal{N}(S_1)$ and $S_3 \in \mathcal{N}(S_1)$, then $S_2 \vee S_3 \in \mathcal{N}(S_1)$ and $S_2 \wedge S_3 \in \mathcal{N}(S_1)$.

Proof. Lemma 2.9 for the stopping time $S_2 \vee S_3$ follows from

$$\{S_2 \vee S_3 > t \geq S_1\} = \{S_2 > t \geq S_1\} \cup \{S_3 > t \geq S_1\} \in \sigma(S_1, X_{S_1}) \vee \mathcal{G}_T^t.$$

The proof of Lemma 2.9 for the stopping time $S_2 \wedge S_3$ is equally simple. $\square$

The next definition will be important in our presentation of the strong Markov property (see Theorems 2.17 and 2.20 below). The σ-algebra $\mathcal{G}_T^{S,\vee}$ in Definition 2.11 is a good candidate to represent the future information about the process X_t after the stopping time S.

Definition 2.11 Let S be a stopping time. Then the σ-algebra $\mathcal{G}_T^{S,\vee}$ is defined as follows: $\mathcal{G}_T^{S,\vee} = \sigma(S \vee \rho, X_{S \vee \rho} : 0 \leq \rho \leq T)$.

The σ-algebra $\mathcal{G}_T^{S,\vee}$ is a special case of the σ-algebras introduced in the following definition.

Definition 2.12 Let $\mathcal{M}$ be a family of stopping times that is closed with respect to the operations $\wedge$ and $\vee$. For $S \in \mathcal{M}$, put

$$\mathcal{M}(S) = \{S' \in \mathcal{M} : S' \geq S\}.$$

Then the σ-algebras $\widetilde{\mathcal{G}}_T^{\mathcal{M}(S)}$ and $\widehat{\mathcal{G}}_T^{\mathcal{M}(S)}$ are defined as follows:

$$\widetilde{\mathcal{G}}_T^{\mathcal{M}(S)} = \sigma(X_{S'} : S' \in \mathcal{M}(S)), \quad \widehat{\mathcal{G}}_T^{\mathcal{M}(S)} = \sigma(S', X_{S'} : S' \in \mathcal{M}(S)). \tag{2.93}$$

It is clear that if S is a stopping time and $\mathcal{M} = \{S \vee \rho : 0 \leq \rho \leq T\}$, then $\mathcal{G}_T^{S,\vee} = \widehat{\mathcal{G}}_T^{\mathcal{M}(S)}$. Another interesting special case of the σ-algebras introduced in Definition 2.12 is as follows.

Definition 2.13 Let S be a stopping time. Then the σ-algebra $\widehat{\mathcal{G}}_T^{\mathcal{N}(S)}$ is defined by

$$\widehat{\mathcal{G}}_T^{\mathcal{N}(S)} = \sigma\left(S', X_{S'} : S' \in \mathcal{N}(S)\right).$$

For the σ-algebra $\widehat{\mathcal{G}}_T^{\mathcal{N}(S)}$, we have $\mathcal{M} = \mathcal{N}(S)$. It follows from Lemma 2.9 that this family is closed under the operations $\wedge$ and $\vee$. It is also clear that $\mathcal{M}(S) = \mathcal{N}(S)$.

The next lemma shows how various σ-algebras representing the future after a stopping time S are related.

Lemma 2.10 *Let S be a stopping time. Then*

$$\mathcal{G}^S \subset \mathcal{G}_T^{S,\vee} \subset \widehat{\mathcal{G}}_T^{\mathcal{N}(S)} \subset \widehat{\mathcal{G}}_T^S. \qquad (2.94)$$

Proof. The first inclusion in (2.94) follows from equality (2.90). Since $S \vee \rho \in \mathcal{N}(S)$ for all $\rho \in [0,T]$, we get the second inclusion in (2.94). The third inclusion is obvious.

This completes the proof of Lemma 2.10. $\square$

It is interesting to notice that for a right-continuous stochastic process X_t on $(\Omega, \mathcal{F})$ and the two-parameter filtration $\mathcal{F}_t^\tau$ generated by the process X_t, the second and third σ-algebras in (2.94) coincide.

Lemma 2.11 *Let X_t be a right-continuous stochastic process on $(\Omega, \mathcal{F})$ with state space E, and let S be a $\mathcal{F}_t^\tau$-stopping time. Then*

$$\mathcal{G}_T^{S,\vee} = \widehat{\mathcal{G}}_T^{\mathcal{N}(S)}. \qquad (2.95)$$

Proof. The inclusion $\mathcal{G}_T^{S,\vee} \subset \widehat{\mathcal{G}}_T^{\mathcal{N}(S)}$ follows from Lemma 2.10. In order to prove the opposite inclusion, let us consider the family $\mathcal{H}_t$, $t \in [0,T]$, of vector spaces of random variables on Ω defined as follows. For every $t \in [0,T]$, the vector space $\mathcal{H}_t$ consists of all bounded random variables F on Ω such that F is measurable with respect to the σ-algebra $\sigma(S, X_S) \vee \sigma(X_\rho : t \leq \rho \leq T)$ and $F\chi_{\{S \leq t\}}$ is measurable with respect to the σ-algebra $\sigma(S, X_S) \vee \sigma(S \vee \rho, X_{S \vee \rho} : t \leq \rho \leq T)$. Then $\mathcal{H}_t$ is closed under the pointwise convergence of uniformly bounded sequences. Moreover,

$\mathcal{H}_t$ contains all random variables of the form

$$f_0(S, X_S) \prod_{j=1}^{n} f_j(\rho_j, X_{\rho_j}),$$

where $f_j : [0,T] \times E \to \mathbb{R}$, $0 \le j \le n$, are bounded Borel functions, and $t \le \rho_1 < \cdots < \rho_n \le T$. By the monotone class theorem, $\mathcal{H}_t$ contains all bounded random variables which are measurable with respect to the σ-algebra $\sigma(S, X_S) \vee \sigma(X_{S \vee \rho} : t \le \rho \le T)$.

Next, let S' be a $\mathcal{F}_t^\tau$-stopping time from the class $\mathcal{N}(S)$ (see Definition 2.9). Then the event $\{S' > t \ge S\}$ belongs to the σ-algebra

$$\sigma(S, X_S) \vee \sigma(X_\rho : t \le \rho \le T).$$

It is not hard to see that the event $\{S' > t \ge S\}$ belongs to the σ-algebra

$$\sigma(S, X_S) \vee \sigma(S \vee \rho, X_{S \vee \rho} : t \le \rho \le T).$$

Hence, the same is true for the event $\{S' > t\} = \{S' > t \ge S\} \cup \{S > t\}$. It follows that the stopping time S' is measurable with respect to the σ-algebra $\sigma(S \vee \rho, X_{S \vee \rho} : 0 \le \rho \le T)$. By Theorem 2.20 (this theorem will be obtained below), we see that the state variable $X_{S'}$ is also measurable with respect to the σ-algebra $\sigma(S \vee \rho, X_{S \vee \rho} : 0 \le \rho \le T)$ (here we use the right-continuity of the process X_t). Since $S' \in \mathcal{N}(S)$ is arbitrary, the inclusion

$$\widehat{\mathcal{G}}_T^{\mathcal{N}(S)} \subset \mathcal{G}_T^{S,\vee}$$

holds.

This completes the proof of Lemma 2.11. □

The next definition concerns the strong Markov property in the case of a family of stopping times.

Definition 2.14 Let $\mathcal{M}$ be a family of stopping times that is closed with respect to the operations $\wedge$ and $\vee$. A stochastic process $(X_t, \mathcal{G}, \mathcal{G}_t, \mathbb{P})$ with state space E is called a strong Markov process with respect to the family $\mathcal{M}$ provided that the following conditions hold:

(1) The process X_t is progressively measurable.
(2) For every pair of stopping times $S_1, S_2 \in \mathcal{M}$ with $S_2 \ge S_1$ and every bounded Borel function f on $[0,T] \times E$, the equality

$$\mathbb{E}\left[f(S_2, X_{S_2}) \mid \mathcal{G}_{S_1}\right] = \mathbb{E}\left[f(S_2, X_{S_2}) \mid \sigma(S_1, X_{S_1})\right] \tag{2.96}$$

holds $\mathbb{P}$-almost surely.

The progressive measurability of the process X_t is assumed in Definition 2.14, because this implies the measurability of X_S with respect to the σ-algebra $\mathcal{G}_S$ (see Lemma 2.8). There are more versions of the strong Markov property with respect to the family $\mathcal{M}$, namely

$$\mathbb{E}\left[f(X_{S_2}) \mid \widetilde{\mathcal{G}}_{S_1}\right] = \mathbb{E}\left[f(X_{S_2}) \mid \sigma(X_{S_1})\right] \quad \mathbb{P}\text{-a.s.} \tag{2.97}$$

for all bounded Borel functions f on E, or

$$\mathbb{E}\left[f(S_2, X_{S_2}) \mid \widehat{\mathcal{G}}_{S_1}\right] = \mathbb{E}\left[f(S_2, X_{S_2}) \mid \sigma(S_1, X_{S_1})\right] \quad \mathbb{P}\text{-a.s.} \tag{2.98}$$

for all bounded Borel functions f on $[0, T] \times E$, or

$$\mathbb{E}\left[f(X_{S_2}) \mid \mathcal{G}_{S_1}\right] = \mathbb{E}\left[f(X_{S_2}) \mid \sigma(S_1, X_{S_1})\right] \quad \mathbb{P}\text{-a.s.} \tag{2.99}$$

for all bounded Borel functions f on E. Recall that it is assumed in equalities (2.97), (2.98), and (2.99) that $S_2 \geq S_1$.

The next theorem provides several equivalent conditions for the validity of the strong Markov property with respect to the family $\mathcal{M}$ (see Definition 2.14). This theorem is related to Lemmas 1.2 and 1.20. Recall that for $S \in \mathcal{M}$, we put $\mathcal{M}(S) = \{S' \in \mathcal{M} : S' \geq S\}$.

Theorem 2.14 *Let $\mathcal{M}$ be a family of stopping times that is closed with respect to $\wedge$ and $\vee$, and let X_s be a progressively measurable stochastic process on $(\Omega, \mathcal{F}, \mathbb{P})$ with state space $(E, \mathcal{E})$. Then the following are equivalent:*

(1) For all pairs $S_1 \in \mathcal{M}$ and $S_2 \in \mathcal{M}$ with $S_2 \geq S_1$ and all bounded Borel functions f on $[0, T] \times E$, the following equality holds:

$$\mathbb{E}\left[f(S_2, X_{S_2}) \mid \mathcal{G}_{S_1}\right] = \mathbb{E}\left[f(S_2, X_{S_2}) \mid \sigma(S_1, X_{S_1})\right] \quad \mathbb{P}\text{-a.s.} \tag{2.100}$$

(2) For any $S \in \mathcal{M}$ and any bounded $\widehat{\mathcal{G}}_T^{\mathcal{M}(S)}$-measurable random variable F, the equality

$$\mathbb{E}\left[F \mid \mathcal{G}_S\right] = \mathbb{E}\left[F \mid \sigma(S, X_S)\right]$$

holds $\mathbb{P}$-a.s.

(3) For any $S \in \mathcal{M}$, any bounded $\mathcal{G}_S$-measurable random variable G, and any bounded real-valued $\widehat{\mathcal{G}}_T^{\mathcal{M}(S)}$-measurable random variable F, the equality

$$\mathbb{E}\left[GF\right] = \mathbb{E}\left[G\mathbb{E}\left[F \mid \sigma(S, X_S)\right]\right]$$

holds.

(4) For any $S \in \mathcal{M}$, $A \in \mathcal{G}_S$, and $B \in \widehat{\mathcal{G}}_T^{\mathcal{M}(S)}$, the equality

$$\mathbb{P}\left[A \cap B \mid \sigma(S, X_S)\right] = \mathbb{P}\left[A \mid \sigma(S, X_S)\right] \mathbb{P}\left[B \mid \sigma(S, X_S)\right]$$

holds $\mathbb{P}$-a.s.

Proof. $(1) \Longrightarrow (2)$.
Suppose that condition (1) in Theorem 2.14 holds. The most important step in the proof of the implication $(1) \Longrightarrow (2)$ is to show that the equality

$$\mathbb{E}\left[F \mid \mathcal{G}_S\right] = \mathbb{E}\left[F \mid \sigma(X_S)\right] \qquad (2.101)$$

holds $\mathbb{P}$-almost surely provided that

$$F = \prod_{j=1}^{m} f_j(\widetilde{S}_j, X_{\widetilde{S}_j}) \qquad (2.102)$$

where for any $1 \leq j \leq m$, f_j is a bounded Borel function on $[0, T] \times E$, and $\widetilde{S}_j$ is a stopping time from $\mathcal{M}(S)$, satisfying $S \leq \widetilde{S}_1 \leq \cdots \leq \widetilde{S}_m \leq T$. We will obtain equality (2.101) using the method of mathematical induction. Since we assumed that condition (1) holds, equality (2.101) is satisfied for $m = 1$. Next, assume that (2.101) is true for a given integer $m \geq 1$ and for all finite sets $\left\{\widetilde{S}_j : 1 \leq j \leq m\right\}$ from $\mathcal{M}(S)$ with $S \leq \widetilde{S}_1 \leq \widetilde{S}_2 \leq \cdots \leq \widetilde{S}_m \leq T$. Let $\widetilde{S}_1 \leq \cdots \leq \widetilde{S}_{m+1}$ be an increasing family of stopping times from $\mathcal{M}(S)$, and put

$$\mathcal{H} = \sigma\left(\mathcal{G}_S, \widetilde{S}_1, X_{\widetilde{S}_1}, \ldots, \widetilde{S}_m, X_{\widetilde{S}_m}\right).$$

Then, using the tower property of conditional expectations, we get

$$\mathbb{E}\left[\prod_{j=1}^{m+1} f_j\left(\widetilde{S}_j, X_{\widetilde{S}_j}\right) \mid \mathcal{G}_S\right]$$

$$= \mathbb{E}\left[\left(\prod_{j=1}^{m} f_j\left(\widetilde{S}_j, X_{\widetilde{S}_j}\right)\right) f_{m+1}\left(\widetilde{S}_{m+1}, X_{\widetilde{S}_{m+1}}\right) \mid \mathcal{G}_S\right]$$

$$= \mathbb{E}\left[\prod_{j=1}^{m} f_j\left(\widetilde{S}_j, X_{\widetilde{S}_j}\right) \mathbb{E}\left[f_{m+1}\left(\widetilde{S}_{m+1}, X_{\widetilde{S}_{m+1}}\right) \mid \mathcal{H}\right] \mid \mathcal{G}_S\right]. \qquad (2.103)$$

It follows from condition (1) in Theorem 2.14 and from the inclusions $\sigma\left(\widetilde{S}_m, X_{\widetilde{S}_m}\right) \subset \mathcal{H} \subset \mathcal{G}_{\widetilde{S}_m}$ that

$$\mathbb{E}\left[f_{m+1}\left(\widetilde{S}_{m+1}, X_{\widetilde{S}_{m+1}}\right) \mid \mathcal{H}\right]$$
$$= \mathbb{E}\left[f_{m+1}\left(\widetilde{S}_{m+1}, X_{\widetilde{S}_{m+1}}\right) \mid \sigma\left(\widetilde{S}_m, X_{\widetilde{S}_m}\right)\right]. \quad (2.104)$$

Therefore, (2.103) gives

$$\mathbb{E}\left[\prod_{j=1}^{m+1} f_j\left(\widetilde{S}_j, X_{\widetilde{S}_j}\right) \mid \mathcal{G}_S\right]$$
$$= \mathbb{E}\left[\prod_{j=1}^{m} f_j\left(\widetilde{S}_j, X_{\widetilde{S}_j}\right) \mathbb{E}\left[f_{m+1}\left(\widetilde{S}_{m+1}, X_{\widetilde{S}_{m+1}}\right) \mid \sigma\left(\widetilde{S}_m, X_{\widetilde{S}_m}\right)\right] \mid \mathcal{G}_S\right]. $$
$$(2.105)$$

Lemma 2.12 *There exists a bounded Borel function g on $[0,T] \times E$ such that*

$$\mathbb{E}\left[f_{m+1}\left(\widetilde{S}_{m+1}, X_{\widetilde{S}_{m+1}}\right) \mid \sigma\left(\widetilde{S}_m, X_{\widetilde{S}_m}\right)\right] = g\left(\widetilde{S}_m, X_{\widetilde{S}_m}\right).$$

Proof. Lemma 2.12 follows from the Radon–Nikodym theorem. Indeed, consider the following Borel measures:

$$\mu_1(B) = \mathbb{E}\left[f_{m+1}\left(\widetilde{S}_{m+1}, X_{\widetilde{S}_{m+1}}\right), \left(\widetilde{S}_m, X_{\widetilde{S}_m}\right) \in B\right]$$

and

$$\mu_2(B) = \mathbb{P}\left[\left(\widetilde{S}_m, X_{\widetilde{S}_m}\right) \in B\right],$$

where B belongs to the Borel σ-algebra of $[0,T] \times E$. It is easy to see that the measure μ_1 is absolutely continuous with respect to the measure μ_2. Denote the Radon–Nikodym derivative of the measure μ_1 with respect to the measure μ_2 by g. Then the equality in the formulation of Lemma 2.12 holds. Note that μ_2 is the image measure of the measure $\mathbb{P}$ under the mapping $\left(\widetilde{S}_m, X_{\widetilde{S}_m}\right)$.

This completes the proof of Lemma 2.12. □

Let us return to the proof of Theorem 2.14. It follows from Lemma 2.12 and equality (2.105) that

$$\mathbb{E}\left[\prod_{j=1}^{m+1} f_j\left(\widetilde{S}_j, X_{\widetilde{S}_j}\right) \mid \mathcal{G}_S\right]$$
$$= \mathbb{E}\left[\prod_{j=1}^{m} f_j\left(\widetilde{S}_j, X_{\widetilde{S}_j}\right) g\left(\widetilde{S}_m, X_{\widetilde{S}_m}\right) \mid \mathcal{G}_S\right]. \qquad (2.106)$$

Applying the induction hypothesis to the right-hand side of (2.106) and using (2.104) and the tower property of conditional expectations, we get

$$\mathbb{E}\left[\prod_{j=1}^{m+1} f_j\left(\widetilde{S}_j, X_{\widetilde{S}_j}\right) \mid \mathcal{G}_S\right] = \mathbb{E}\left[\prod_{j=1}^{m} f_j\left(\widetilde{S}_j, X_{\widetilde{S}_j}\right) g\left(\widetilde{S}_m, X_{\widetilde{S}_m}\right) \mid \sigma(S, X_S)\right]$$
$$\mathbb{E}\left[\prod_{j=1}^{m} f_j\left(\widetilde{S}_j, X_{\widetilde{S}_j}\right) \mathbb{E}\left[f_{m+1}\left(\widetilde{S}_{m+1}, X_{\widetilde{S}_{m+1}}\right) \mid \sigma\left(\widetilde{S}_m, X_{\widetilde{S}_m}\right)\right] \mid \sigma(S, X_S)\right]$$
$$= \mathbb{E}\left[\prod_{j=1}^{m} f_j\left(\widetilde{S}_j, X_{\widetilde{S}_j}\right) \mathbb{E}\left[f_{m+1}\left(\widetilde{S}_{m+1}, X_{\widetilde{S}_{m+1}}\right) \mid \mathcal{H}\right] \mid \sigma(S, X_S)\right]$$
$$= \mathbb{E}\left[\mathbb{E}\left[\prod_{j=1}^{m+1} f_j\left(\widetilde{S}_j, X_{\widetilde{S}_j}\right) \mid \mathcal{H}\right] \mid \sigma(S, X_S)\right]$$
$$= \mathbb{E}\left[\prod_{j=1}^{m+1} f_j\left(\widetilde{S}_j, X_{\widetilde{S}_j}\right) \mid \sigma(S, X_S)\right]. \qquad (2.107)$$

This establishes condition (2) in Theorem 2.14 in the case where the random variable F is given by (2.102).

Our next goal is to prove the equality in (2) in the case where

$$F = \chi_{\left\{\left(\widetilde{S}_1, X_{\widetilde{S}_1}\right) \in B_1\right\}} \times \cdots \times \chi_{\left\{\left(\widetilde{S}_m, X_{\widetilde{S}_m}\right) \in B_m\right\}}. \qquad (2.108)$$

In (2.108), $B_1, \ldots, B_m$ are Borel subsets of $[0, T] \times E$ and $\widetilde{S}_1, \ldots, \widetilde{S}_m$ are stopping times from the class $\mathcal{M}(S)$. The sequence of stopping times $\widetilde{S}_i$, $1 \leq i \leq m$, is not necessarily increasing. For $1 \leq k \leq m$, define an increasing sequence of stopping times $S'_k \in \mathcal{M}(S)$ by

$$S'_k = \min\left\{\widetilde{S}_{j_1} \vee \cdots \vee \widetilde{S}_{j_k} : 1 \leq j_1 < \cdots < j_k \leq m\right\}.$$

Let π be a permutation of the set $\{1, \ldots, m\}$, and define the event A_π by

$$A_\pi = \left\{ \left(S'_{\pi(1)}, X_{S'_{\pi(1)}} \right) \in B_1, \ldots, \left(S'_{\pi(m)}, X_{S'_{\pi(m)}} \right) \in B_m \right\}.$$

Then it is not hard to see that

$$\left\{ \left(\widetilde{S}_1, X_{\widetilde{S}_1} \right) \in B_1, \ldots, \left(\widetilde{S}_m, X_{\widetilde{S}_m} \right) \in B_m \right\} = \bigcup_\pi A_\pi.$$

Next, using the inclusion-exclusion principle, we get

$$F = \chi_{\left\{ \left(\widetilde{S}_1, X_{\widetilde{S}_1} \right) \in B_1 \right\}} \times \cdots \times \chi_{\left\{ \left(\widetilde{S}_m, X_{\widetilde{S}_m} \right) \in B_m \right\}}$$
$$= \sum_\pi \chi_{A_\pi} - \sum_{\pi, \pi' : \pi \neq \pi'} \chi_{A_\pi \cap A_{\pi'}}$$
$$+ \sum_{\pi, \pi', \pi'' : \pi \neq \pi', \pi \neq \pi'', \pi' \neq \pi''} \chi_{A_\pi \cap A_{\pi'} \cap A_{\pi''}} - \cdots . \quad (2.109)$$

Since S'_k is an increasing sequence of stopping times, (2.107) and (2.109) give

$$\mathbb{E}\left[F \mid \mathcal{G}_S \right] = \mathbb{E}\left[F \mid \sigma\left(S, X_S \right) \right],$$

where F is defined by

$$F = \chi_{\left\{ \left(\widetilde{S}_1, X_{\widetilde{S}_1} \right) \in B_1 \right\}} \times \cdots \times \chi_{\left\{ \left(\widetilde{S}_m, X_{\widetilde{S}_m} \right) \in B_m \right\}}.$$

It follows from

$$G = \int_0^\infty \chi_{\{G^+ \geq \lambda\}} d\lambda - \int_0^\infty \chi_{\{G^- \geq \lambda\}} d\lambda \quad (2.110)$$

that condition (2) holds for all functions F such that

$$F = \prod_{j=1}^m f_j \left(\widetilde{S}_j, X_{\widetilde{S}_j} \right).$$

Finally, using the monotone class theorem, we obtain condition (2) as it is formulated in Theorem 2.14.
(2) $\implies$ (3).
This implication follows from the properties of conditional expectations.
(3) $\implies$ (4).
Suppose that condition (3) in Theorem 2.14 holds, and let $A \in \mathcal{G}_S$, $B \in$

$\widehat{\mathcal{G}}_T^{\mathcal{M}(S)}$, and $D \in \sigma(S, X_S)$. Then, applying condition (3) with $G = \chi_A$ and $F = \chi_B \chi_D$, we obtain

$$\mathbb{E}\left[\chi_D \chi_A \chi_B\right] = \mathbb{E}\left[\chi_A \mathbb{E}\left[\chi_D \chi_B \mid \sigma(S, X_S)\right]\right]$$
$$= \mathbb{E}\left[\chi_D \chi_A \mathbb{E}\left[\chi_B \mid \sigma(S, X_S)\right]\right]. \qquad (2.111)$$

Therefore,

$$\mathbb{E}\left[\chi_A \chi_B \mid \sigma(S, X_S)\right] = \mathbb{E}\left[\chi_A \mathbb{E}\left[\chi_B \mid \sigma(S, X_S)\right] \mid \sigma(S, X_S)\right]$$
$$= \mathbb{P}\left[A \mid \sigma(S, X_S)\right] \mathbb{P}\left[B \mid \sigma(S, X_S)\right]. \qquad (2.112)$$

This implies condition (4) in Theorem 2.14.
(4) $\implies$ (1).
Suppose that condition (4) in Theorem 2.14 holds. Then it is not hard to show that (2.111) is valid. Let $S_1 \in \mathcal{M}$ and $S_2 \in \mathcal{M}$ be stopping times such as in condition (1). Taking $D = \Omega$ in (2.111) and using (2.110) with $G = f(S_2, X_{S_2})$, we obtain

$$\mathbb{E}\left[\chi_A f(S_2, X_{S_2})\right] = \mathbb{E}\left[\chi_A \mathbb{E}\left[f(S_2, X_{S_2}) \mid \sigma(S_1, X_{S_1})\right]\right].$$

It follows from the definition of conditional expectations that

$$\mathbb{E}\left[f(S_2, X_{S_2}) \mid \mathcal{G}_{S_1}\right] = \mathbb{E}\left[f(S_2, X_{S_2}) \mid \sigma(S_1, X_{S_1})\right].$$

This implies condition (1) in Theorem 2.14.
The proof of Theorem 2.14 is thus completed. $\square$

The next theorem concerns the strong Markov property with respect to the σ-algebra $\widetilde{\mathcal{G}}_S$. Theorem 2.15 below contains an extra condition (condition (2)) in comparison with Theorem 2.14. The reason why we did not include this condition in the formulation of Theorem 2.14 is that we do not know whether $\widetilde{\mathcal{G}}_S = \mathcal{G}_S$.

Theorem 2.15 *Let $\mathcal{M}$ be a family of stopping times that is closed with respect to $\wedge$ and $\vee$, and let X_s be a progressively measurable stochastic process on $(\Omega, \mathcal{F}, \mathbb{P})$ with state space $(E, \mathcal{E})$. Then the following are equivalent:*

(1) For all stopping times $S_1 \in \mathcal{M}$ and $S_2 \in \mathcal{M}$ with $S_1 \leq S_2$,

$$\mathbb{E}\left[f(X_{S_2}) \mid \widetilde{\mathcal{G}}_{S_1}\right] = \mathbb{E}\left[f(X_{S_2}) \mid \sigma(X_{S_1})\right] \quad \mathbb{P}\text{-a.s.}$$

(2) For all stopping times $S \in \mathcal{M}$ and $\widetilde{S} \in \mathcal{M}$ with $S \leq \widetilde{S}$, any finite set of stopping times S_i, $1 \leq i \leq n$, such that $S_i \leq S$ for all i, and any bounded Borel function f on E, the equality

$$\mathbb{E}\left[f\left(X_{\widetilde{S}}\right) \mid \sigma\left(X_{S_1}, \ldots, X_{S_n}, X_S\right)\right] = \mathbb{E}\left[f\left(X_{\widetilde{S}}\right) \mid \sigma\left(X_S\right)\right]$$

holds $\mathbb{P}$-a.s..

(3) For any stopping time $S \in \mathcal{M}$ and any bounded $\widetilde{\mathcal{G}}_T^{\mathcal{M}(S)}$-measurable random variable F, the equality

$$\mathbb{E}\left[F \mid \widetilde{\mathcal{G}}_S\right] = \mathbb{E}\left[F \mid \sigma(X_S)\right]$$

holds $\mathbb{P}$-a.s.

(4) For any stopping time $S \in \mathcal{M}$, any bounded $\widetilde{\mathcal{G}}_S$-measurable random variable G, and any bounded $\widetilde{\mathcal{G}}_T^{\mathcal{M}(S)}$-measurable random variable F, the equality

$$\mathbb{E}[GF] = \mathbb{E}\left[G\mathbb{E}\left[F \mid \sigma(X_S)\right]\right]$$

holds.

(5) For any stopping time $S \in \mathcal{M}$, the equality

$$\mathbb{P}\left[A \cap B \mid \sigma(X_S)\right] = \mathbb{P}\left[A \mid \sigma(X_S)\right]\mathbb{P}\left[B \mid \sigma(X_S)\right]$$

holds $\mathbb{P}$-a.s. for all $A \in \widetilde{\mathcal{G}}_S$ and $B \in \widetilde{\mathcal{G}}_T^{\mathcal{M}(S)}$.

Proof. We will only prove the implications (1) $\implies$ (2) and (2) $\implies$ (3). The remaining implications can be obtained as in Theorem 2.14.

(1) $\implies$ (2).

Suppose that condition (1) in Theorem 2.15 holds, and let S, $\widetilde{S}$, $S_1, \ldots, S_n$ be such as in condition (2). It is not hard to see that condition (2) follows from condition (1) applied to the pair S and $\widetilde{S}$ and from the following inclusions:

$$\sigma\left(X_S\right) \subset \sigma\left(X_{S_1}, \ldots, X_{S_n}, X_S\right) \subset \widetilde{\mathcal{G}}_S.$$

(2) $\implies$ (3).

Suppose that condition (2) in Theorem 2.15 holds. Then, reasoning as in the proof of the implication (1) $\implies$ (2) in Theorem 2.14, we see that the equality

$$\mathbb{E}\left[F \mid \sigma\left(X_{S_1}, \ldots, X_{S_n}, X_S\right)\right] = \mathbb{E}\left[F \mid \sigma(X_S)\right] \qquad (2.113)$$

holds for any $\widetilde{\mathcal{G}}_T^{\mathcal{M}(S)}$-measurable random variable F. Next, replacing the σ-algebra $\sigma(X_{S_1}, \ldots, X_{S_n}, X_S)$ by the σ-algebra $\widetilde{\mathcal{G}}_S$ in formula (2.113) and using the monotone class theorem, we get condition (3) in Theorem 2.15.

This completes the proof of Theorem 2.15. □

Remark 2.7 By using the inclusion-exclusion principle as we have already done in the proof of the implication (1) $\implies$ (2) in Theorem 2.14, we can show that the following condition is equivalent to conditions (1)-(5) in Theorem 2.15:

(2') For any pair of stopping times $S \in \mathcal{M}$ and $\widetilde{S} \in \mathcal{M}$ with $S \leq \widetilde{S}$, any increasing finite sequence of stopping times $\{S_i : 1 \leq i \leq n\}$ such that $S_i \leq S$ for $1 \leq i \leq n$, and any bounded Borel function f on E, the equality

$$\mathbb{E}\left[f\left(X_{\widetilde{S}}\right) \mid \sigma(X_{S_1}, \ldots, X_{S_n}, X_S)\right] = \mathbb{E}\left[f\left(X_{\widetilde{S}}\right) \mid \sigma(X_S)\right]$$

holds $\mathbb{P}$-a.s.

An assertion, similar to Theorem 2.15, holds for the σ-algebras $\widehat{\mathcal{G}}_S$ and $\widehat{\mathcal{G}}_T^{\mathcal{M}(S)}$. The proof of this fact is similar to the proof of Theorems 2.14 and 2.15, and we leave it as an exercise for the reader.

Theorem 2.16 *Let $\mathcal{M}$ be a family of stopping times that is closed with respect to $\wedge$ and $\vee$, and let X_s be a progressively measurable stochastic process on $(\Omega, \mathcal{F}, \mathbb{P})$ with state space $(E, \mathcal{E})$. Then the following are equivalent:*

(1) For all pairs of stopping times $S_1 \in \mathcal{M}$ and $S_2 \in \mathcal{M}$ with $S_2 \geq S_1$, the following equality holds:

$$\mathbb{E}\left[f(S_2, X_{S_2}) \mid \widehat{\mathcal{G}}_{S_1}\right] = \mathbb{E}\left[f(S_2, X_{S_2}) \mid \sigma(S_1, X_{S_1})\right] \quad \mathbb{P}\text{-a.s.}$$

for all bounded Borel functions f on $[0, T] \times E$.

(2) For any pair of stopping times $S \in \mathcal{M}$ and $\widetilde{S} \in \mathcal{M}$ with $S \leq \widetilde{S}$, any finite set of stopping times $\{S_i : 1 \leq i \leq n\}$ such that $S_i \leq S$ for $1 \leq i \leq n$, and all bounded Borel functions f on $[0, T] \times E$, the equality

$$\mathbb{E}\left[f\left(\widetilde{S}, X_{\widetilde{S}}\right) \mid \sigma(S_1, X_{S_1}, \ldots, S_n, X_{S_n}, S, X_S)\right]$$
$$= \mathbb{E}\left[f\left(\widetilde{S}, X_{\widetilde{S}}\right) \mid \sigma(S, X_S)\right]$$

holds $\mathbb{P}$-a.s..

(3) Let $S \in \mathcal{M}$. Then for any bounded $\widehat{\mathcal{G}}_T^{\mathcal{M}(S)}$-measurable random variable F, the equality

$$\mathbb{E}\left[F \mid \widehat{\mathcal{G}}_S\right] = \mathbb{E}\left[F \mid \sigma(S, X_S)\right]$$

holds $\mathbb{P}$-a.s.

(4) For any $S \in \mathcal{M}$, any bounded $\widehat{\mathcal{G}}_S$-measurable random variable G, and any bounded real-valued $\widehat{\mathcal{G}}_T^{\mathcal{M}(S)}$-measurable random variable F, the equality

$$\mathbb{E}\left[GF\right] = \mathbb{E}\left[G\mathbb{E}\left[F \mid \sigma\left(S, X_S\right)\right]\right]$$

holds.

(5) For any $S \in \mathcal{M}$, $A \in \widehat{\mathcal{G}}_S$, and $B \in \widehat{\mathcal{G}}_T^{\mathcal{M}(S)}$, the equality

$$\mathbb{P}\left[A \cap B \mid \sigma(S, X_S)\right] = \mathbb{P}\left[A \mid \sigma(S, X_S)\right]\mathbb{P}\left[B \mid \sigma(S, X_S)\right]$$

holds $\mathbb{P}$-a.s.

Remark 2.8 Arguing as in the proof of the implication (1) $\implies$ (2) in Theorem 2.14, we can show that the following condition is equivalent to conditions (1)-(5) in Theorem 2.16:

(2') For all pairs of stopping times $S \in \mathcal{M}$ and $\widetilde{S} \in \mathcal{M}$ with $S \leq \widetilde{S}$, all increasing finite sequences of stopping times $\{S_i : 1 \leq i \leq n\}$ such that $S_i \leq S$ for $1 \leq i \leq n$, and all bounded Borel functions f on $[0, T] \times E$, the equality

$$\mathbb{E}\left[f\left(\widetilde{S}, X_{\widetilde{S}}\right) \mid \sigma\left(S_1, X_{S_1}, \ldots, S_n, X_{S_n}, S, X_S\right)\right] = \mathbb{E}\left[f\left(\widetilde{S}, X_{\widetilde{S}}\right) \mid \sigma\left(S, X_S\right)\right]$$

holds $\mathbb{P}$-a.s.

We conclude the present section by the definition of the strong Markov property of a stochastic process. It is based on condition (2) in Theorem 2.14 and on our choice of the future σ-algebra.

Definition 2.15 A stochastic process $(X_t, \mathcal{G}, \mathcal{G}_t, \mathbb{P})$ with state space E is called a strong Markov process if for every stopping time S and every bounded $\mathcal{G}_T^{S,\vee}$-measurable random variable F, the equality

$$\mathbb{E}\left[F \mid \mathcal{G}_S\right] = \mathbb{E}\left[F \mid \sigma\left(S, X_S\right)\right]$$

holds $\mathbb{P}$-almost surely.

2.7 Strong Markov Property with Respect to Families of Measures

This section is a continuation of Section 2.6. It is devoted to strong Markov processes with respect to families of stopping times and families of measures. Let $(X_t, \mathcal{G}_t^\tau, \mathbb{P}_{\tau,x})$ be a non-homogeneous Markov process on $(\Omega, \mathcal{F})$ with state space $(E, \mathcal{E})$. Fix $\tau \in [0, T]$, and consider a pair (S_1, S_2) of $\mathcal{G}_t^\tau$-stopping times with $\tau \leq S_1 \leq S_2 \leq T$. One of the possible ways to define the strong Markov property with respect to the pair (S_1, S_2) and the family of measures $\{\mathbb{P}_{s,y} : \tau \leq s \leq T, y \in E\}$ is the following:

$$\mathbb{E}_{\tau,x}\left[f(S_2, X_{S_2}) \mid \mathcal{G}_{S_1}^\tau\right] = \mathbb{E}_{S_1, X_{S_1}}\left[f(S_2, X_{S_2})\right] \quad \mathbb{P}_{\tau,x}\text{-a.s.} \qquad (2.114)$$

for every bounded Borel function f on $[\tau, T] \times E$. However, in order for the expressions in equality (2.114) to be well-defined, one has to impose certain restrictions on the process X_t and the stopping time S_2. Let us first assume that X_t is an $\mathcal{G}_t^\tau$-progressively measurable process (see Definition 1.15). Then, since S_1 is a stopping time with $\tau \leq S_1 \leq T$, the random variable X_S is $\mathcal{G}_{S_1}^\tau$-measurable. Indeed, by the progressive measurability of the process X_t, the function $(t, \omega) \mapsto X_t(\omega)$ is $\mathcal{B}_{[\tau,T]} \otimes \mathcal{G}_T^\tau / \mathcal{E}$-measurable. In addition, the stopping time S_1 is $\mathcal{G}_{S_1}^\tau / \mathcal{B}_{[\tau,T]}$-measurable. Since the composition of two measurable functions is measurable, the function X_{S_1} is $\mathcal{G}_{S_1}^\tau / \mathcal{E}$-measurable (see Lemma 2.8 for more details). If equality (2.114) holds, then the expression on the left-hand side of (2.114) is $\mathcal{G}_{S_1}^\tau$-measurable. Hence, we need a condition guaranteeing the $\mathcal{G}_{S_1}^\tau$-measurability of the right-hand side of (2.114). It looks natural to require the measurability of S_1 and X_{S_1} with respect to the σ-algebra $\mathcal{G}_{S_1}^\tau$. As we have already established, this property follows from the progressive measurability of the process X_t. However, this is not enough. It is necessary to impose more restrictions on S_2 and X_t in order for the expression on the right-hand side of (2.114) to be well-defined. We will see below that under these restrictions, the right-hand side of equality (2.114) is $\sigma(S_1, X_{S_1})$-measurable. This is one of the important features of the strong Markov property.

Let P be a transition probability function, and let $(X_t, \mathcal{G}_t^\tau, \mathbb{P}_{\tau,x})$ be a corresponding Markov process. We will impose several restrictions on P, X_t, and S_2 in order the expression on the right-hand side of (2.114) to be defined and $\sigma(S_1, X_{S_1})$-measurable. These restrictions include a measurability condition for P, the right-continuity of the process X_t, and the $\mathcal{G}_T^{S_1,\vee}$-measurability of the stopping time S_2. Recall that the right-continuity of a stochastic process implies its progressive measurability (see Theorem 1.5), and the measurability of S_2 with respect to the σ-algebra $\mathcal{G}_T^{S_1,\vee}$ implies the measurability of X_{S_2} with respect to the same σ-algebra (see Theorem 2.17 below). Now we are finally ready to define the strong Markov property with respect to a pair of stopping times.

Definition 2.16 Let P be a transition probability function, and let $(X_t, \mathcal{G}_t^\tau, \mathbb{P}_{\tau,x})$ be a corresponding adapted Markov process. Fix $\tau \in [0, T]$, and suppose that S_1 and S_2 are $\mathcal{G}_t^\tau$-stopping times with $\tau \leq S_1 \leq S_2 \leq T$. Then the process X_t is called a strong Markov process with respect to the pair of stopping times (S_1, S_2) and the family of measures $\{\mathbb{P}_{s,y} : \tau \leq s \leq T,\ y \in E\}$ provided that the following conditions hold:

(1) The process X_t is right-continuous.
(2) For every $B \in \mathcal{E}$, the function $(\tau, x, t) \mapsto P(\tau, x; t, B)$ is $\mathcal{B}_{[0,T]} \otimes \mathcal{E} \otimes \mathcal{B}_{[0,T]}$-measurable.
(3) The stopping time S_2 is $\mathcal{G}_T^{S_1,\vee}$-measurable.
(4) The equality

$$\mathbb{E}_{\tau,x}\left[f(S_2, X_{S_2}) \mid \mathcal{G}_{S_1}^\tau\right] = \mathbb{E}_{S_1, X_{S_1}}\left[f(S_2, X_{S_2})\right]\ \ \mathbb{P}_{\tau,x}\text{-a.s.} \qquad (2.115)$$

holds for all bounded Borel functions f on $[\tau, T] \times E$.

Recall that our choice for the σ-algebra representing the future after the stopping time S_1 is the σ-algebra $\mathcal{G}_T^{S_1,\vee}$. Hence, condition (3) in Definition 2.16 can be interpreted as follows: the stopping time S_2 resides in the future after the stopping time S_1.

It remains to show that under the restrictions described in conditions (1)-(3) in Definition 2.16 the function on the right-hand side of (2.115) makes sense. Moreover, we will establish that this function is measurable with respect to the σ-algebra $\sigma(S_1, X_{S_1})$. We will prove this fact for more general functions of the form $\mathbb{E}_{S_1, X_{S_1}}[F]$, where F is a $\mathcal{G}_T^{S_1,\vee}$-measurable random variable. Note that by condition (3) in Definition 2.16, S_2 is $\mathcal{G}_T^{S_1,\vee}$-measurable. Moreover, using condition (1) in Definition 2.16 and Theorem 2.17 below, we see that X_{S_2} is $\mathcal{G}_T^{S_1,\vee}$-measurable. Therefore, $f(S_2, X_{S_2})$ is $\mathcal{G}_T^{S_1,\vee}$-measurable for any bounded Borel function f on $[\tau, T] \times E$.

Lemma 2.13 Let P be a transition probability function, and let $(X_t, \mathcal{G}_t^\tau, \mathbb{P}_{\tau,x})$ be a corresponding Markov process. Fix $\tau \in [0, T]$ and $x \in E$, and suppose that S is a $\mathcal{G}_t^\tau$-stopping time with $\tau \leq S \leq T$ and F is a bounded $\mathcal{G}_T^{S,\vee}$-measurable random variable. Suppose also that condition (2) in Definition 2.16 holds. Then the expression $\mathbb{E}_{S,X_S}[F]$ is well-defined, and moreover, there exists a bounded Borel function g on $[0,T] \times E$ such that

$$\mathbb{E}_{S,X_S}[F] = g(S, X_S). \tag{2.116}$$

Remark 2.9 It follows from (2.116) that the random variable $\mathbb{E}_{S,X_S}[F]$ is $\sigma(S, X_S)$-measurable.

Proof. Let $\mathcal{H}$ be the vector space of all bounded $\mathcal{G}_T^{S,\vee}$-measurable random variables F satisfying the following conditions:

(i) $\mathbb{E}_{S,X_S}[F]$ exists $\mathbb{P}_{\tau,x}$-almost surely.
(ii) There exists a bounded Borel function g on $[0,T] \times E$ such that the equality in (2.116) holds.

It is not hard to see that $\mathcal{H}$ contains the constant functions and is closed under the convergence of uniformly bounded nonnegative nondecreasing sequences of its elements. Moreover, for any random variable F of the form $F = f(S \vee \rho, X_{S \vee \rho})$, where f is a bounded Borel function on $[\tau, T] \times E$ and $\tau \leq \rho \leq T$, we have $F \in \mathcal{H}$. Indeed, the random variable F can be rewritten in the following form:

$$\mathbb{E}_{S(\omega), X_{S(\omega)}(\omega)}[F] = \mathbb{E}_{S(\omega), X_{S(\omega)}(\omega)}\left[f\left(S(\omega) \vee \rho, X_{S(\omega) \vee \rho}(\cdot)\right)\right]$$
$$= Y(S(\omega), S(\omega) \vee \rho) f(S(\omega) \vee \rho) (X_{S(\omega)}(\omega))$$
$$= \int_E f(S(\omega) \vee \rho, y) P(S(\omega), X_{S(\omega)}(\omega); S(\omega) \vee \rho, dy). \tag{2.117}$$

Next, put

$$g(s, z) = \int_E f(s \vee \rho, y) P(s, z; s \vee \rho, dy).$$

Then it follows from (2.117) and condition (2) in Definition 2.16 that $F \in \mathcal{H}$.

Let F be a random variable defined by

$$F = \prod_{j=1}^{n+1} f_j\left(S \vee \rho_j, X_{S \vee \rho_j}\right), \quad n \geq 0,$$

where $\tau \leq \rho_1 < \rho_2 < \cdots < \rho_n < \rho_{n+1} \leq T$, and for every j with $1 \leq j \leq n+1$, f_j is a bounded Borel function on $[\tau, T] \times E$. Our next goal is to prove that any such random variable F belongs to $\mathcal{H}$. We will use the method of mathematical induction in the proof. For $n = 0$, the assertion above follows from (2.117). By the Markov property for constant times, we have

$$\mathbb{E}_{S(\omega),X_{S(\omega)}(\omega)}\left[\prod_{j=1}^{n+1} f_j\left(S(\omega) \vee \rho_j, X_{S(\omega)\vee\rho_j}(\cdot)\right)\right]$$

$$= \mathbb{E}_{S(\omega),X_{S(\omega)}(\omega)}\left[\prod_{j=1}^{n} f_j\left(S(\omega) \vee \rho_j, X_{S(\omega)\vee\rho_j}(\cdot)\right)\right.$$

$$\mathbb{E}_{S(\omega)\vee\rho_n,X_{S(\omega)\vee\rho_n}(\omega)}\left[f_{n+1}\left(S(\omega) \vee \rho_{n+1}, X_{S(\omega)\vee\rho_{n+1}}(\cdot)\right)\right]\Big]$$

$$= \mathbb{E}_{S(\omega),X_{S(\omega)}(\omega)}\left[\prod_{j=1}^{n} g_j\left(S(\omega) \vee \rho_j, X_{S(\omega)\vee\rho_j}(\cdot)\right)\right], \tag{2.118}$$

where $g_j = f_j$ for $1 \leq j \leq n-1$ and

$$g_n(s,y) = f_n(s,y)\mathbb{E}_{s,y}\left[f_{n+1}\left(s \vee \rho_{n+1}, X_{s\vee\rho_{n+1}}(\cdot)\right)\right].$$

Next, using condition (2) in Definition 2.16, we see that the function g_n is a bounded Borel function. Taking into account (2.118) and the fact that $F \in \mathcal{H}$ for $n = 0$, we see that induction gives $F \in \mathcal{H}$ for all $n \geq 0$. By the monotone class theorem for bounded functions, we conclude that the space $\mathcal{H}$ contains all bounded $\mathcal{G}_T^{S,\vee}$-measurable random variables.

This completes the proof of Lemma 2.13. □

The next theorem has already been used in the beginning of the present section. In our opinion, Theorem 2.17 has an independent interest.

Theorem 2.17 *Let X_t, $\tau \leq t \leq T$, be a right-continuous stochastic process with state space E, and let (S_1, S_2) be a pair of $\mathcal{G}_t^\tau$-stopping times with $\tau \leq S_1 \leq S_2 \leq T$. Suppose that S_2 is measurable with respect to the σ-algebra $\mathcal{G}_T^{S_1,\vee}$. Then the state variable X_{S_2} is measurable with respect to the same σ-algebra.*

Proof. Define a family of random variables by

$$S_{2,n} = S_1 + \frac{T-S_1}{2^n}\left\lceil\frac{2^n(S_2-S_1)}{T-S_1}\right\rceil, \quad n \geq 1.$$

We will need the following lemma in the proof of Theorem 2.17.

Lemma 2.14 *For every $n \geq 1$, the random variable $S_{2,n}$ is a stopping time. Moreover,*

$$S_2 \leq S_{2,n+1} \leq S_{2,n} \leq S_2 + \frac{T}{2^n}. \qquad (2.119)$$

Proof. It is not difficult to see that (2.119) follows from the definition of the function $x \mapsto \lceil x \rceil$ (see Subsection 1.1.1). The fact that $S_{2,n}$ is a stopping time can be established as follows. For every $\tau \leq t \leq T$, we have

$$\{S_{2,n} \leq t\} = \bigcup_{k=0}^{2^n} \left\{ \left\lceil \frac{2^n(S_2 - S_1)}{T - S_1} \right\rceil = k \right\} \cap \{S_{2,n} \leq t\}$$

$$= \{S_1 = S_2 \leq t\} \cup \bigcup_{k=1}^{2^n} \left\{ S_1 + \frac{T - S_1}{2^n}(k-1) < S_2 \leq S_1 + \frac{T - S_1}{2^n} k \leq t \right\}. \qquad (2.120)$$

It is also true that

$$\{S_2 \leq t\} \setminus \{S_1 = S_2 \leq t\} = \{S_1 < S_2 \leq t\}$$
$$= \bigcup_{r \in \mathbb{Q} \cap [\tau, t]} \{S_1 \leq r\} \cap \{r < S_2 \leq t\}. \qquad (2.121)$$

Now it is not hard to see that $\{S_1 = S_2 \leq t\} \in \mathcal{G}_t^\tau$.

For every $k \geq 1$, we have

$$\left\{ S_1 + \frac{T - S_1}{2^n}(k-1) < S_2 \leq S_1 + \frac{T - S_1}{2^n} k \leq t \right\}$$
$$= \left\{ S_1 + \frac{T - S_1}{2^n}(k-1) < S_2 \leq t \right\} \cap \left\{ S_2 \leq S_1 + \frac{T - S_1}{2^n} k \leq t \right\}$$
$$= \left\{ S_1 + \frac{T - S_1}{2^n}(k-1) < S_2 \leq t \right\} \cap \left[\{S_2 \leq t\} \right.$$
$$\left. \cap \left\{ S_1 + \frac{T - S_1}{2^n} k \leq t \right\} \setminus \left\{ S_1 + \frac{T - S_1}{2^n} k < S_2 \leq t \right\} \right]. \qquad (2.122)$$

We will next prove that for every $k \geq 0$, the random variable $S_1 + \frac{T - S_1}{2^n} k$ is a stopping time. We have already established this fact for $k = 0$. The case where $k = 2^n$ is also simple. In the case where $1 \leq k < 2^n$, we have

$$\left\{ S_1 + \frac{T - S_1}{2^n} k \leq t \right\} = \left\{ S_1 \leq \left(\frac{t 2^n - kT}{2^n - k} \right) \wedge t \right\} \in \mathcal{G}_t.$$

This implies the assertion above.

Let us go back to the proof of Lemma 2.14. By taking into account the fact that the random variable $S_1 + \dfrac{T - S_1}{2^n} k$ is a stopping time and reasoning as in the proof of (2.121), we see that the event on the right-hand side of (2.122) belongs to the σ-algebra $\mathcal{G}_t$. Now it is clear that Lemma 2.14 can be obtained from (2.120), (2.121), and (2.122). □

It follows from the right-continuity of the process X_t and from (2.119) that in order to establish the $\mathcal{G}_T^{S_1,\vee}$-measurability of the state variable X_{S_2}, it suffices to prove that the state variables $X_{S_{2,n}}$, $n \geq 1$, are $\mathcal{G}_T^{S_1,\vee}$-measurable. Consider the following events:

$$\left\{ S_{2,n} = S_1 + \frac{T - S_1}{2^n} k \right\} = \left\{ S_{2,n} = \left(1 - \frac{k}{2^n}\right) S_1 + \frac{k}{2^n} T \right\} \quad (2.123)$$

where $0 \leq k \leq 2^n$. It is not hard to see that the events in (2.123) belong to the σ-algebra $\mathcal{G}_T^{S_1,\vee}$. On these events, the state variables $X_{S_{2,n}}$ coincide with

$$X_{S_1 + \frac{T-S_1}{2^n} k} = X_{(1-\frac{k}{2^n})S_1 + \frac{k}{2^n} T}.$$

Consider the following approximating sequence for S_1:

$$S_{1,m} = \tau + \frac{T - \tau}{2^m} \left\lceil \frac{2^m (S_1 - \tau)}{T - \tau} \right\rceil.$$

Since the process X_t is right-continuous, it suffices to prove that the state variable $X_{(1-\frac{k}{2^n})S_{1,m} + \frac{k}{2^n} T}$ is $\mathcal{G}_T^{S_1,\vee}$-measurable.

Let $0 \leq \ell \leq 2^m$. Then it follows from

$$\left(1 - \frac{\ell}{2^m}\right) \tau + \frac{\ell}{2^m} T \geq S_1$$

that on the event

$$\left\{ S_{1,m} = \left(1 - \frac{\ell}{2^m}\right) \tau + \frac{\ell}{2^m} T \right\}, \quad (2.124)$$

the equality

$$X_{(1-\frac{k}{2^n})S_{1,m} + \frac{k}{2^n} T} = X_{\rho(k,\ell)}$$

holds. Here we take into account that $\rho(k,\ell) \geq S_1$. Therefore, on the event in (2.124) we have

$$X_{(1-\frac{k}{2^n})S_{1,m} + \frac{k}{2^n} T} = X_{S_1 \vee \rho(k,\ell)}. \quad (2.125)$$

Now it is not hard to see that the $\mathcal{G}_T^{S_1,\vee}$-measurability of the state variable X_{S_2} follows from the $\mathcal{G}_T^{S_1,\vee}$-measurability of the events in (2.123) and (2.124) together with equality (2.125).

This completes the proof of Theorem 2.17. □

The next definition is motivated by condition (3) in Definition (2.16).

Definition 2.17 Let X_t be a stochastic process, and let $\tau \in [0,T]$. The relation $\preceq$ is defined on the set $\mathcal{S}_T^\tau \times \mathcal{S}_T^\tau$ as follows:

$$S_1 \preceq S_2 \iff S_1 \leq S_2 \text{ and } S_2 \text{ is } \mathcal{G}_T^{S_1,\vee}\text{-measurable.} \qquad (2.126)$$

Lemma 2.15 *Suppose that X_t is a right-continuous process, and let $\tau \in [0,T]$. Then the relation $\preceq$ defined in (2.126) is a partial ordering on the set $\mathcal{S}_T^\tau \times \mathcal{S}_T^\tau$.*

Proof. The reflexivity and antisymmetry of the relation $\preceq$ are clear. Next let $S_1, S_2, S_3 \in \mathcal{S}_T^\tau \times \mathcal{S}_T^\tau$ with $S_1 \preceq S_2 \preceq S_3$. Then, it is clear that $S_1 \leq S_3$. Since S_2 is $\mathcal{G}_T^{S_1,\vee}$-measurable, $S_2 \vee \alpha$ is also $\mathcal{G}_T^{S_1,\vee}$-measurable for all $\alpha \in [\tau, T]$. It follows from the right-continuity of X_t that the function $(S_2 \vee \alpha, X_{S_2 \vee \alpha})$ is $\mathcal{G}_T^{S_1,\vee}$-measurable (apply Theorem 2.17 to $S_2 \vee \alpha$ and S_1). Therefore, the σ-algebra $\mathcal{G}_T^{S_2,\vee}$ is contained in the σ-algebra $\mathcal{G}_T^{S_1,\vee}$. Now the measurability of S_3 with respect to the σ-algebra $\mathcal{G}_T^{S_2,\vee}$ implies that S_3 is also $\mathcal{G}_T^{S_1,\vee}$-measurable.

This completes the proof of Lemma 2.15. □

Let P be a transition probability function, and let $(X_t, \mathcal{G}_t^\tau, \mathbb{P}_{\tau,x})$ be a corresponding Markov process on Ω with state space E. Suppose that for every $\tau \in [0,T]$, a family $\mathbb{P}_{\tau,x}$, $x \in E$, of probability measures is defined on the measurable space $(\Omega, \mathcal{G}_T^\tau)$. Put

$$\mathcal{M}_T^0 = \bigcup_{\tau \in [0,T]} \mathcal{S}_T^\tau.$$

Here the symbol $\mathcal{S}_T^\tau$ stands for the family of all $\mathcal{G}_t^\tau$-stopping times where $\tau \leq t \leq T$. Let $\mathcal{M}$ be a subfamily of the family $\mathcal{M}_T^0$. For every $\tau \in [0,T]$, put $\mathcal{M}(\tau) = \mathcal{M} \cap \mathcal{S}_T^\tau$.

Definition 2.18 A family $\mathcal{M} \subset \mathcal{M}_T^0$ is called an admissible family of stopping times provided that the following two conditions hold:

(i) For every $\tau \in [0,T]$, the family $\mathcal{M}(\tau)$ is closed with respect to the operations $\vee$ and $\wedge$.

(ii) For every $\tau \in [0,T]$, the conditions $S_1 \in \mathcal{M}(\tau)$, $S_2 \in \mathcal{M}(\tau)$, and $S_1 \leq S_2$ imply the condition $S_1 \preceq S_2$.

The next definition is a generalization of Definition 2.16.

Definition 2.19 Let P be a transition probability function, and let $(X_t, \mathcal{G}_t^\tau, \mathbb{P}_{\tau,x})$ be a corresponding Markov process. Suppose that $\mathcal{M}$ is an admissible family of stopping times. Then the process X_t is called a strong Markov process with respect to the family of stopping times $\mathcal{M}$ and the family of measures $\{\mathbb{P}_{\tau,x} : 0 \leq \tau \leq T,\ x \in E\}$ provided that the following conditions hold:

(1) The process X_t is right-continuous.
(2) For every $B \in \mathcal{E}$, the function $(\tau, x, t) \mapsto P(\tau, x; t, B)$ is $\mathcal{B}_{[0,T]} \otimes \mathcal{E} \otimes \mathcal{B}_{[0,T]}$-measurable.
(3) The equality

$$\mathbb{E}_{\tau,x}\left[f(S_2, X_{S_2}) \mid \mathcal{G}_{S_1}^\tau\right] = \mathbb{E}_{S_1, X_{S_1}}\left[f(S_2, X_{S_2})\right] \quad \mathbb{P}_{\tau,x}\text{-a.s.} \quad (2.127)$$

holds for all $\tau \in [0,T]$, $x \in E$, $S_1 \in \mathcal{M}(\tau)$, $S_2 \in \mathcal{M}(\tau)$ with $S_1 \leq S_2$, and all bounded Borel functions f on $[\tau, T] \times E$.

Lemma 2.16 Let X_t be a strong Markov process with respect to an admissible family of stopping times $\mathcal{M}$ and the family of measures $\mathbb{P}_{\tau,x}$. Then formulae (2.96) and (2.98) hold for every measure $\mathbb{P}_{\tau,x}$ and all $\mathcal{G}_t^\tau$-stopping times $S_1 \in \mathcal{M}(\tau)$ and $S_2 \in \mathcal{M}(\tau)$ with $\tau \leq S_1 \leq S_2 \leq T$.

Proof. Let X_t be a strong Markov process with respect to the family $\mathcal{M}$ and the family $\mathbb{P}_{\tau,x}$, and let S_1 and S_2 be stopping times such as in the formulation of Lemma 2.16. Then the random variable $F = f(S_2, X_{S_2})$ is $\mathcal{G}_T^{S_1, \vee}$-measurable (see Definition 2.18 and Theorem 2.17). Therefore, equality (2.127), Remark 2.9, and the properties of conditional expectations give

$$\begin{aligned}
\mathbb{E}_{S_1, X_{S_1}}\left[f(S_2, X_{S_2})\right] &= \mathbb{E}_{\tau,x}\left[f(S_2, X_{S_2}) \mid \mathcal{G}_{S_1}^\tau\right] \\
&= \mathbb{E}_{\tau,x}\left[\mathbb{E}_{\tau,x}\left[f(S_2, X_{S_2}) \mid \mathcal{G}_{S_1}^\tau\right] \mid \sigma(S_1, X_{S_1})\right] \\
&= \mathbb{E}_{\tau,x}\left[f(S_2, X_{S_2}) \mid \sigma(S_1, X_{S_1})\right].
\end{aligned}$$

This implies condition (2.96). Condition (2.98) follows from (2.96), since $\sigma(S_1, X_{S_1}) \subset \widehat{\mathcal{G}}_{S_1}^\tau \subset \mathcal{G}_{S_1}^\tau$.

This completes the proof of Lemma 2.16. □

The next theorems provide equivalent conditions for the validity of the strong Markov property with respect to a family of stopping times $\mathcal{M}$ and the family of measures $\mathbb{P}_{\tau,x}$. Recall that if $\mathcal{M}$ is a family of stopping times, then the family $\mathcal{M}(S)$ is defined by $\mathcal{M}(S) = \{S' \in \mathcal{M}(\tau) : S' \geq S\}$ and the symbol $\widehat{\mathcal{G}}_T^{\mathcal{M}(S)}$ stands for the σ-algebra given by

$$\widehat{\mathcal{G}}_T^{\mathcal{M}(S)} = \sigma\left(S', X_{S'} : S' \in \mathcal{M}(S)\right).$$

Theorem 2.18 *Let P be a transition probability function, and let $(X_t, \mathcal{G}_t^\tau, \mathbb{P}_{\tau,x})$ be an adapted Markov process with P as its transition function. Suppose that $\mathcal{M}$ is an admissible family of stopping times, and assume that conditions (1) and (2) in Definition 2.16 hold. Then for every $\tau \in [0,T]$ and $x \in E$, the following are equivalent:*

(1) For all $\mathcal{G}_t^\tau$-stopping times $S_1 \in \mathcal{M}(\tau)$ and $S_2 \in \mathcal{M}(\tau)$ with $\tau \leq S_1 \leq S_2 \leq T$ and all bounded Borel functions f on $[\tau, T] \times E$, the equality

$$\mathbb{E}_{\tau,x}\left[f(S_2, X_{S_2}) \mid \mathcal{G}_{S_1}^\tau\right] = \mathbb{E}_{S_1, X_{S_1}}\left[f(S_2, X_{S_2})\right]$$

holds $\mathbb{P}_{\tau,x}$-a.s.

(2) For all stopping times $S \in \mathcal{M}(\tau)$ and all bounded real-valued $\widehat{\mathcal{G}}_T^{\mathcal{M}(S)}$-measurable random variables F, the equality

$$\mathbb{E}_{\tau,x}\left[F \mid \mathcal{G}_S^\tau\right] = \mathbb{E}_{S, X_S}\left[F\right]$$

holds $\mathbb{P}_{\tau,x}$-a.s.

(3) For any stopping time $S \in \mathcal{M}(\tau)$, any bounded $\mathcal{G}_S^\tau$-measurable random variable G, and any bounded $\widehat{\mathcal{G}}_T^{\mathcal{M}(S)}$-measurable random variable F, the equality

$$\mathbb{E}_{\tau,x}[GF] = \mathbb{E}_{\tau,x}[G\mathbb{E}_{S, X_S}[F]]$$

holds.

(4) For any $A \in \mathcal{G}_S^\tau$, $B \in \widehat{\mathcal{G}}_T^{\mathcal{M}(S)}$, and $S \in \mathcal{M}(\tau)$, the equality

$$\mathbb{P}_{\tau,x}\left[A \cap B \mid \sigma(S, X_S)\right] = \mathbb{P}_{\tau,x}\left[A \mid \sigma(S, X_S)\right] \mathbb{P}_{S, X_S}[B]$$

holds $\mathbb{P}_{\tau,x}$-a.s.

Proof. We will only prove the implications (1) $\Longrightarrow$ (2) and (4) $\Longrightarrow$ (1). The remaining implications can be obtained as in Theorem 2.14.
(1) $\Longrightarrow$ (2).

Suppose that condition (1) in Theorem 2.18 holds. It will be first shown that the equality

$$\mathbb{E}_{\tau,x}\left[F \mid \mathcal{G}_S^\tau\right] = \mathbb{E}_{S,X_S}[F] \qquad (2.128)$$

holds for all random variables F given by

$$F = \prod_{j=1}^m f_j\left(\widetilde{S}_j, X_{\widetilde{S}_j}\right),$$

where for every $1 \leq j \leq m$, f_j is a bounded Borel function on $[0,T] \times E$, and $\widetilde{S}_j \in \mathcal{M}(S)$ is a stopping time such that $S \leq \widetilde{S}_1 \leq \cdots \leq \widetilde{S}_m \leq T$. We will use the method of mathematical induction in the proof of equality (2.128).

It follows from condition (1) in Theorem 2.18 that (2.128) holds for $m = 1$. Next, let $m \geq 1$ and assume that (2.128) is true for all $1 \leq k \leq m$ and for all finite subsets $\left\{\widetilde{S}_j : 1 \leq j \leq k\right\}$ of $\mathcal{M}(S)$ with

$$S \leq \widetilde{S}_1 \leq \widetilde{S}_2 \leq \cdots \leq \widetilde{S}_k \leq T.$$

For any family of stopping times $\widetilde{S}_1 \leq \cdots \leq \widetilde{S}_{m+1}$ with $m + 1$ elements contained in the class $\mathcal{M}(S)$, denote

$$\mathcal{H} = \sigma\left(\mathcal{G}_S^\tau, \widetilde{S}_1, X_{\widetilde{S}_1}, \ldots, \widetilde{S}_m, X_{\widetilde{S}_m}\right).$$

Then, using the tower property of conditional expectations, we get

$$\mathbb{E}_{\tau,x}\left[\prod_{j=1}^{m+1} f_j\left(\widetilde{S}_j, X_{\widetilde{S}_j}\right) \mid \mathcal{G}_S^\tau\right]$$

$$= \mathbb{E}_{\tau,x}\left[\left(\prod_{j=1}^m f_j\left(\widetilde{S}_j, X_{\widetilde{S}_j}\right)\right) f_{m+1}\left(\widetilde{S}_{m+1}, X_{\widetilde{S}_{m+1}}\right) \mid \mathcal{G}_S^\tau\right]$$

$$= \mathbb{E}_{\tau,x}\left[\prod_{j=1}^m f_j\left(\widetilde{S}_j, X_{\widetilde{S}_j}\right) \mathbb{E}_{\tau,x}\left[f_{m+1}\left(\widetilde{S}_{m+1}, X_{\widetilde{S}_{m+1}}\right) \mid \mathcal{H}\right] \mid \mathcal{G}_S^\tau\right]. \quad (2.129)$$

Now it follows from condition (1) and from the inclusions

$$\sigma\left(\widetilde{S}_m, X_{\widetilde{S}_m}\right) \subset \mathcal{H} \subset \mathcal{G}_{\widetilde{S}_m}^\tau$$

that

$$\mathbb{E}_{\tau,x}\left[f_{m+1}\left(\widetilde{S}_{m+1}, X_{\widetilde{S}_{m+1}}\right) \mid \mathcal{H}\right]$$
$$= \mathbb{E}_{\widetilde{S}_m, X_{\widetilde{S}_m}}\left[f_{m+1}\left(\widetilde{S}_{m+1}, X_{\widetilde{S}_{m+1}}\right)\right]. \tag{2.130}$$

Next, we obtain from (2.129) and (2.130) that

$$\mathbb{E}_{\tau,x}\left[\prod_{j=1}^{m+1} f_j\left(\widetilde{S}_j, X_{\widetilde{S}_j}\right) \mid \mathcal{G}_S^\tau\right]$$
$$= \mathbb{E}_{\tau,x}\left[\prod_{j=1}^{m} f_j\left(\widetilde{S}_j, X_{\widetilde{S}_j}\right) \mathbb{E}_{\widetilde{S}_m, X_{\widetilde{S}_m}}\left[f_{m+1}\left(\widetilde{S}_{m+1}, X_{\widetilde{S}_{m+1}}\right)\right] \mid \mathcal{G}_S^\tau\right]. \tag{2.131}$$

Since $\mathcal{M}$ is an admissible family, the stopping time $\widetilde{S}_{m+1}$ is $\mathcal{G}_T^{\widetilde{S}_m,\vee}$- measurable. Moreover, Theorem 2.17 implies that $X_{\widetilde{S}_{m+1}}$ is $\mathcal{G}_T^{\widetilde{S}_m,\vee}$- measurable. It follows from Lemma 2.13 that there exists a bounded Borel measurable function g on $[\tau, T] \times E$ such that

$$g\left(\widetilde{S}_m, X_{\widetilde{S}_m}\right) = \mathbb{E}_{\widetilde{S}_m, X_{\widetilde{S}_m}}\left[f_{m+1}\left(\widetilde{S}_{m+1}, X_{\widetilde{S}_{m+1}}\right)\right]. \tag{2.132}$$

Therefore, (2.131) and (2.132) give

$$\mathbb{E}_{\tau,x}\left[\prod_{j=1}^{m+1} f_j\left(\widetilde{S}_j, X_{\widetilde{S}_j}\right) \mid \mathcal{G}_S^\tau\right]$$
$$= \mathbb{E}_{\tau,x}\left[\prod_{j=1}^{m} f_j\left(\widetilde{S}_j, X_{\widetilde{S}_j}\right) g\left(\widetilde{S}_m, X_{\widetilde{S}_m}\right) \mid \mathcal{G}_S^\tau\right]. \tag{2.133}$$

By equality (2.130), the tower property of conditional expectations, and the induction hypothesis applied to the right-hand side of (2.133), we have

$$\mathbb{E}_{\tau,x}\left[\prod_{j=1}^{m+1} f_j\left(\widetilde{S}_j, X_{\widetilde{S}_j}\right) \mid \mathcal{G}_S^\tau\right] = \mathbb{E}_{S,X_S}\left[\prod_{j=1}^{m} f_j\left(\widetilde{S}_j, X_{\widetilde{S}_j}\right) g\left(\widetilde{S}_m, X_{\widetilde{S}_m}\right)\right]$$
$$= \mathbb{E}_{S,X_S}\left[\prod_{j=1}^{m} f_j\left(\widetilde{S}_j, X_{\widetilde{S}_j}\right) \mathbb{E}_{\widetilde{S}_m, X_{\widetilde{S}_m}}\left[f_{m+1}\left(\widetilde{S}_{m+1}, X_{\widetilde{S}_{m+1}}\right)\right]\right]$$

$$= \mathbb{E}_{S,X_S}\left[\prod_{j=1}^{m} f_j\left(\widetilde{S}_j, X_{\widetilde{S}_j}\right) \mathbb{E}_{S,X_S}\left[f_{m+1}\left(\widetilde{S}_{m+1}, X_{\widetilde{S}_{m+1}}\right) \mid \mathcal{H}\right]\right]$$

$$= \mathbb{E}_{S,X_S}\left[\mathbb{E}_{S,X_S}\left[\prod_{j=1}^{m+1} f_j\left(\widetilde{S}_j, X_{\widetilde{S}_j}\right) \mid \mathcal{H}\right]\right]$$

$$= \mathbb{E}_{S,X_S}\left[\prod_{j=1}^{m+1} f_j\left(\widetilde{S}_j, X_{\widetilde{S}_j}\right)\right]. \qquad (2.134)$$

It follows from (2.134) that equality (2.128) holds for any random variable F given by

$$F = \prod_{j=1}^{m} f_j\left(\widetilde{S}_j, X_{\widetilde{S}_j}\right),$$

where f_j, $1 \leq j \leq m$, is a finite sequence of bounded Borel functions on $[\tau, T] \times E$, and $\widetilde{S}_j \in \mathcal{M}(S)$, $1 \leq j \leq m$, is a finite sequence of stopping times satisfying $\tau \leq S \leq \widetilde{S}_1 \leq \cdots \leq \widetilde{S}_m \leq T$.

Our next goal is to prove the equality in condition (2) in Theorem 2.18 for a function F given by

$$F = \chi_{\left\{\left(\widetilde{S}_1, X_{\widetilde{S}_1}\right) \in B_1\right\}} \times \cdots \times \chi_{\left\{\left(\widetilde{S}_m, X_{\widetilde{S}_m}\right) \in B_m\right\}},$$

where $B_1, \ldots, B_m$ are Borel subsets of $[\tau, T] \times E$, and $\widetilde{S}_1, \ldots, \widetilde{S}_m$ are stopping times from the class $\mathcal{M}(S)$ satisfying $\widetilde{S}_j \geq S$ for all $1 \leq j \leq m$. Define an increasing sequence of stopping times S'_k by

$$S'_k = \min\left\{\widetilde{S}_{j_1} \vee \cdots \vee \widetilde{S}_{j_k} : 1 \leq j_1 < \cdots < j_k \leq m\right\}, \quad 1 \leq k \leq m.$$

It is clear that $S'_k \in \mathcal{M}(S)$.

For any permutation π of the set $\{1, \ldots, m\}$, put

$$A_\pi = \left\{\left(S'_{\pi(1)}, X_{S'_{\pi(1)}}\right) \in B_1, \ldots, \left(S'_{\pi(m)}, X_{S'_{\pi(m)}}\right) \in B_m\right\}.$$

Then we have

$$\left\{\left(\widetilde{S}_1, X_{\widetilde{S}_1}\right) \in B_1, \ldots, \left(\widetilde{S}_m, X_{\widetilde{S}_m}\right) \in B_m\right\} = \bigcup_\pi A_\pi.$$

Next, using the inclusion-exclusion principle, we get

$$F = \chi_{\left\{\left(\widetilde{S}_1, X_{\widetilde{S}_1}\right) \in B_1\right\}} \times \cdots \times \chi_{\left\{\left(\widetilde{S}_m, X_{\widetilde{S}_m}\right) \in B_m\right\}}$$

$$= \sum_{\pi} \chi_{A_\pi} - \sum_{\pi,\pi':\pi\neq\pi'} \chi_{A_\pi \cap A_{\pi'}}$$
$$+ \sum_{\pi,\pi',\pi'':\pi\neq\pi',\pi\neq\pi'',\pi'\neq\pi''} \chi_{A_\pi \cap A_{\pi'} \cap A_{\pi''}} - \cdots. \qquad (2.135)$$

Since S'_k is an increasing sequence of stopping times from the class $\mathcal{M}(S)$, equalities (2.134) and (2.135) imply that

$$\mathbb{E}_{\tau,x}\left[F \mid \mathcal{G}_S\right] = \mathbb{E}_{S,X_S}[F].$$

By formula (2.110), the previous equality holds for all functions F which can be represented as follows:

$$F = \prod_{j=1}^{m} f_j\left(\widetilde{S}_j, X_{\widetilde{S}_j}\right).$$

Now using the monotone class theorem, we see that condition (2) in Theorem 2.18 holds.

(4) $\implies$ (1).

Suppose that condition (4) in Theorem 2.18 holds, and let $A \in \mathcal{G}^\tau_{S_1}$, $B \in \widehat{\mathcal{G}}^{\mathcal{M}(S_1)}_T$, $S_1 \in \mathcal{M}(\tau)$, and $S_2 \in \mathcal{M}(\tau)$ with $\tau \leq S_1 \leq S_2 \leq T$. Then, using condition (4) in Theorem 2.18 with $S = S_1$, we get

$$\mathbb{P}_{\tau,x}\left[A \cap B \mid \sigma(S_1, X_{S_1})\right] = \mathbb{E}_{\tau,x}\left[\chi_A \mathbb{P}_{S_1,X_{S_1}}[B] \mid \sigma(S_1, X_{S_1})\right]$$

$\mathbb{P}_{\tau,x}$-a.s. Taking the expectation $\mathbb{E}_{\tau,x}$ in the previous equality, we see that

$$\mathbb{P}_{\tau,x}[B \cap A] = \mathbb{E}_{\tau,x}\left[\mathbb{P}_{S_1,X_{S_1}}[B] \chi_A\right]$$

for all $A \in \mathcal{G}^\tau_{S_1}$ and $B \in \widehat{\mathcal{G}}^{\mathcal{M}(S_1)}_T$. It follows that

$$\mathbb{E}_{\tau,x}\left[f(S_2, X_{S_2}) \chi_A\right] = \mathbb{E}_{\tau,x}\left[\mathbb{E}_{S_1,X_{S_1}}\left[f(S_2, X_{S_2})\right] \chi_A\right],$$

and hence

$$\mathbb{E}_{\tau,x}\left[f(S_2, X_{S_2}) \mid \mathcal{G}^\tau_{S_1}\right] = \mathbb{E}_{S_1,X_{S_1}}\left[f(S_2, X_{S_2})\right] \quad \mathbb{P}_{\tau,x}\text{-a.s.}$$

This establishes condition (1) in Theorem 2.18.

The proof of Theorem 2.18 is thus completed. □

The next theorem is similar to Theorems 2.14 and 2.18. Its proof is omitted.

Theorem 2.19 *Let P be a transition probability function, and let $(X_t, \mathcal{G}_t^\tau, \mathbb{P}_{\tau,x})$ be an adapted Markov process with P as its transition function. Suppose that $\mathcal{M}$ is an admissible family of stopping times, and assume that conditions (1) and (2) in Definition 2.16 hold. Then, for every $\tau \in [0, T]$ and $x \in E$, the following are equivalent:*

(1) For all $\mathcal{G}_t^\tau$-stopping times $S_1 \in \mathcal{M}(\tau)$ and $S_2 \in \mathcal{M}(\tau)$ with $\tau \leq S_1 \leq S_2 \leq T$, and all bounded Borel functions f on $[\tau, T] \times E$, the equality

$$\mathbb{E}_{\tau,x}\left[f(S_2, X_{S_2}) \mid \widehat{\mathcal{G}}_{S_1}^\tau\right] = \mathbb{E}_{S_1, X_{S_1}} f(S_2, X_{S_2})$$

holds $\mathbb{P}_{\tau,x}$-a.s.

(2) For all stopping times $S \in \mathcal{M}(\tau)$ and $\widetilde{S} \in \mathcal{M}(S)$, all finite families of stopping times $\{S_i : 1 \leq i \leq n\}$ such that $\tau \leq S_i \leq S$ for $1 \leq i \leq n$, and all bounded Borel functions f on $[\tau, T] \times E$, the equality

$$\mathbb{E}_{\tau,x}\left[f\left(\widetilde{S}, X_{\widetilde{S}}\right) \mid \sigma(S_1, X_{S_1}, \ldots, S_n, X_{S_n}, S, X_S)\right]$$
$$= \mathbb{E}_{S, X_S}\left[f\left(\widetilde{S}, X_{\widetilde{S}}\right)\right]$$

holds $\mathbb{P}_{\tau,x}$-a.s.

(3) For any stopping time $S \in \mathcal{M}(\tau)$ and any bounded $\widehat{\mathcal{G}}_T^{\mathcal{M}(S)}$-measurable random variable F, the equality

$$\mathbb{E}_{\tau,x}\left[F \mid \widehat{\mathcal{G}}_S^\tau\right] = \mathbb{E}_{S, X_S}[F]$$

holds $\mathbb{P}_{\tau,x}$-a.s.

(4) For any stopping time $S \in \mathcal{M}(\tau)$, any bounded $\widehat{\mathcal{G}}_S^\tau$-measurable random variable G, and any bounded real-valued $\widehat{\mathcal{G}}_T^{\mathcal{M}(S)}$-measurable random variable F, the equality

$$\mathbb{E}_{\tau,x}[GF] = \mathbb{E}_{\tau,x}[G\mathbb{E}_{S, X_S}[F]]$$

holds.

(5) For $A \in \widehat{\mathcal{G}}_S^\tau$, $B \in \widehat{\mathcal{G}}_T^{\mathcal{M}(S)}$, and any stopping time $S \in \mathcal{M}(\tau)$, the equality

$$\mathbb{P}_{\tau,x}\left[A \cap B \mid \sigma(S, X_S)\right] = \mathbb{P}_{\tau,x}\left[A \mid \sigma(S, X_S)\right] \mathbb{P}_{S, X_S}[B]$$

holds $\mathbb{P}_{\tau,x}$-a.s.

Finally, we are ready to define the strong Markov property of a stochastic process with respect to a family of measures.

Definition 2.20 Let P be a transition probability function, and let $(X_t, \mathcal{G}_t^\tau, \mathbb{P}_{\tau,x})$ be an adapted Markov process associated with P. It is said that the process X_t is a strong Markov process with respect to the family of measures $\mathbb{P}_{\tau,x}$ if for every $\tau \in [0,T]$, $x \in E$, every $\mathcal{G}_t^\tau$-stopping time S with $\tau \leq S \leq T$, and every $\mathcal{G}_T^{S,\vee}$-measurable random variable F, the equality

$$\mathbb{E}_{\tau,x}\left[F \mid \mathcal{G}_S^\tau\right] = \mathbb{E}_{S,X_S}[F]$$

holds $\mathbb{P}_{\tau,x}$-a.s.

The next theorem states that under certain restrictions on the transition probability function P, the process X_t is a strong Markov process with respect to the family of measures $\mathbb{P}_{\tau,x}$. Let P be a transition probability function, and let $(X_t, \mathcal{G}_t^\tau, \mathbb{P}_{\tau,x})$ be an adapted Markov process associated with P. Then we have

$$\mathbb{E}_{\tau,x}\left[f(X_{t_2}) \mid \mathcal{G}_{t_1}^\tau\right] = \mathbb{E}_{t_1,X_{t_1}}[f(X_{t_2})] \tag{2.136}$$

for all $\tau \leq t_1 < t_2 \leq T$, $(\tau, x) \in [0,T] \times E$, and all bounded Borel functions f on E. By the monotone class theorem, the equality in (2.136) is equivalent to the following equality:

$$\mathbb{E}_{\tau,x}\left[F \mid \mathcal{G}_{t_1}^\tau\right] = \mathbb{E}_{t_1,X_{t_1}}[F] \tag{2.137}$$

where F is any bounded $\mathcal{G}_T^{t_1}$-measurable random variable.

Let Y be the free backward propagator associated with P. It is defined for all $0 \leq \tau \leq t \leq T$ and $f \in C_0$ by the following formula:

$$Y(\tau, t) f(x) = \int_E f(y) P(\tau, x; t, dy).$$

Recall that for a given stopping time S, the symbol $\mathcal{G}_T^{S,\vee}$ stands for the σ-algebra $\sigma(S \vee \rho, X_{S \vee \rho} : 0 \leq \rho \leq T)$ (see Definition 2.11). It is clear that for a stopping time S with $\tau \leq S \leq T$,

$$\sigma(S \vee \rho, X_{S \vee \rho} : 0 \leq \rho \leq T) = \sigma(S \vee \rho, X_{S \vee \rho} : \tau \leq \rho \leq T).$$

Theorem 2.20 Let P be a transition probability function, and let $(X_t, \mathcal{G}_t^\tau, \mathbb{P}_{\tau,x})$ be an adapted Markov process associated with P. Suppose that the following conditions hold:

(1) The process X_t is right-continuous.
(2) For every $B \in \mathcal{E}$, the function $(\tau, x, t) \mapsto P(\tau, x; t, B)$ is $\mathcal{B}_{[0,T]} \otimes \mathcal{E} \otimes \mathcal{B}_{[0,T]}$-measurable.

(3) For any function $f \in C_0$ and $t \in (0,T]$, the function $(\tau,x) \mapsto Y(\tau,t) f(x)$ is continuous from the right in τ on the interval $[0,t)$ and continuous in x on E.

Then X_t is a strong Markov process with respect to the family $\{\mathbb{P}_{\tau,x}\}$.

Remark 2.10 Condition (3) in Theorem 2.20 is equivalent to the following condition:

$$\lim_{s\downarrow s_0, x\to x_0} \mathbb{E}_{s,x}[f(X_t)] = \mathbb{E}_{s_0,x_0}[f(X_t)]$$

for all $x_0 \in E$, $\tau \leq s_0 < t$, and $f \in C_0$. It follows from conditions (1) and (3) in the formulation of Theorem 2.20 that

$$\lim_{s\downarrow s_0} \mathbb{E}_{s,X_s}[f(X_t)] = \mathbb{E}_{s_0,X_{s_0}}[f(X_t)], \quad \mathbb{P}_{\tau,x}\text{-a.s.} \tag{2.138}$$

for all $x \in E$, $f \in C_0$, and $0 \leq \tau \leq s_0 < t \leq T$.

Proof. Let $\tau \in [0,T]$, $x \in E$, and let S be a $\mathcal{G}_t^\tau$-stopping time with $\tau \leq S \leq T$. Our goal is to prove that the equality

$$\mathbb{E}_{\tau,x}[F \mid \mathcal{G}_S^\tau] = \mathbb{E}_{S,X_S}[F] \tag{2.139}$$

holds $\mathbb{P}_{\tau,x}$-almost surely for any bounded $\mathcal{G}_T^{S,\vee}$-measurable random variable F. We will derive equality (2.139) from the following assertion. For all $\tau \in [0,T]$, $x \in E$, all bounded Borel functions f on $[\tau,T] \times E$, and all $\tau \leq \rho_1 < \rho_2 \leq T$,

$$\mathbb{E}_{\tau,x}\left[f(S \vee \rho_2, X_{S\vee\rho_2}) \mid \mathcal{G}_{S\vee\rho_1}^\tau\right]$$
$$= \mathbb{E}_{S\vee\rho_1, X_{S\vee\rho_1}} f(S \vee \rho_2, X_{S\vee\rho_2}) \quad \mathbb{P}_{\tau,x}\text{-a.s.} \tag{2.140}$$

Indeed, if (2.140) holds, then the validity of (1) $\Longrightarrow$ (2) in Theorem 2.18, where the family $\mathcal{M}$ is defined by $\mathcal{M} = \{S \vee \rho : \tau \leq \rho \leq T\}$, implies equality (2.139).

It remains to prove equality (2.140). Let $\tau \leq \rho_1 < \rho_2 \leq T$, and put $\widetilde{S} = S \vee \rho_1$. Then $\widetilde{S} \vee \rho_2 = S \vee \rho_2$. Consider the following approximations from above of the stopping time $\widetilde{S}$:

$$\widetilde{S}_m = \tau + \frac{T-\tau}{2^m}\left\lceil \frac{2^m(\widetilde{S}-\tau)}{T-\tau} \right\rceil, \quad m \geq 1.$$

Next we will prove that for all $m > n$ and all functions $f \in C_0([\tau, T] \times E)$,

$$\mathbb{E}_{\tau,x}\left[f\left(\widetilde{S}_n \vee \rho_2, X_{\widetilde{S}_n \vee \rho_2}\right) \mid \mathcal{G}^{\tau}_{\widetilde{S}_m}\right]$$
$$= \mathbb{E}_{\widetilde{S}_m, X_{\widetilde{S}_m}} f\left(\widetilde{S}_n \vee \rho_2, X_{\widetilde{S}_n \vee \rho_2}\right) \quad \mathbb{P}_{\tau,x}\text{-a.s.} \qquad (2.141)$$

Let us first show that (2.141) implies (2.140). Assume that (2.141) holds, and let $\omega \in \Omega$. Then the right-hand side of (2.141) evaluated at ω is equal to

$$\mathbb{E}_{\widetilde{S}_m(\omega), X_{\widetilde{S}_m(\omega)}(\omega)} f\left(\widetilde{S}_n(\omega) \vee \rho_2, X_{\widetilde{S}_n(\omega) \vee \rho_2}(\cdot)\right)$$
$$= Y\left(\widetilde{S}_m(\omega), \widetilde{S}_n(\omega)\right) f\left(\widetilde{S}_n(\omega) \vee \rho_2\right)\left(X_{\widetilde{S}_n(\omega) \vee \rho_2}\right), \qquad (2.142)$$

where the function $f\left(\widetilde{S}_n(\omega) \vee \rho_2\right)$ is defined by

$$f\left(\widetilde{S}_n(\omega) \vee \rho_2\right)(y) = f\left(\widetilde{S}_n(\omega) \vee \rho_2, y\right), \quad y \in E.$$

By the right-continuity of the process X_t, condition (2) in Theorem 2.20, and equality (2.142), we see that

$$\lim_{m \to \infty} \mathbb{E}_{\widetilde{S}_m(\omega), X_{\widetilde{S}_m(\omega)}(\omega)} f\left(\widetilde{S}_n(\omega) \vee \rho_2, X_{\widetilde{S}_n(\omega) \vee \rho_2}(\cdot)\right)$$
$$= \mathbb{E}_{\widetilde{S}(\omega), X_{\widetilde{S}(\omega)}(\omega)} f\left(\widetilde{S}_n(\omega) \vee \rho_2, X_{\widetilde{S}_n(\omega) \vee \rho_2}(\cdot)\right) \qquad (2.143)$$

for $\mathbb{P}_{\tau,x}$-almost all $\omega \in \Omega$. Let $A \in \mathcal{G}^{\tau}_{\widetilde{S}}$. Then $A \in \mathcal{G}^{\tau}_{\widetilde{S}_m}$ for all $m > n$, and hence (2.141) gives

$$\mathbb{E}_{\tau,x}\left[f\left(\widetilde{S}_n \vee \rho_2, X_{\widetilde{S}_n \vee \rho_2}\right), A\right]$$
$$= \mathbb{E}_{\tau,x}\left[\mathbb{E}_{\widetilde{S}_m, X_{\widetilde{S}_m}}\left[f\left(\widetilde{S}_n \vee \rho_2, X_{\widetilde{S}_n \vee \rho_2}\right)\right], A\right]. \qquad (2.144)$$

Passing to the limit as $m \to \infty$ in (2.144) and using (2.143), we get

$$\mathbb{E}_{\tau,x}\left[f\left(\widetilde{S}_n \vee \rho_2, X_{\widetilde{S}_n \vee \rho_2}\right), A\right]$$
$$= \mathbb{E}_{\tau,x}\left[\mathbb{E}_{\widetilde{S}, X_{\widetilde{S}}}\left[f\left(\widetilde{S}_n \vee \rho_2, X_{\widetilde{S}_n \vee \rho_2}\right), A\right]\right] \qquad (2.145)$$

for all $f \in C_0([\tau, T] \times E)$. Now taking into account the continuity of the function f and the right-continuity of the process X_t, we see that the dominated convergence theorem can be applied in (2.145). Therefore,

$$\mathbb{E}_{\tau,x}\left[f\left(\widetilde{S} \vee \rho_2, X_{\widetilde{S} \vee \rho_2}\right), A\right]$$

$$= \mathbb{E}_{\tau,x}\left[\mathbb{E}_{\widetilde{S},X_{\widetilde{S}}}\left[f\left(\widetilde{S}\vee\rho_2, X_{\widetilde{S}\vee\rho_2}\right),A\right]\right]. \qquad (2.146)$$

The random variable $\widetilde{S}$ is $\mathcal{G}_{\widetilde{S}}^\tau$-measurable (use the definition of the σ-algebra $\mathcal{G}_{\widetilde{S}}^\tau$). Moreover, the state variable $X_{\widetilde{S}}$ is also $\mathcal{G}_{\widetilde{S}}^\tau$-measurable. This follows from Lemma 2.8 and from the fact that the right-continuity of a stochastic process implies its progressive measurability. Hence, the random variable $\mathbb{E}_{\widetilde{S},X_{\widetilde{S}}}\left[f\left(\widetilde{S}\vee\rho_2, X_{\widetilde{S}\vee\rho_2}\right)\right]$ is $\mathcal{G}_{\widetilde{S}}^\tau$-measurable. Now, using the definition of conditional expectations and equality (2.146), we see that equality (2.140) holds for all $f\in C_0\left([\tau,T]\times E\right)$. By Urysohn's Lemma and the monotone class theorem, equality (2.140) holds for all bounded Borel functions. This establishes the implication (2.141) $\Longrightarrow$ (2.140).

Next we will prove equality (2.141). The following lemma will be needed in the proof.

Lemma 2.17 *Let n be a given nonnegative integer. Then for every integer m with $m\geq n$ and every integer k with $0\leq k\leq 2^m$, there exists a unique integer $\ell_{k,m}^{(n)}$ such that*

$$k\leq 2^{m-n}\ell_{k,m}^{(n)}\leq 2^{m-n}+k-1.$$

Proof. It is easy to see that Lemma 2.17 holds for $m=n$. Now suppose that $m\neq n$. Then the inequalities in the formulation of Lemma 2.17 are equivalent to the inequalities

$$k2^{n-m}\leq \ell_{k,m}^{(n)}\leq k2^{n-m}+1-2^{n-m}. \qquad (2.147)$$

It is clear that the interval $[k2^{n-m}, k2^{n-m}+1-2^{n-m}]$ contains at most one integer. We will show that

$$\ell_{k,m}^{(n)}=\lfloor k2^{n-m}+1-2^{n-m}\rfloor=\lceil 2^{n-m}k\rceil. \qquad (2.148)$$

Let $\ell=\lceil 2^{n-m}k\rceil$. Then $2^{n-m}k\leq \ell<2^{n-m}k+1$, and hence $k\leq 2^{m-n}\ell<k+2^{m-n}$. It follows that $k\leq 2^{m-n}\ell\leq k+2^{m-n}-1$. Therefore,

$$k2^{n-m}\leq \ell\leq k2^{n-m}+1-2^{n-m}$$

and $\ell\leq \lfloor k2^{n-m}+1-2^{n-m}\rfloor$. Combining the previous equality with the facts that $\ell=\lceil 2^{n-m}k\rceil$ and there is at most one integer between $k2^{n-m}$ and $k2^{n-m}+1-2^{n-m}$, we see that (2.147) holds.

This completes the proof of Lemma 2.17. $\square$

We are finally ready to finish the proof of Theorem 2.20. Note that it only remains to establish equality (2.141). For $n \geq 1$, $m \geq n$, and $0 \leq k \leq 2^m$, choose $\ell_{k,m}^{(n)}$ as in Lemma 2.17. Then for $\widetilde{S}_m$ defined by

$$\widetilde{S}_m = \tau + \frac{T-\tau}{2^m}\left\lceil \frac{2^m\left(\widetilde{S}-\tau\right)}{T-\tau}\right\rceil,$$

we have

$$\left\{\widetilde{S}_m = \tau + \frac{T-\tau}{2^m}k\right\} \subset \left\{\widetilde{S}_n = \tau + \frac{T-\tau}{2^n}\ell_{k,m}^{(n)}\right\}. \tag{2.149}$$

Indeed, the event on the left-hand side of (2.149) is the same as the event $\left\{k-1 \leq \dfrac{2^m\left(\widetilde{S}-\tau\right)}{T-\tau} < k\right\}$, while the event on the right-hand side of (2.149) coincides with the event $\left\{\ell_{k,m}^{(n)} - 1 \leq \dfrac{2^n\left(\widetilde{S}-\tau\right)}{T-\tau} < \ell_{k,m}^{(n)}\right\}$. Now we see that (2.149) follows from Lemma 2.17. Next put

$$t(k,m) = \tau + \frac{T-\tau}{2^m}k = \left(1 - \frac{k}{2^m}\right)\tau + \frac{k}{2^m}T$$

and

$$t(n,k,m) = \tau + \frac{T-\tau}{2^n}\ell_{k,m}^{(n)} = \left(1 - \frac{\ell_{k,m}^{(n)}}{2^n}\right)\tau + \frac{\ell_{k,m}^{(n)}}{2^n}T.$$

Then, it follows from the properties of the number $\ell_{k,m}^{(n)}$ that $t(k,m) \leq t(n,k,m)$. Therefore, (2.149) gives

$$\left\{\widetilde{S}_m = t(k,m)\right\} \subset \left\{\widetilde{S}_n = t(n,k,m)\right\}. \tag{2.150}$$

Let $A \in \mathcal{G}_{\widetilde{S}_m}^{\tau}$. Then, using (2.150), we get

$$\mathbb{E}_{\tau,x}\left[f\left(\widetilde{S}_n \vee \rho_2, X_{\widetilde{S}_n \vee \rho_2}\right), A\right]$$

$$= \sum_{k=0}^{2^m} \mathbb{E}_{\tau,x}\left[f\left(\widetilde{S}_n \vee \rho_2, X_{\widetilde{S}_n \vee \rho_2}\right), A \cap \left\{\widetilde{S}_m = t(k,m)\right\}\right]$$

$$= \sum_{k=0}^{2^m} \mathbb{E}_{\tau,x} \left[f\left(t(n,k,m) \vee \rho_2, X_{t(n,k,m) \vee \rho_2}\right), A \cap \left\{\widetilde{S}_m = t(k,m)\right\}\right]$$

$$= \sum_{k=0}^{2^m} \mathbb{E}_{\tau,x} \left[\mathbb{E}_{\tau,x} \left[f\left(t(n,k,m) \vee \rho_2, X_{t(n,k,m) \vee \rho_2}\right), \right.\right.$$
$$\left.\left. A \cap \left\{\widetilde{S}_m = t(k,m)\right\} \mid \mathcal{G}^\tau_{t(k,m)}\right]\right]. \tag{2.151}$$

It follows from the definition of the σ-algebra $\mathcal{G}^\tau_{\widetilde{S}_m}$ that

$$A \cap \left\{\widetilde{S}_m = t(k,m)\right\} \in \mathcal{G}^\tau_{t(k,m)}.$$

Therefore, (2.151) gives

$$\mathbb{E}_{\tau,x}\left[f\left(\widetilde{S}_n \vee \rho_2, X_{\widetilde{S}_n \vee \rho_2}\right), A\right]$$
$$= \sum_{k=0}^{2^m} \mathbb{E}_{\tau,x}\left[\mathbb{E}_{\tau,x}\left[f\left(t(n,k,m) \vee \rho_2, X_{t(n,k,m) \vee \rho_2}\right) \mid \mathcal{G}^\tau_{t(k,m)}\right],\right.$$
$$\left. A \cap \left\{\widetilde{S}_m = t(k,m)\right\}\right]. \tag{2.152}$$

Next, applying the Markov property for constant times and using (2.150) in (2.152), we obtain

$$\mathbb{E}_{\tau,x}\left[f\left(\widetilde{S}_n \vee \rho_2, X_{\widetilde{S}_n \vee \rho_2}\right), A\right]$$
$$= \mathbb{E}_{\tau,x}\left[\sum_{k=0}^{2^m} \mathbb{E}_{t(k,m), X_{t(k,m)}}\left[f\left(t(n,k,m) \vee \rho_2, X_{t(n,k,m) \vee \rho_2}\right)\right],\right.$$
$$\left. \chi_{\left\{\widetilde{S}_m = t(k,m)\right\}}, A\right]. \tag{2.153}$$

Since the random variable

$$\sum_{k=0}^{2^m} \mathbb{E}_{t(k,m), X_{t(k,m)}}\left[f\left(t(n,k,m) \vee \rho_2, X_{t(n,k,m) \vee \rho_2}\right)\right] \chi_{\left\{\widetilde{S}_m = t(k,m)\right\}}$$

is $\mathcal{G}^\tau_{\widetilde{S}_m}$-measurable, inclusion (2.150) and equality (2.153) give

$$\mathbb{E}_{\tau,x}\left[f\left(\widetilde{S}_n \vee \rho_2, X_{\widetilde{S}_n \vee \rho_2}\right) \mid \mathcal{G}^\tau_{\widetilde{S}_m}\right]$$
$$= \sum_{k=0}^{2^m} \mathbb{E}_{t(k,m), X_{t(k,m)}}\left[f\left(t(n,k,m) \vee \rho_2, X_{t(n,k,m) \vee \rho_2}\right)\right] \chi_{\left\{\widetilde{S}_m = t(k,m)\right\}}$$

$$= \sum_{k=0}^{2^m} \mathbb{E}_{\widetilde{S}_m, X_{\widetilde{S}_m}} \left[f \left(\widetilde{S}_n \vee \rho_2, X_{\widetilde{S}_n \vee \rho_2} \right) \right] \chi_{\{\widetilde{S}_m = t(k,m)\}}$$

$$= \mathbb{E}_{\widetilde{S}_m, X_{\widetilde{S}_m}} \left[f \left(\widetilde{S}_n \vee \rho_2, X_{\widetilde{S}_n \vee \rho_2} \right) \right].$$

This implies equality (2.141).

The proof of Theorem 2.20 is thus completed. □

Remark 2.11 Fix $\tau \in [0,T]$ and $x \in E$, and denote by $\bar{\mathcal{G}}_T^{S,\vee}$ the completion of the σ-algebra $\mathcal{G}_T^{S,\vee}$ with respect to the measure $\mathbb{P}_{\tau,x}$. Arguing as in the proof of Theorem 2.20, we see that the $\mathcal{G}_T^{S,\vee}$-measurability assumption for the random variable F in the formulation of the strong Markov property can be replaced by the $\bar{\mathcal{G}}_T^{S_1,\vee}$-measurability of F. This means that under the conditions in Theorem 2.20, the $\bar{\mathcal{G}}_T^{S,\vee}$-measurability of the random variable F implies the strong Markov property in the following form:

$$\mathbb{E}_{\tau,x} \left[F \mid \bar{\mathcal{G}}_{S_1}^{\tau} \right] = \mathbb{E}_{S_1, X_{S_1}} [F]$$

$\mathbb{P}_{\tau,x}$-almost surely.

The next assertion follows from Theorem 2.20.

Corollary 2.1 *Let P be a transition probability function, and let $(X_t, \mathcal{G}_t^{\tau}, \mathbb{P}_{\tau,x})$ be an adapted Markov process associated with P. Suppose that the following conditions hold:*

(1) The process X_t is right-continuous.
(2) For every $B \in \mathcal{E}$, the function $(\tau, x, t) \mapsto P(\tau, x; t, B)$ is $\mathcal{B}_{[0,T]} \otimes \mathcal{E} \otimes \mathcal{B}_{[0,T]}$-measurable.
(3) For every $f \in C_0$ and $t \in (0,T]$, the function $(\tau, x) \mapsto Y(\tau, t) f(x)$ is continuous from the right in τ on the interval $[0,t)$ and continuous in $x \in E$.

Let $\mathcal{M}$ be an admissible family of stopping times. Then the process X_t is a strong Markov process with respect to the families $\mathcal{M}$ and $\{\mathbb{P}_{\tau,x}\}$.

Next we will give examples of families of stopping times which can be used as the families $\mathcal{M}(\tau)$ in Corollary 2.1. More examples will be given in Section 2.10.

Example 2.3 Let P be a transition probability function, and let $(X_t, \mathcal{G}_t^{\tau}, \mathbb{P}_{\tau,x})$ be a corresponding Markov process. Assume that the process X_t is right-continuous, and fix $\tau \in [0,T]$. For a given $\mathcal{G}_t^{\tau}$-stopping time S with $\tau \leq S \leq T$, consider the following family of stopping times:

$\mathcal{M} = \mathcal{M}(\tau) = \{(S + \rho) \wedge T : \rho \geq 0\}$. It is not difficult to prove that the family $\mathcal{M}$ satisfies the following condition: for all $S_1 \in \mathcal{M}$ and $S_2 \in \mathcal{M}$ with $\tau \leq S_1 \leq S_2 \leq T$, the stopping time S_2 is measurable with respect to the σ-algebra $\sigma(S_1) \subset \mathcal{G}_T^{S_1,\vee}$. In the next example, we will consider a more general case.

Example 2.4 In this example, the family $\mathcal{M} = \mathcal{M}(\tau)$ consists of certain functions of a given stopping time. Let $\tau \in [0,T]$, and let S be a $\mathcal{G}_t^\tau$-stopping time with $\tau \leq S \leq T$. Suppose that the process X_t is right-continuous, and consider the family Φ of all functions $\varphi : [\tau, T] \to [\tau, T]$ satisfying the following conditions:

(1) φ is continuous on $[\tau, T]$.
(2) There exists a point $u(\varphi) \in [\tau, T]$ such that the function φ is strictly increasing on the interval $[\tau, u(\varphi)]$ and $\varphi(u) = T$ for all $u \in [u(\varphi), T]$.
(3) $\varphi(u) \geq u$ for all $u \in [0, T]$.

Let $\mathcal{M} = \mathcal{M}(\tau) = \{\varphi(S) : \varphi \in \Phi\}$. We will first prove that any random variable $\varphi(S) \in \Phi$ is a $\mathcal{G}_t^\tau$-stopping time. Then we will establish that for all $S_1 \in \mathcal{M}$ and $S_2 \in \mathcal{M}$ with $\tau \leq S_1 \leq S_2 \leq T$, the stopping time S_2 is measurable with respect to the σ-algebra $\sigma(S_1) \subset \mathcal{G}_T^{S_1,\vee}$.

Let us first notice that if $\varphi \in \Phi$, then the inverse function φ^{-1} exists on the interval $[\varphi(\tau), T]$ and maps $[\varphi(\tau), T]$ onto $[\tau, u(\varphi)]$. Moreover,

$$\varphi^{-1}(\varphi(t)) = t \quad \text{for all } t \in [\tau, u(\varphi)]. \tag{2.154}$$

In order to show that any random variable $\varphi(S)$ with $\varphi \in \Phi$ is a $\mathcal{G}_t^\tau$-stopping time, we need to prove that for every $t \in [\tau, T]$, the event $\{\varphi(S) \leq t\}$ belongs to the σ-algebra $\mathcal{G}_t^\tau$. If $t = T$, then

$$\{\varphi(S) \leq t\} = \Omega \in \mathcal{G}_t^\tau.$$

If $\tau \leq t < \varphi(\tau)$, then

$$\{\varphi(S) \leq t\} = \emptyset \in \mathcal{G}_t^\tau.$$

Finally, if $\varphi(\tau) \leq t < T$, then

$$\{\varphi(S) \leq t\} = \{\varphi(\tau) \leq \varphi(S) \leq t\} = \{S \in [\tau, \varphi^{-1}(t)]\} \in \mathcal{G}_{\varphi^{-1}(t)}^\tau \subset \mathcal{G}_t^\tau.$$

Therefore, $\varphi(S)$ is a $\mathcal{G}_t^\tau$-stopping time.

Next, let $\varphi_1(S) \in \Phi$ and $\varphi_2(S) \in \Phi$ be such that $\varphi_1(S) \leq \varphi_2(S)$. Then

$$\varphi_1(r) \leq \varphi_2(r) \quad \text{for all } r \in S(\Omega). \tag{2.155}$$

Consider the following function on the interval $[\varphi_1(\tau), T]$:

$$\Psi(r) = \begin{cases} \varphi_2\left(\varphi_1^{-1}(r)\right) & \text{if } r \in [\varphi_1(\tau), T), \\ T & \text{if } r = T. \end{cases} \quad (2.156)$$

Then Ψ is a Borel function on $[\varphi_1(\tau), T]$. Moreover,

$$\varphi_2(S(\omega)) = \Psi\left(\varphi_1(S(\omega))\right) \quad \text{for all } \omega \in \Omega. \quad (2.157)$$

Indeed, if $S(\omega) \in [\tau, u(\varphi_1))$, then (2.157) follows from (2.154) and (2.156). If $S(\omega) \in [u(\varphi_1), T]$, then $\varphi_1(S(\omega)) = T$, and (2.156) implies the equality $\Psi(\varphi_1(S(\omega))) = T$. On the other hand, (2.155) gives $\varphi_2(S(\omega)) = T$. This establishes (2.157). It follows from (2.157) that the random variable $\varphi_2(S)$ is measurable with respect to the σ-algebra $\sigma(\varphi(S_1))$. Since

$$\sigma\left(\varphi(S_1)\right) \subset \mathcal{G}_T^{\varphi(S_1), \vee},$$

the stopping time $\varphi_2(S)$ is $\mathcal{G}_T^{\varphi(S_1), \vee}$-measurable.

2.8 Feller-Dynkin Propagators and Completions of σ-Algebras

In this section we continue our study of Markov processes associated with Feller-Dynkin propagators. Let us first recall the construction of the process $\left(X_t, \mathcal{F}_{t+}^\tau, \mathbb{P}_{\tau, x}\right)$ in Theorem 2.13. For a given transition probability function P, we consider the standard realization $\widetilde{X}_t$ of a corresponding Markov process on the space $\left(E^{[0,T]}, \widetilde{\mathcal{F}}_T^\tau, \mathbb{P}_{s,x}\right)$. Then, by restricting the process $\widetilde{X}_t$ to the space Ω of all right-continuous E-valued functions on $[0, T]$ having left limits, we get the process X_t that is stochastically equivalent to the process $\widetilde{X}_t$. It follows from Theorem 2.13, Lemma 2.6, and Lemma 2.7 that if the free backward propagator Y associated with P is a strongly continuous backward Feller-Dynkin propagator, then X_t is a Markov process with respect to the filtration $\mathcal{F}_{t+}^\tau$. This means that

$$\mathbb{E}_{s,x}\left[f(X_u) | \mathcal{F}_{t+}^s\right] = \mathbb{E}_{t, X_t}\left[f(X_u)\right] \quad (2.158)$$

$\mathbb{P}_{s,x}$-a.s. for all $0 \leq s \leq t \leq u \leq T$ and all bounded Borel functions f on E. Since X_t is a Markov process with respect to the filtration $\mathcal{F}_t^\tau$, we have

$$\mathbb{E}_{s,x}\left[f(X_u) | \mathcal{F}_{t+}^s\right] = \mathbb{E}_{s,x}\left[f(X_u) | \mathcal{F}_t^s\right] \quad (2.159)$$

$\mathbb{P}_{s,x}$-a.s. for all $0 \leq s \leq t \leq u \leq T$ and all bounded Borel functions f on E. It follows from (2.158) and (2.159) that

$$\mathbb{E}_{s,x}\left[f(X_u)|\mathcal{F}_{t+}^\tau\right] = \mathbb{E}_{s,x}\left[f(X_u)|\mathcal{F}_t^\tau\right] = \mathbb{E}_{t,X_t}\left[f(X_u)\right] \qquad (2.160)$$

$\mathbb{P}_{s,x}$-a.s. for all $0 \leq s \leq \tau \leq t \leq u \leq T$, $x \in E$, and all bounded Borel functions f on E. In (2.160), the measure $\mathbb{P}_{s,x}$ is restricted to the σ-algebra $[\mathcal{F}_T^\tau]^{V_\tau}$. Recall that the symbol $[\mathcal{F}_T^\tau]^{V_\tau}$ stands for the completion of the σ-algebra $\mathcal{F}_T^\tau$ with respect to the family of measures defined by

$$V_\tau = \{\mathbb{P}_{s,x} : 0 \leq s \leq \tau, x \in E\}. \qquad (2.161)$$

We will also need the σ-algebras $\bar{\mathcal{F}}_t^\tau$ and $\bar{\mathcal{F}}_{t+}^\tau$ which are the completions of the σ-algebras $\mathcal{F}_t^\tau$ and $\mathcal{F}_{t+}^\tau$ with respect to the family of measures in (2.161) (see Section 1.7).

Next, suppose that $\tau \leq u \leq t$. Since the random variable $f(X_u)$ is measurable with respect to the σ-algebras $\mathcal{F}_t^\tau$ and $\mathcal{F}_{t+}$, equality (2.160) holds for $0 \leq s \leq \tau \leq t \leq T$, $\tau \leq u \leq T$, and $x \in E$. Let $F = \prod_{k=1}^n f_k(X_{u_k})$, where

$$\tau \leq u_1 < \ldots < u_i \leq t < u_{i+1} < \cdots < u_n \leq T,$$

and let f_k be bounded Borel functions on E. Then, using (2.160) and the equivalence (1) $\Longleftrightarrow$ (3) in Theorem 2.16, we get

$$\mathbb{E}_{s,x}\left[F|\mathcal{F}_{t+}^\tau\right] = \prod_{k=1}^i f_k(X_{u_k}) \mathbb{E}_{s,x}\left[\prod_{k=i+1}^n f_k(X_{u_k})|\mathcal{F}_{t+}^\tau\right]$$

$$= \prod_{k=1}^i f_k(X_{u_k}) \mathbb{E}_{s,x}\left[\prod_{k=i+1}^n f_k(X_{u_k})|\mathcal{F}_t^\tau\right] = \mathbb{E}_{s,x}\left[F|\mathcal{F}_t^\tau\right]. \qquad (2.162)$$

By the monotone class theorem, (2.162) implies that the equality

$$\mathbb{E}_{s,x}\left[F|\mathcal{F}_{t+}^\tau\right] = \mathbb{E}_{s,x}\left[F|\mathcal{F}_t^\tau\right] \qquad (2.163)$$

holds $\mathbb{P}_{s,x}$-a.s. for all $0 \leq s \leq \tau \leq t \leq T$ and all $\mathcal{F}_T^\tau$-measurable random variable F. It is assumed in (2.163) that the measure $\mathbb{P}_{s,x}$ is restricted to the σ-algebra $[\mathcal{F}_T^\tau]^{V_\tau}$.

Lemma 2.18 *Let $(X_t, \mathcal{F}_{t+}^\tau, \mathbb{P}_{\tau,x})$ be the process constructed in the proof of Theorem 2.13. Then for all $0 \leq \tau \leq t \leq T$, the σ-algebras $\bar{\mathcal{F}}_t^\tau$ and $\bar{\mathcal{F}}_{t+}^\tau$ coincide.*

Proof. The inclusion

$$\bar{\mathcal{F}}_t^\tau \subset \bar{\mathcal{F}}_{t+}^\tau \tag{2.164}$$

has already been established (see (1.47)). Next we will prove the opposite inclusion. Let $A \in \mathcal{F}_{t+}^\tau$. Then equality (2.163) with $(s,x) \in [0,\tau] \times E$ and $F = \chi_A$ shows that the function χ_A is equal to an $\mathcal{F}_t^\tau$-measurable function almost surely with respect to the measure $\mathbb{P}_{s,x}$ restricted to the σ-algebra $[\mathcal{F}_T^\tau]^{V_\tau}$. It follows that for every pair $(s,x) \in [0,\tau] \times E$, there exists a set $A_{s,x} \in \mathcal{F}_t^\tau$ such that $A \triangle A_{s,x} \in [\mathcal{F}_T^\tau]^{\mathbb{P}_{s,x}}$ and $\mathbb{P}_{s,x}(A \triangle A_{s,x}) = 0$. This means that for all $(s,x) \in [0,\tau] \times E$, the set A belongs to the completion $[\mathcal{F}_t^\tau]^{\mathbb{P}_{s,x}}$ of the σ-algebra $\mathcal{F}_t^\tau$ in the σ-algebra $[\mathcal{F}T^\tau]^{\mathbb{P}_{s,x}}$. It is not hard to prove that for all $(s,x) \in [0,\tau] \times E$, we have

$$\mathcal{F}_{t+}^\tau \subset \left[\bar{\mathcal{F}}_t^\tau\right]^{\mathbb{P}_{s,x}}.$$

Therefore,

$$\left[\bar{\mathcal{F}}_{t+}^\tau\right]^{\mathbb{P}_{s,x}} \subset \left[\bar{\mathcal{F}}_t^\tau\right]^{\mathbb{P}_{s,x}}. \tag{2.165}$$

By intersecting the sets in (2.165) with respect to all $(s,x) \in [0,\tau] \times E$, we obtain the inclusion

$$\bar{\mathcal{F}}_{t+}^\tau \subset \bar{\mathcal{F}}_t^\tau. \tag{2.166}$$

Now it is clear that Lemma 2.18 follows from (2.164) and (2.166). □

2.9 Feller-Dynkin Propagators and Standard Processes

It is natural to expect that under certain restrictions on a transition probability function P, the class of all Markov processes associated with P contains a process with special properties. For instance, the process X_t constructed in the proof of Theorem 2.13 is such a process. In this section we continue our discussion of the behavior of this process. Let us recall that the process X_t is defined on the space Ω of all E-valued functions on the interval $[0,T]$ which are right-continuous on the interval $[0,T)$ and have left limits on the interval $(0,T]$. Define a family $\{\mathcal{G}_t^\tau\}$ of σ-algebras by

$$\mathcal{G}_t^\tau = \bar{\mathcal{F}}_t^\tau, \quad 0 \leq \tau \leq t \leq T, \tag{2.167}$$

where $\bar{\mathcal{F}}_t^\tau$ is the completion of the σ-algebra $\mathcal{F}_t^\tau$ with respect to the family of measures $V_\tau = \{\mathbb{P}_{s,x} : 0 \leq s \leq \tau, x \in E\}$. The family $\{\mathcal{G}_t^\tau\}$ is a two-parameter filtration. It is not hard to see that

$$\mathcal{F}_t^\tau \subset \mathcal{G}_t^\tau \subset [\mathcal{F}_T^\tau]^{V_\tau}. \qquad (2.168)$$

For all $\tau \in [0,T]$, the process X_t, $\tau \le t \le T$, is adapted to the filtration $\{\mathcal{G}_t^\tau\}$. It is also clear that for every pair $(\tau,x) \in [0,T] \times E$ and $t \ge \tau$, the measure $\mathbb{P}_{\tau,x}$ can be extended to the σ-algebra $\mathcal{G}_t^\tau$. By (1.48) and Lemma 2.18, we see that $\mathcal{G}_t^\tau = \mathcal{G}_{t+}^\tau$, $0 \le \tau \le t \le T$. Moreover, since the σ-algebra $\mathcal{G}_t^\tau$ is V_τ-complete, we have $\mathcal{G}_t^\tau = \bar{\mathcal{G}}_t^\tau$, $0 \le \tau \le t \le T$.

In the remaining part of the present section, we will restrict ourselves to the study of the behavior of the process

$$(X_t, \mathcal{G}_t^\tau, \mathbb{P}_{\tau,x}),\ 0 \le \tau \le t \le T,\ x \in E. \qquad (2.169)$$

It follows from (2.160) that the process $(X_t, \mathcal{G}_t^\tau, \mathbb{P}_{\tau,x})$ is a Markov process, that is,

$$\mathbb{E}_{s,x}\left[f(X_u)\,|\,\mathcal{G}_t^\tau\right] = \mathbb{E}_{t,X_t}\left[f(X_u)\right]$$

for all $0 \le s \le \tau \le t \le u \le T$, $x \in E$, and all bounded Borel functions f on E. Moreover, using Corollary 2.1, we see that X_t is a strong Markov process. Recall that this means the following. Let $\mathcal{M}$ be an admissible family of stopping times. Then for all $\tau \in [0,T]$, $x \in E$, $S_1 \in \mathcal{M}(\tau)$, and $S_2 \in \mathcal{M}(\tau)$ with $\tau \le S_1 \le S_2 \le T$, the equality

$$\mathbb{E}_{\tau,x}\left[f(S_2, X_{S_2})\,|\,\mathcal{G}_{S_1}^\tau\right] = \mathbb{E}_{S_1, X_{S_1}}\left[f(S_2, X_{S_2})\right] \qquad (2.170)$$

holds $\mathbb{P}_{\tau,x}$-almost surely for any bounded Borel function f on $[\tau,T] \times E$.

The process in (2.169) has left limits but is not necessarily left-continuous. However, it satisfies a weaker condition which is called the quasi left-continuity. The next definition explains what the quasi left-continuity means for a general stochastic process.

Definition 2.21 A stochastic process $\left(\widetilde{X}_t, \widetilde{\mathcal{G}}_t^\tau, \widetilde{\mathbb{P}}_{\tau,x}\right)$ adapted to the filtration $\widetilde{\mathcal{G}}_t^\tau$ is called quasi left-continuous provided that for all $(\tau,x) \in [0,T] \times E$, all $\widetilde{\mathcal{G}}_t^\tau$-stopping times S, and all non-decreasing sequences S_n of $\widetilde{\mathcal{G}}_t^\tau$-stopping times such that $\tau \le S_n \le S \le T$ and $S = \lim_{n \to \infty} S_n$, the equality $\lim_{n \to \infty} \widetilde{X}_{S_n} = \widetilde{X}_S$ holds $\widetilde{\mathbb{P}}_{\tau,x}$-a.s.

Theorem 2.21 *Let P be a transition probability function such that the corresponding free backward propagator Y is a strongly continuous backward Feller-Dynkin propagator. Then the process $(X_t, \mathcal{G}_t^\tau, \mathbb{P}_{\tau,x})$ in (2.169) is quasi left-continuous.*

Proof. Let S_n and S be as in Definition 2.21. It follows from Theorem 2.13 that the process X_t has left limits on $(\tau, T]$. Set $X_{S-} = \lim_{n \to \infty} X_{S_n}$. Then Corollary 2.1 and Example 2.3 show that for every $n \geq 1$, the process X_t is a strong Markov process in the sense of equality (2.170). Here the family of stopping times $\mathcal{M}$ is defined by

$$\mathcal{M} = \{(S_n + \rho) \wedge T : 0 \leq \rho \leq T\}.$$

It follows that

$$\mathbb{E}_{\tau,x}\left[f\left(X_{(S_n+\rho)\wedge T}\right) | \mathcal{G}_{S_n}^\tau\right] = \mathbb{E}_{S_n, X_{S_n}}\left[f\left(X_{(S_n+\rho)\wedge T}\right)\right] \qquad (2.171)$$

$\mathbb{P}_{\tau,x}$-a.s. for all $\rho \geq 0$, $n \geq 1$, and all bounded Borel functions f on E. Now let f and g be functions from the space C_0. Then, using the right-continuity of the process X_t and equality (2.171), and taking into account the restrictions on the backward free propagator Y in Theorem 2.13, we get

$$\mathbb{E}_{\tau,x}\left[f(X_{S-}) g(X_S)\right] = \lim_{\rho \downarrow 0} \lim_{n \to \infty} \mathbb{E}_{\tau,x}\left[f(X_{S_n}) g\left(X_{(S_n+\rho)\wedge T}\right)\right]$$

$$= \lim_{\rho \downarrow 0} \lim_{n \to \infty} \mathbb{E}_{\tau,x}\left[f(X_{S_n}) \mathbb{E}_{\tau,x}\left[g\left(X_{(S_n+\rho)\wedge T}\right) | \mathcal{G}_{S_n}^\tau\right]\right]$$

$$= \lim_{\rho \downarrow 0} \lim_{n \to \infty} \mathbb{E}_{\tau,x}\left[f(X_{S_n}) \mathbb{E}_{S_n, X_{S_n}}\left[g\left(X_{(S_n+\rho)\wedge T}\right)\right]\right]$$

$$= \lim_{\rho \downarrow 0} \lim_{n \to \infty} \mathbb{E}_{\tau,x}\left[f(X_{S_n}) Y(S_n, (S_n+\rho)\wedge T) g(X_{S_n})\right]$$

$$= \lim_{\rho \downarrow 0} \mathbb{E}_{\tau,x}\left[f(X_{S-}) Y(S, (S+\rho)\wedge T) g(X_{S-})\right]$$

$$= \mathbb{E}_{\tau,x}\left[f(X_{S-}) g(X_{S-})\right]. \qquad (2.172)$$

Taking $f(x) = 1$, $x \in E$, in (2.172) and replacing g by g^2, we get

$$\mathbb{E}_{\tau,x}\left[g^2(X_S)\right] = \mathbb{E}_{\tau,x}\left[g^2(X_{S-})\right]$$

for all $g \in C_0$. Moreover, taking $f = g$ in (2.172), we obtain

$$\mathbb{E}_{\tau,x}\left[g(X_{S-}) g(X_S)\right] = \mathbb{E}_{\tau,x}\left[g^2(X_{S-})\right]$$

for all $g \in C_0$. It follows that

$$\mathbb{E}_{\tau,x}\left[(g(X_{S-}) - g(X_S))^2\right] = \mathbb{E}_{\tau,x}\left[g^2(X_{S-})\right] - 2\mathbb{E}_{\tau,x}\left[g(X_{S-}) g(X_S)\right]$$
$$+ \mathbb{E}_{\tau,x}\left[g^2(X_S)\right] = 0$$

for all $g \in C_0$. Hence,

$$g(X_{S-}) = g(X_S) \tag{2.173}$$

$\mathbb{P}_{\tau,x}$-a.s. for all $g \in C_0$. Since the space C_0 is separable, the exceptional set in (2.173) can be chosen independently of g. Since the functions from C_0 separate points, we have $X_{S-} = X_S$ $\mathbb{P}_{\tau,x}$-a.s.
This completes the proof of Theorem 2.21. □

The next definition concerns general adapted stochastic processes $(X_t, \mathcal{G}_t^\tau, \mathbb{P}_{\tau,x})$. It is based on the properties of the process in (2.169).

Definition 2.22 An adapted stochastic process $(X_t, \mathcal{G}_t^\tau, \mathbb{P}_{\tau,x})$ is called a standard process provided that

(1) The process X_t is right-continuous and has left limits.
(2) $\mathcal{G}_t^\tau = \mathcal{G}_{t+}^\tau = \overline{\mathcal{G}}_t^\tau$.
(3) For every admissible family $\mathcal{M}$ of stopping times, the process X_t is a strong Markov process with respect to $\mathcal{M}$ and $\{\mathbb{P}_{\tau,x}\}$.
(4) The process X_t is quasi left-continuous.

Remark 2.12 Let $(X_t, \mathcal{G}_t^\tau, \mathbb{P}_{\tau,x})$ be a standard process. Recall that for fixed $\tau \in [0, T]$ and $x \in E$, the symbol $\bar{\mathcal{G}}_T^{S,\vee}$ stands for the completion of the σ-algebra $\mathcal{G}_T^{S,\vee} = \sigma(S \vee \rho, X_{S \vee \rho} : 0 \leq \rho \leq T)$ with respect to the measure $\mathbb{P}_{\tau,x}$. Then, for all pairs of stopping times (S_1, S_2) such that S_2 is $\bar{\mathcal{G}}_T^{S_1,\vee}$-measurable, all $\tau \leq S_1 \leq S_2 \leq T$, and all bounded Borel functions f on $[\tau, T] \times E$, the equality

$$\mathbb{E}_{\tau,x}\left[f(S_2, X_{S_2}) \mid \bar{\mathcal{G}}_{S_1}^\tau\right] = \mathbb{E}_{S_1, X_{S_1}}[f(S_2, X_{S_2})]$$

holds $\mathbb{P}_{\tau,x}$-almost surely (see Remark 2.11).

Now we are ready to formulate one of the main results of this chapter.

Theorem 2.22 *Let P be a transition probability function such that the corresponding free backward propagator Y is a strongly continuous backward Feller-Dynkin propagator. Then there exists a standard process $(X_t, \mathcal{G}_t^\tau, \mathbb{P}_{\tau,x})$ with P as its transition function.*

Proof. Since the stochastic process in (2.169) satisfies conditions (1)-(4) in Definition 2.22, it is a standard process. It is also clear that this process has P as its transition function.
This completes the proof of Theorem 2.22. □

2.10 Hitting Times and Standard Processes

In this section we study how fast a Markov process reaches a Borel subset of the state space.

Definition 2.23 Let $(X_t, \mathcal{G}_t^\tau, \mathbb{P}_{\tau,x})$ be a Markov process on Ω with state space E, and let B be a Borel subset of E. Let $\tau \in [0, T)$, and suppose that $S: \Omega \to [\tau, T]$ is a $\mathcal{G}_t^\tau$-stopping time. For the process X_t, the entry time of the set B after time S is defined by

$$D_B^S = \begin{cases} \inf\{t: t \geq S, \, X_t \in B\} & \text{on } \bigcup_{\tau \leq t < T} \{S \leq t, \, X_t \in B\}, \\ T & \text{elsewhere.} \end{cases} \quad (2.174)$$

The pseudo-hitting time of the set B after time S is defined by

$$\widetilde{D}_B^S = \begin{cases} \inf\{t: t \geq S, \, X_t \in B\} & \text{on } \bigcup_{\tau < t < T} \{S \leq t, \, X_t \in B\}, \\ T & \text{elsewhere.} \end{cases} \quad (2.175)$$

Finally, the hitting time of the set B after time S is defined by

$$T_B^S = \begin{cases} \inf\{t: t > S, \, X_t \in B\} & \text{on } \bigcup_{\tau \leq t < T} \{S < t, \, X_t \in B\}, \\ T & \text{elsewhere.} \end{cases} \quad (2.176)$$

It is not hard to prove that

$$\bigcup_{t: \tau \leq t < T} \{S \leq t, \, X_t \in B\} = \bigcup_{t: \tau \leq t < T} \{S \vee t \leq t, \, X_{S \vee t} \in B\}$$

and

$$\bigcup_{t: \tau < t < T} \{S \leq t, \, X_t \in B\} = \bigcup_{t: \tau < t < T} \{S \vee t \leq t, \, X_{S \vee t} \in B\}.$$

We also have $D_B^S \leq \widetilde{D}_B^S \leq T_B^S$. Next we will show that the following equalities hold:

$$T_B^S = \inf_{\varepsilon > 0} \left\{ D_B^{(\varepsilon + S) \wedge T} \right\} = \inf_{r \in \mathbb{Q}^+} \left\{ D_B^{(r + S) \wedge T} \right\}. \quad (2.177)$$

Indeed, the first equality in (2.177) can be obtained using the inclusion

$$\{t \geq (\varepsilon + S) \wedge T, X_t \in B\} \subset \{t > S, X_t \in B\}$$

and the fact that for every $t \in [\tau, T)$ and $\omega \in \{S < t, X_t \in B\}$ there exists $\varepsilon > 0$ dependent on ω and such that $\omega \in \{(\varepsilon + S) \wedge T \le t, X_t \in B\}$. The second equality in (2.177) follows from the monotonicity of the entry time D_B^S with respect to S.

Our next goal is to prove that under certain restrictions on the process $(X_t, \mathcal{G}_t^\tau, \mathbb{P}_{\tau,x})$, the entry time D_B^S, the pseudo-hitting time $\widetilde{D}_B^S$, and the hitting time T_B^S are stopping times. Throughout the present section, the symbols $\mathcal{K}$ and $\mathcal{O}$ will stand for the family of all compact subsets and the family of all open subsets of the space E, respectively.

The Choquet capacitability theorem will be used in the proof of the fact that D_B^S, $\widetilde{D}_B^S$, and T_B^S are stopping times. We will restrict ourselves to positive capacities and the pavement of the space E by compact subsets. For more general cases of the Choquet theorem, we refer the reader to [Doob (2001); Meyer (1966)] (more references can be found in Section 2.11).

Definition 2.24 A function I from the class $\mathcal{P}(E)$ of all subsets of E into the extended real half-line $\overline{\mathbb{R}}_+$ is called a Choquet capacity provided that

(i) If $A_1 \in \mathcal{P}(E)$ and $A_2 \in \mathcal{P}(E)$ are such that $A_1 \subset A_2$, then
$$I(A_1) \le I(A_2).$$

(ii) If $A_n \in \mathcal{P}(E)$, $n \ge 1$, and $A \in \mathcal{P}(E)$ are such that $A_n \uparrow A$, then
$$I(A_n) \to I(A) \quad \text{as } n \to \infty.$$

(iii) If $K_n \in \mathcal{K}$, $n \ge 1$, and $K \in \mathcal{K}$ are such that $K_n \downarrow K$, then
$$I(K_n) \to I(K) \quad \text{as } n \to \infty.$$

Definition 2.25 A function $\varphi : \mathcal{K} \to [0, \infty)$ is called strongly subadditive provided that the following conditions hold:

(i) If $K_1 \in \mathcal{K}$ and $K_2 \in \mathcal{K}$ are such that $K_1 \subset K_2$, then $\varphi(K_1) \le \varphi(K_2)$.
(ii) If $K_1 \in \mathcal{K}$ and $K_2 \in \mathcal{K}$, then
$$\varphi(K_1 \cup K_2) + \varphi(K_1 \cap K_2) \le \varphi(K_1) + \varphi(K_2). \tag{2.178}$$

The following construction allows one to define a Choquet capacity starting with a strongly subadditive function. Let φ be a strongly subadditive function satisfying the following additional condition:

(iii) For all $K \in \mathcal{K}$ and $\varepsilon > 0$, there exists $G \in \mathcal{O}$ such that $K \subset G$ and $\varphi(K') \leq \varphi(K) + \varepsilon$ for all compact subsets K' of G.

For any $G \in \mathcal{O}$, put

$$I^*(G) = \sup_{K \in \mathcal{K}; K \subset G} \varphi(K), \qquad (2.179)$$

and define a set function $I : \mathcal{P}(E) \to \bar{\mathbb{R}}_+$ by

$$I(A) = \inf_{G \in \mathcal{O}; A \subset G} I^*(G), \qquad A \in \mathcal{P}(E). \qquad (2.180)$$

It is known that the function I is a Choquet capacity. It is clear that for any $G \in \mathcal{O}$, we have $I(G) = I^*(G)$. Moreover, it is not hard to see that for any $K \in \mathcal{K}$, $\varphi(K) = I(K)$.

Definition 2.26 Let $\varphi : \mathcal{K} \to [0, \infty)$ be a strongly subadditive function satisfying condition (iii), and let I be the Choquet capacity generated by φ (see formulas (2.179) and (2.180)). A subset B of E is said to be I-capacitable if the following equality holds:

$$I(B) = \sup \{\varphi(K) : K \subset B, K \in \mathcal{K}\}. \qquad (2.181)$$

Now we are ready to formulate the Choquet capacitability theorem (see, e.g., [Doob (2001); Dellacherie and Meyer (1978); Meyer (1966)]).

Theorem 2.23 *Let $\varphi : \mathcal{K} \to [0, \infty)$ be a strongly subadditive function satisfying condition (iii), and let I be the Choquet capacity generated by φ. Then every analytic subset of E, and in particular, every Borel subset of E is I-capacitable.*

The definition of analytic sets can be found in [Doob (2001); Dellacherie and Meyer (1978)]. We will only need the Choquet capacitability theorem for Borel sets.

The symbol $P(E)$ will stand for the collection of all Borel probability measures on the space E. For $B \in \mathcal{E}$ and $\mu \in P(E)$, we put

$$\mathbb{P}_{\tau,\mu}(B) = \int \mathbb{P}_{\tau,x}(B) d\mu(x).$$

For instance, if $\mu = \delta_x$ is the Dirac measure concentrated at $x \in E$, then $\mathbb{P}_{\tau,\delta_x} = \mathbb{P}_{\tau,x}$.

Lemma 2.19 *Let $\tau \in [0, T]$, and let $(X_t, \mathcal{G}_t^\tau, \mathbb{P}_{\tau,x})$ be an adapted, right-continuous, and quasi left-continuous stochastic process. Suppose that S is a $\mathcal{G}_t^\tau$-stopping time such that $\tau \leq S \leq T$. Then, for any $t \in [\tau, T]$ and*

$\mu \in P(E)$, the following functions are strongly subadditive on $\mathcal{K}$ and satisfy condition (iii):
$$K \mapsto \mathbb{P}_{\tau,\mu}\left[D_K^S \leq t\right], \quad \text{and} \quad K \mapsto \mathbb{P}_{\tau,\mu}\left[\widetilde{D}_K^S \leq t\right], \quad K \in \mathcal{K}. \quad (2.182)$$

Proof. We will check conditions (i) and (ii) in Definition 2.25 and also condition (iii) for the set functions in (2.182). Let $K_1 \in \mathcal{K}$ and $K_2 \in \mathcal{K}$ be such that $K_1 \subset K_2$. Then $D_{K_1}^S \geq D_{K_2}^S$, and hence
$$\mathbb{P}_{\tau,\mu}\left[D_{K_1}^S \leq t\right] \leq \mathbb{P}_{\tau,\mu}\left[D_{K_2}^S \leq t\right].$$

This proves condition (i) for the function $K \mapsto \mathbb{P}_{\tau,\mu}\left[D_K^S \leq t\right]$. The proof of (i) for the second function in (2.182) is similar.

In order to prove condition (iii) for the function $K \mapsto \mathbb{P}_{\tau,\mu}\left[D_K^S \leq t\right]$, we need part (a) of Lemma 2.22 (this lemma will be obtained below). More precisely, let $K \in \mathcal{K}$ and $G_n \in \mathcal{O}$, $n \in \mathbb{N}$, be such as in Lemma 2.22. Then by part (a) of Lemma 2.22 (note that part (a) of Lemma 2.22 holds under the restrictions in Lemma 2.19), we get

$$\begin{aligned}\mathbb{P}_{\tau,\mu}\left[D_K^S \leq t\right] &\leq \inf_{G \in \mathcal{O}: G \supset K} \sup_{K' \in \mathcal{K}: K' \subset G} \mathbb{P}_{\tau,\mu}\left[D_{K'}^S \leq t\right] \\ &\leq \inf_{n \in \mathbb{N}} \sup_{K' \in \mathcal{K}: K' \subset G_n} \mathbb{P}_{\tau,\mu}\left[D_{K'}^S \leq t\right] \\ &\leq \inf_{n \in \mathbb{N}} \mathbb{P}_{\tau,\mu}\left[D_{G_n}^S \leq t\right] = \mathbb{P}_{\tau,\mu}\left[D_K^S \leq t\right]. \quad (2.183)\end{aligned}$$

It follows from (2.183) that
$$\mathbb{P}_{\tau,\mu}\left[D_K^S \leq t\right] = \inf_{G \in \mathcal{O}: G \supset K} \sup_{K' \in \mathcal{K}, K' \subset G} \mathbb{P}_{\tau,\mu}\left[D_{K'}^S \leq t\right]. \quad (2.184)$$

Now it is clear that the equality in (2.184) implies property (iii) for the function $K \mapsto \mathbb{P}_{\tau,\mu}\left[D_K^S \leq t\right]$. The proof of (iii) for the function $K \mapsto \mathbb{P}_{\tau,\mu}\left[\widetilde{D}_K^S \leq t\right]$ is similar. Here we use part (d) of Lemma 2.22 (note that part (d) of Lemma 2.22 holds under the restrictions in Lemma 2.19).

Next we will prove that the function $K \mapsto \mathbb{P}_{\tau,\mu}\left[D_K^S \leq t\right]$ satisfies condition (ii). In the proof, the following simple facts will be used: for all Borel subsets B_1 and B_2 of E,
$$D_{B_1 \cup B_2}^S = D_{B_1}^S \wedge D_{B_2}^S, \quad (2.185)$$
and
$$D_{B_1 \cap B_2}^S \geq D_{B_1}^S \vee D_{B_2}^S. \quad (2.186)$$

By (2.185) and (2.186) with $K_1 \in \mathcal{K}$ and $K_2 \in \mathcal{K}$ instead of B_1 and B_2, respectively, we get

$$\{D^S_{K_1 \cup K_2} \leq t\} \setminus \{D^S_{K_2} \leq t\} = (\{D^S_{K_1} \leq t\} \cup \{D^S_{K_2} \leq t\}) \setminus \{D^S_{K_2} \leq t\}$$
$$= \{D^S_{K_1} \leq t\} \setminus \{D^S_{K_2} \leq t\} = \{D^S_{K_1} \leq t\} \setminus (\{D^S_{K_1} \leq t\} \cap \{D^S_{K_2} \leq t\})$$
$$= \{D^S_{K_1} \leq t\} \setminus \{D^S_{K_1} \vee D^S_{K_2} \leq t\} \subset \{D^S_{K_1} \leq t\} \setminus \{D^S_{K_1 \cap K_2} \leq t\}. \tag{2.187}$$

It follows from (2.187) that

$$\mathbb{P}_{\tau,\mu}\left[D^S_{K_1 \cup K_2} \leq t\right] + \mathbb{P}_{\tau,\mu}\left[D^S_{K_1 \cap K_2} \leq t\right]$$
$$\leq \mathbb{P}_{\tau,\mu}\left[D^S_{K_1} \leq t\right] + \mathbb{P}_{\tau,\mu}\left[D^S_{K_2} \leq t\right]. \tag{2.188}$$

Now it is clear that (2.188) implies condition (ii) for the function $K \mapsto \mathbb{P}_{\tau,\mu}\left[D^S_K \leq t\right]$. The proof of condition (ii) for the second function in Lemma 2.19 is similar.

This completes the proof of Lemma 2.19. □

The next theorem states that under certain restrictions, the entry time D^S_B, the pseudo-hitting time $\widetilde{D}^S_B$, and the hitting time T^S_B are stopping times. Recall that we denoted by $\overline{\mathcal{G}}^\tau_t$ the completion $[\mathcal{G}^\tau_t]^{V_\tau}$ of the σ-algebra $\mathcal{G}^\tau_t$ with respect to the family of measures $V_\tau = \{\mathbb{P}_{s,x} : 0 \leq s \leq \tau, x \in E\}$ (see Section 1.7).

Theorem 2.24 *Let $(X_t, \mathcal{G}^\tau_t, \mathbb{P}_{\tau,x})$ be an adapted stochastic process satisfying the following conditions:*

(i) The process X_t is right-continuous and quasi left-continuous.
(ii) $\mathcal{G}^\tau_t = \mathcal{G}^\tau_{t+} = \overline{\mathcal{G}}^\tau_t$ for $0 \leq \tau \leq t \leq T$.

Then for every $\tau \in [0,T)$ and every $\mathcal{G}^\tau_t$-stopping time $S : \Omega \to [\tau, T]$, the random variables D^S_B, $\widetilde{D}^S_B$, and T^S_B are $\mathcal{G}^\tau_t$-stopping times.

Proof. We will first prove Theorem 2.24 assuming that it holds for all open and all compact subsets of E. The validity of Theorem 2.24 for such sets will be established in Lemmas 2.20 and 2.21 below.

Let B be a Borel subset of E, and suppose that we have already shown that for any $\varepsilon \geq 0$ the random time $D^{(\varepsilon+S) \wedge T}_B$ is an $\overline{\mathcal{G}}^\tau_{t+}$-stopping time. Since

$$T^S_B = \inf_{\varepsilon > 0, \varepsilon \in \mathbb{Q}^+} D^{(\varepsilon+S) \wedge T}_B$$

(see (2.177)), we see that T_B^S is a $\bar{\mathcal{G}}_{t+}^\tau$-stopping time. Therefore, in order to prove that T_B^S is a $\bar{\mathcal{G}}_{t+}^\tau$-stopping time, it suffices to show that for every Borel subset B of E, the random time D_B^S is a $\bar{\mathcal{G}}_{t+}^\tau$-stopping time. Since the process X_t is continuous from the right, it is enough to prove the previous assertion with S replaced by $(\varepsilon + S) \wedge T$.

Fix $t \in [\tau, T)$, $\mu \in P(E)$, and $B \in \mathcal{E}$. By Lemma 2.19 and the Choquet capacitability theorem, the set B is capacitable with respect to the capacity I associated with the strongly subadditive function $K \mapsto \mathbb{P}_{\tau,\mu}\left[D_K^S \leq t\right]$. Therefore, there exists an increasing sequence $K_n \in \mathcal{K}$, $n \in \mathbb{N}$, and a decreasing sequence $G_n \in \mathcal{O}$, $n \in \mathbb{N}$, such that

$$K_n \subset K_{n+1} \subset B \subset G_{n+1} \subset G_n, \quad n \in \mathbb{N},$$

and

$$\sup_{n \in \mathbb{N}} \mathbb{P}_{\tau,\mu}\left[D_{K_n}^S \leq t\right] = \inf_{n \in \mathbb{N}} \mathbb{P}_{\tau,\mu}\left[D_{G_n}^S \leq t\right].$$

Put

$$\Lambda_1^{\tau,\mu,S} = \bigcup_{n \in \mathbb{N}} \{D_{K_n}^S \leq t\} \quad \text{and} \quad \Lambda_2^{\tau,\mu,S} = \bigcap_{n \in \mathbb{N}} \{D_{G_n}^S \leq t\}. \tag{2.189}$$

Then Lemma 2.20 implies $\Lambda_2^{\tau,\mu,S} \in \bar{\mathcal{G}}_{t+}^\tau$, and Lemma 2.21 gives $\Lambda_1^{\tau,\mu,S} \in \bar{\mathcal{G}}_{t+}^\tau$. Moreover, we have

$$\Lambda_1^{\tau,\mu,S} \subset \{D_B^S \leq t\} \subset \Lambda_2^{\tau,\mu,S}, \tag{2.190}$$

and

$$\mathbb{P}_{\tau,\mu}\left[\Lambda_2^{\tau,\mu,S}\right] = \inf_{n \in \mathbb{N}} \mathbb{P}_{\tau,\mu}\left[D_{G_n}^S \leq t\right]$$
$$= \sup_{n \in \mathbb{N}} \mathbb{P}_{\tau,\mu}\left[D_{K_n}^S \leq t\right] = \mathbb{P}_{\tau,\mu}\left[\Lambda_1^{\tau,\mu,S}\right]. \tag{2.191}$$

It follows from (2.190) and (2.191) that

$$\mathbb{P}_{\tau,\mu}\left[\Lambda_2^{\tau,\mu,S} \setminus \Lambda_1^{\tau,\mu,S}\right] = 0.$$

Then, using (2.190) again, we see that the event $\{D_B^S \leq t\}$ belongs to the σ-algebra $\bar{\mathcal{G}}_{t+}^\tau$. Therefore, the random time D_B^S is a $\bar{\mathcal{G}}_{t+}^\tau$-stopping time. As we have already observed, it also follows that the random time T_B^S is a $\bar{\mathcal{G}}_{t+}^\tau$-stopping time.

A similar argument with D_B^S replaced by $\widetilde{D}_B^S$ shows that the random time $\widetilde{D}_B^S$, $B \in \mathcal{E}$, is a $\bar{\mathcal{G}}_{t+}^\tau$-stopping times.

This completes the proof of the fact that Theorem 2.24 for all open and all closed sets implies the general case. □

Next we will prove that Theorem 2.24 holds for all open sets.

Lemma 2.20 *Let* $S: \Omega \to [\tau, T]$ *be a* $\bar{\mathcal{G}}_{t+}^\tau$*-stopping time, and let* $G \in \mathcal{O}$. *Then the random times* D_G^S, $\widetilde{D}_G^S$, *and* T_G^S *are* $\bar{\mathcal{G}}_{t+}^\tau$*-stopping times.*

Proof. It is not hard to see that

$$\{D_G^S \leq t\} = \bigcap_{m \in \mathbb{N}} \left\{ D_G^S < t + \frac{1}{m} \right\} = \bigcap_{m \in \mathbb{N}} \bigcup_{\tau \leq \rho < t + \frac{1}{m}, \rho \in \mathbb{Q}^+} \{S \leq \rho, X_\rho \in G\}. \tag{2.192}$$

The last event in (2.192) belongs to the σ-algebra $\bar{\mathcal{G}}_{t+}^\tau$, and hence the random time D_G^S is a $\bar{\mathcal{G}}_{t+}^\tau$-stopping time. The fact that $\widetilde{D}_G^S$ is a $\bar{\mathcal{G}}_{t+}^\tau$-stopping time follows from

$$\{\widetilde{D}_G^S \leq t\} = \bigcap_{m \in \mathbb{N}} \left\{ D_G^S < t + \frac{1}{m} \right\} = \bigcap_{m \in \mathbb{N}} \bigcup_{\rho \in (\tau, t + \frac{1}{m}) \cap \mathbb{Q}^+} \{S \leq \rho, X_\rho \in G\}.$$

The equality (2.177) with G instead of B implies that T_G^S is a $\bar{\mathcal{G}}_{t+}^\tau$-stopping time. □

Finally, we will establish that Theorem 2.24 holds for all closed sets.

Lemma 2.21 *Let* $S: \Omega \to [\tau, T]$ *be a* $\bar{\mathcal{G}}_{t+}^\tau$*-stopping time, and let* $K \in \mathcal{K}$. *Then the random times* D_K^S, $\widetilde{D}_K^S$ *and* T_K^S *are* $\bar{\mathcal{G}}_{t+}^\tau$*-stopping times.*

Proof. Let K be a compact subset of E, and let G_n, $n \in \mathbb{N}$, be a sequence of open subsets of E satisfying the following conditions: $K \subset \overline{G}_{n+1} \subset G_n$ and $\bigcap_{n \in \mathbb{N}} G_n = K$. Then, every random time $D_{G_n}^S$ is a $\bar{\mathcal{G}}_{t+}^\tau$-stopping time (see Lemma 2.20), and for every $\mu \in P(E)$ the sequence of random times $D_{G_n}^S$, $n \in \mathbb{N}$, increases $\mathbb{P}_{\tau,\mu}$-almost surely to D_K^S. This implies that the random time T_K^S is a $\bar{\mathcal{G}}_{t+}^\tau$-stopping time. It is not hard to see that the equality (2.177) with K instead of B implies that T_K^S is a $\bar{\mathcal{G}}_{t+}^\tau$-stopping time. Our next goal is to show that the sequence $D_{G_n}^S$, $n \in \mathbb{N}$, is $\mathbb{P}_{\tau,\mu}$-almost surely convergent. Put $D_K = \sup_n D_{G_n}^S$. Since $D_{G_n}^S \leq D_{G_{n+1}}^S \leq D_K^S$, we have $D_K \leq D_K^S$. By Lemma 2.20, the random times $D_{G_n}^S$, $n \in \mathbb{N}$, are $\bar{\mathcal{G}}_{t+}^\tau$-stopping times. Therefore, the quasi left-continuity of the process X_t, $t \in [0, T]$, implies that

$$\lim_{n \to \infty} X_{D_{G_n}^S} = X_{D_K} \quad \mathbb{P}_{\tau,\mu}\text{-a.s.}$$

Moreover,
$$X_{D_K} \in \bigcap_n \overline{G}_n = K \quad \mathbb{P}_{\tau,\mu}\text{-a.s.}$$

It follows from $D_K^S \geq S$ that $D_K^S \leq D_K$ $\mathbb{P}_{\tau,\mu}$-almost surely, and hence $D_K^S = D_K$ $\mathbb{P}_{\tau,\mu}$-almost surely. This proves that $D_{G_n}^S \to D_K^S$ $\mathbb{P}_{\tau,\mu}$-almost surely as $n \to \infty$.

In order to finish the proof of Lemma 2.21, we will establish that for every $\mu \in P(E)$, the sequence of random times $\widetilde{D}_{G_n}^S$ increases $\mathbb{P}_{\tau,\mu}$-almost surely to $\widetilde{D}_K^S$. Put $\widetilde{D}_K = \sup_n \widetilde{D}_{G_n}^S$. It follows from $\widetilde{D}_{G_n}^S \leq \widetilde{D}_{G_{n+1}}^S \leq \widetilde{D}_K^S$ that $\widetilde{D}_K \leq \widetilde{D}_K^S$. Since the process X_t, $t \in [0,T]$, is quasi left-continuous,
$$\lim_{n \to \infty} X_{\widetilde{D}_{G_n}^S} = X_{\widetilde{D}_K} \quad \mathbb{P}_{\tau,\mu}\text{-a.s.}$$

Therefore
$$X_{\widetilde{D}_K} \in \bigcap_n \overline{G}_n = K \quad \mathbb{P}_{\tau,\mu}\text{-a.s.}$$

It follows from $\widetilde{D}_K^S \geq S$ that $\widetilde{D}_K^S \leq \widetilde{D}_K$ $\mathbb{P}_{\tau,\mu}$-almost surely, and hence $\widetilde{D}_K^S = \widetilde{D}_K$ $\mathbb{P}_{\tau,\mu}$-almost surely. This equality shows that the random time $\widetilde{D}_K^S$ is a $\bar{\mathcal{G}}_{t+}^\tau$-stopping time.

This completes the proof of Lemma 2.21. □

Let us return to the study of standard processes (see Definition 2.22). It was established in Section 2.9 that if P is a transition probability function such that the backward free propagator Y associated with P is a strongly continuous backward Feller-Dynkin propagator, then there exists a standard Markov process $(X_t, \mathcal{G}_t^\tau, \mathbb{P}_{\tau,x})$ with P as its transition function (see Theorem 2.22). By Theorem 2.24, for a standard process $(X_t, \mathcal{G}_t^\tau, \mathbb{P}_{\tau,x})$, the random variables D_B^S, $\widetilde{D}_B^S$, and T_B^S are $\mathcal{G}_t^\tau$-stopping times. Since a standard process is always a strong Markov process in the sense of condition (3) in Definition 2.22, the stopping times in Examples 2.3 and 2.4 can be used in the formulation of the strong Markov property of the process X_t. Our next goal is to construct more examples of such families of stopping times.

Let $(X_t, \mathcal{G}_t^\tau, \mathbb{P}_{\tau,x})$ be a Markov process and let $\tau \in [0,T]$. Suppose that S is a $\mathcal{G}_t^\tau$-stopping time with $\tau \leq S \leq T$. Fix a measure $\nu \in P(E)$, and denote by $\bar{\mathcal{G}}_T^{S,\vee}$ the completion of the σ-algebra $\mathcal{G}_T^{S,\vee} = \sigma\left(S \vee \rho, X_{S \vee \rho} : 0 \leq \rho \leq T\right)$ with respect to the measure ν. The next theorem will allow us to construct more examples of families of stopping times which can be used in the formulation of the strong Markov property with respect to families of measures.

Theorem 2.25 *Let $(X_t, \mathcal{G}_t^\tau, \mathbb{P}_{\tau,x})$ be a standard process, and let $B \in \mathcal{E}$. Then the stopping times D_B^S, $\widetilde{D}_B^S$, and T_B^S are measurable with respect to the σ-algebra $\bar{\mathcal{G}}_T^{S,\vee}$.*

Proof. The proof of Theorem 2.25 is based on the following two lemmas.

Lemma 2.22 *Let $K \in \mathcal{K}$, $\tau \in [0, T)$, and suppose that $G_n \in \mathcal{O}$, $n \in \mathbb{N}$, is a sequence such that $K \subset \overline{G}_{n+1} \subset G_n$ and $\bigcap_{n \in \mathbb{N}} G_n = K$. Then the following assertions hold:*

(a) *For every $\mu \in P(E)$, the sequence of stopping times $D_{G_n}^S$ increases and tends to D_K^S $\mathbb{P}_{\tau,\mu}$-almost surely.*
(b) *For every $t \in [\tau, T]$, the events $\{D_{G_n}^S \leq t\}$, $n \in \mathbb{N}$, are $\mathcal{G}_T^{S,\vee}$-measurable, and the event $\{D_K^S \leq t\}$ is $\bar{\mathcal{G}}_T^{S,\vee}$-measurable.*
(c) *For every $t \in [\tau, T]$, the events $\{T_{G_n}^S \leq t\}$, $n \in \mathbb{N}$, are $\mathcal{G}_T^{S,\vee}$-measurable, and the event $\{T_K^S \leq t\}$ is $\bar{\mathcal{G}}_T^{S,\vee}$-measurable.*
(d) *For every $\mu \in P(E)$, the sequence of stopping times $\widetilde{D}_{G_n}^S$ increases and tends to $\widetilde{D}_K^S$ $\mathbb{P}_{\tau,\mu}$-almost surely.*
(e) *For every $t \in [\tau, T]$, the events $\{\widetilde{D}_{G_n}^S \leq t\}$, $n \in \mathbb{N}$, are $\mathcal{G}_T^{S,\vee}$-measurable, and the event $\{\widetilde{D}_K^S \leq t\}$ is $\bar{\mathcal{G}}_T^{S,\vee}$-measurable.*

Proof. (a) Fix $\mu \in P(E)$, and let $K \in \mathcal{K}$ and $G_n \in \mathcal{O}$, $n \in \mathbb{N}$, be as in assertion (a) in the formulation of Lemma 2.22. Put $D_K = \sup_n D_{G_n}^S$. Since $D_{G_n}^S \leq D_{G_{n+1}}^S \leq D_K^S$, we have $D_K \leq D_K^S$. Moreover, D_K is a stopping time. By the quasi left-continuity of the process X_t,

$$\lim_{n \to \infty} X_{D_{G_n}^S} = X_{D_K} \quad \mathbb{P}_{\tau,\mu}\text{-a.s.}$$

Therefore,

$$X_{D_K} \in \bigcap_n \overline{G}_n = K \quad \mathbb{P}_{\tau,\mu}\text{-a.s.}$$

Since $D_K^S \geq S$, we have $D_K^S \leq D_K$ $\mathbb{P}_{\tau,\mu}$-a.s., and hence $D_K^S = D_K$ $\mathbb{P}_{\tau,\mu}$-a.s.

(b) Fix $t \in [\tau, T)$ and $n \in \mathbb{N}$. By the right-continuity of paths, we have

$$\{D_{G_n}^S \leq t\} = \bigcap_{m \in \mathbb{N}} \left\{D_{G_n}^S < t + \frac{1}{m}\right\} = \bigcap_{m \in \mathbb{N}} \bigcup_{\rho \in [\tau, t + \frac{1}{m})} \{S \leq \rho, X_\rho \in G_n\}$$

$$= \bigcap_{m \in \mathbb{N}} \bigcup_{\rho \in [\tau, t + \frac{1}{m}) \cap \mathbb{Q}^+} \{S \vee \rho \leq \rho, X_{S \vee \rho} \in G_n\}. \qquad (2.193)$$

It follows that
$$\{D_{G_n}^S \leq t\} \in \mathcal{G}_T^{S,\vee}, \ 0 \leq t \leq T.$$

Next, using assertion (a), we see that the events $\{D_K^S \leq t\}$ and $\bigcap_{n \in \mathbb{N}} \{D_{G_n}^S \leq t\}$ coincide $\mathbb{P}_{\tau,\mu}$-almost surely. It follows that $\{D_K^S \leq t\} \in \bar{\mathcal{G}}_T^{S,\vee}$. This proves assertion (b).

(c) Since the sets G_n are open and the process X_t is right-continuous, the hitting times $T_{G_n}^S$ and the entry times $D_{G_n}^S$ coincide. Hence, the first part of assertion (c) follows from assertion (b). In order to prove the second part of (c), we reason as follows. By assertion (b), for every $r \in \mathcal{Q}^+$, the stopping time $D_K^{(r+S)\wedge T}$ is $\bar{\mathcal{G}}_T^{(r+S)\wedge T,\vee}$-measurable. Our next goal is to prove that for every $\varepsilon > 0$,
$$\mathcal{G}_T^{(\varepsilon+S)\wedge T,\vee} \subset \mathcal{G}_T^{S,\vee}. \tag{2.194}$$

Fix $\varepsilon > 0$, $\rho \in [\tau, T]$, and put $S_1 = S$ and $S_2 = ((\varepsilon+S)\wedge T)\vee\rho$. Observe that for $t \in [0, T]$, the events $\{((\varepsilon + S) \wedge T) \vee \rho \leq t\}$ and $\{S \vee (\rho - \varepsilon) \leq t - \varepsilon\}$ coincide. Therefore, S_2 is $\mathcal{G}_T^{S,\vee}$-measurable. Since the process X_t is right-continuous, it follows from Theorem 2.17 that X_{S_2} is $\mathcal{G}_T^{S,\vee}$-measurable. This implies inclusion (2.194). It follows that
$$\bar{\mathcal{G}}_T^{(\varepsilon+S)\wedge T,\vee} \subset \bar{\mathcal{G}}_T^{S,\vee}, \tag{2.195}$$

and we see that the stopping times $D_K^{(\varepsilon+S)\wedge T}$, $\varepsilon \geq 0$, are $\bar{\mathcal{G}}_T^{S,\vee}$-measurable. Since the family $D_K^{(\varepsilon+S)\wedge T}$, $\varepsilon > 0$, decreases to T_K^S, the hitting time T_K^S is $\bar{\mathcal{G}}_T^{S,\vee}$-measurable as well.

(d) Fix $\mu \in P(E)$, and let $K \in \mathcal{K}$ and $G_n \in \mathcal{O}$, $n \in \mathbb{N}$, be as in assertion (a). Put $\widetilde{D}_K = \sup_n \widetilde{D}_{G_n}^S$. Since
$$\widetilde{D}_{G_n}^S \leq \widetilde{D}_{G_{n+1}}^S \leq \widetilde{D}_K^S,$$

we have $\widetilde{D}_K \leq \widetilde{D}_K^S$. It follows from the quasi left-continuity of the process X_t that
$$\lim_{n \to \infty} X_{\widetilde{D}_{G_n}^S} = X_{\widetilde{D}_K} \ \mathbb{P}_{\tau,\mu}\text{-almost surely.}$$

Therefore,
$$X_{\widetilde{D}_K} \in \bigcap_n \overline{G}_n = K \ \mathbb{P}_{\tau,\mu}\text{-almost surely.}$$

Now $\widetilde{D}_K^S \geq S$ implies that $\widetilde{D}_K^S \leq \widetilde{D}_K$ $\mathbb{P}_{\tau,\mu}$-almost surely, and hence $\widetilde{D}_K^S = \widetilde{D}_K$ $\mathbb{P}_{\tau,\mu}$-almost surely.

(e) Fix $t \in [\tau, T)$ and $n \in \mathbb{N}$. By the right-continuity of paths,

$$\left\{\widetilde{D}_{G_n}^S \leq t\right\} = \bigcap_{m \in \mathbb{N}} \left\{D_{G_n}^S < t + \frac{1}{m}\right\} = \bigcap_{m \in \mathbb{N}} \bigcup_{\rho \in (\tau, t + \frac{1}{m})} \{S \leq \rho, X_\rho \in G_n\}$$

$$= \bigcap_{m \in \mathbb{N}} \bigcup_{\rho \in (\tau, t + \frac{1}{m}) \cap \mathbb{Q}^+} \{S \vee \rho \leq \rho, X_{S \vee \rho} \in G_n\}. \quad (2.196)$$

It follows that $\left\{\widetilde{D}_{G_n}^S \leq t\right\} \in \mathcal{G}_T^{S,\vee}$. Next, using assertion (d), we see that the events $\left\{\widetilde{D}_K^S \leq t\right\}$ and $\bigcap_{n \in \mathbb{N}} \left\{\widetilde{D}_{G_n}^S \leq t\right\}$ coincide $\mathbb{P}_{\tau,\mu}$-almost surely. Therefore, $\left\{\widetilde{D}_K^S \leq t\right\} \in \bar{\mathcal{G}}_T^{S,\vee}$. This proves assertion (e). □

Let us return to the proof of Theorem 2.25. We will first prove that for any Borel set B, the entry time D_B^S is measurable with respect to the σ-algebra $\bar{\mathcal{G}}_T^{S,\vee}$. The same assertion holds for the hitting time T_B^S. Indeed, if D_B^S is $\bar{\mathcal{G}}_T^{S,\vee}$-measurable for all stopping times S, then for every $\varepsilon > 0$, the stopping time $D_B^{(\varepsilon+S) \wedge T}$ is measurable with respect to the σ-algebra $\bar{\mathcal{G}}_T^{(\varepsilon+S) \wedge T, \vee}$. By (2.195), $D_B^{(\varepsilon+S) \wedge T}$ is $\bar{\mathcal{G}}_T^{S,\vee}$-measurable. Now (2.177) implies the $\bar{\mathcal{G}}_T^{S,\vee}$-measurability of T_B^S.

Fix $t \in [\tau, T)$, $\mu \in P(E)$, and $B \in \mathcal{E}$. By Lemma 2.19, the set B is capacitable with respect to the capacity $K \mapsto \mathbb{P}_{\tau,\mu}[D_K^S \leq t]$. Therefore, there exists an increasing sequence $K_n \in \mathcal{K}$, $n \in \mathbb{N}$, and a decreasing sequence $G_n \in \mathcal{O}$, $n \in \mathbb{N}$, such that

$$K_n \subset K_{n+1} \subset B \subset G_{n+1} \subset G_n, \quad n \in \mathbb{N},$$

and

$$\sup_{n \in \mathbb{N}} \mathbb{P}_{\tau,\mu}\left[D_{K_n}^S \leq t\right] = \inf_{n \in \mathbb{N}} \mathbb{P}_{\tau,\mu}\left[D_{G_n}^S \leq t\right].$$

Consider the following events:

$$\Lambda_1^{\tau,\mu,S} = \bigcup_{n \in \mathbb{N}} \{D_{K_n}^S \leq t\} \quad \text{and} \quad \Lambda_2^{\tau,\mu,S} = \bigcap_{n \in \mathbb{N}} \{D_{G_n}^S \leq t\}. \quad (2.197)$$

These events are $\bar{\mathcal{G}}_T^{S,\vee}$-measurable. Moreover, we have

$$\Lambda_1^{\tau,\mu,S} \subset \{D_B^S \leq t\} \subset \Lambda_2^{\tau,\mu,S} \quad (2.198)$$

and

$$\mathbb{P}_{\tau,\mu}\left[\Lambda_2^{\tau,\mu,S}\right] = \inf_{n \in \mathbb{N}} \mathbb{P}_{\tau,\mu}\left[D_{G_n}^S \leq t\right]$$

$$= \sup_{n \in \mathbb{N}} \mathbb{P}_{\tau,\mu} \left[D_{K_n}^S \leq t \right] = \mathbb{P}_{\tau,\mu} \left[\Lambda_1^{\tau,\mu,S} \leq t \right]. \qquad (2.199)$$

Now (2.198) and (2.199) give $\mathbb{P}_{\tau,\mu} \left[\Lambda_2^{\tau,\mu,S} \setminus \Lambda_1^{\tau,\mu,S} \right] = 0$. It follows from (2.198) that the event $\{D_B^S \leq t\}$ is measurable with respect to the σ-algebra $\bar{\mathcal{G}}_T^{S,\vee}$. This establishes the $\bar{\mathcal{G}}_T^{S,\vee}$-measurability of the entry time D_B^S and the hitting time T_B^S. The proof of Theorem 2.25 for the pseudo-hitting time $\widetilde{D}_B^S$ is similar to that for the entry time D_B^S.

The proof of Theorem 2.25 is thus completed. □

Corollary 2.2 *Let $(X_t, \mathcal{G}_t^\tau, \mathbb{P}_{\tau,x})$ be a standard process, and let A and B be Borel subsets of E with $B \subset A$. Then the entry time D_B^τ is measurable with respect to the σ-algebra $\bar{\mathcal{G}}_T^{D_A^\tau,\vee}$. Moreover, the hitting time T_B^τ is measurable with respect to the σ-algebra $\bar{\mathcal{G}}_T^{T_A^\tau,\vee}$.*

Proof. By Theorem 2.25, it suffices to show that the equalities

$$D_B^{D_A^\tau} = D_B^\tau \text{ and } \widetilde{D}_B^{T_A^\tau} = T_B^\tau \qquad (2.200)$$

hold $\mathbb{P}_{\tau,\mu}$-almost surely for all $\mu \in P(E)$. The first equality in (2.200) follows from

$$\bigcup_{\tau \leq s < T} \{D_A^\tau \leq s, X_s \in B\} = \bigcup_{\tau \leq s < T} \{X_s \in B\},$$

while the second equality in (2.200) can be obtained from

$$\bigcup_{\tau < s < T} \{T_A^\tau \leq s, X_s \in B\} = \bigcup_{\tau < s < T} \{X_s \in B\}.$$

This proves Corollary 2.2. □

Corollary 2.2 implies that the families $\{D_A^\tau : A \in \mathcal{E}\}$ and $\{T_A^\tau : A \in \mathcal{E}\}$ can be used in the definition of the strong Markov property in the case of standard processes. The next theorem states that the strong Markov property holds for entry times and hitting times of comparable Borel subsets.

Theorem 2.26 *Let $(X_t, \mathcal{G}_t^\tau, \mathbb{P}_{\tau,x})$ be a standard process, and fix $\tau \in [0,T]$. Let A and B be Borel subsets of E such that $B \subset A$, and let $f : [\tau,T] \times E \to \mathbb{R}$ be a bounded Borel function. Then the equalities*

$$\mathbb{E}_{\tau,x} \left[f\left(D_B^\tau, X_{D_B^\tau} \right) \mid \mathcal{G}_{D_A^\tau}^\tau \right] = \mathbb{E}_{D_A^\tau, X_{D_A^\tau}} \left[f\left(D_B^\tau, X_{D_B^\tau} \right) \right]$$

and

$$\mathbb{E}_{\tau,x}\left[f\left(T_B^\tau, X_{T_B^\tau}\right) \mid \mathcal{G}_{T_A^\tau}\right] = \mathbb{E}_{T_A^\tau, X_{T_A^\tau}}\left[f\left(T_B^\tau, X_{T_B^\tau}\right)\right]$$

hold $\mathbb{P}_{\tau,x}$-almost surely.

Proof. Theorem 2.26 follows from Corollary 2.2 and Remark 2.12. □

2.11 Notes and Comments

(a) Propagators (evolution families) are two-parameter generalizations of semigroups. However, propagator theory is not yet as complete as semigroup theory. We refer the reader to [Pazy (1983); Goldstein (1985); Engel and Nagel (2000); Demuth and van Casteren (2000); van Casteren (2002)] for more information on semigroup theory. Various results concerning propagators can be found in [Pazy (1983); Nagel (1995); Nickel (1997); Nagel and Nikel (2002); Schnaubelt (2000/2001); Schnaubelt (2000); Liskevich, Vogt, and Voigt (2005)]. Under certain restrictions, propagators generate solutions to Cauchy problems for non-autonomous evolution equations. Such results go back to Sobolevskii (see [Sobolevskii (1961)]) and Tanabe (see [Tanabe (1960a); Tanabe (1960b); Tanabe (1997)]). Important discoveries in the theory of non-autonomous evolution equations were made by Acquistapace and Terreni (see the survey [Acquistapace (1993)] and the references therein).

(b) Theorem 2.1 was first formulated without proof in [Gulisashvili (2004a)]. The proof can be found in [Gulisashvili (2004b); Gulisashvili (2004c)]. This theorem was also obtained independently but later in [Liskevich, Vogt, and Voigt (2005)].

(c) Howland semigroups were introduced in [Howland (1974)] (see [Chicone and Latushkin (1999)] for more information on Howland semigroups).

(d) The Feller property and the Feller-Dynkin property are discussed in [Rogers and Williams (2000a); Rogers and Williams (2000b); Revuz and Yor (1991); Demuth and van Casteren (2000); van Casteren (2002)].

(e) Our presentation of the strong Markov property of non-homogeneous processes has certain similarities with that in [Dynkin (1994); Dynkin (2002)]. For more information on the strong Markov property of non-homogeneous stochastic processes see [Dynkin (1973); Kuznetsov (1982)].

(f) Time-homogeneous standard processes are discussed in [Blumenthal and Getoor (1968)].
(g) For the proof of Choquet's capacitability theorem see [Bourbaki (1956); Dynkin (1965); Meyer (1966)].

Chapter 3

Non-Autonomous Kato Classes of Measures

3.1 Additive and Multiplicative Functionals

In this section we study two-parameter additive and multiplicative functionals. We will first give examples of additive and multiplicative functionals of Markov processes. In these examples, the functionals are generated by a Borel function V on $[0, T] \times E$. More complicated examples, where the functionals are associated with time-dependent measures, will be discussed in Sections 3.9 and 3.10. The additive and multiplicative functionals considered in the present section and in Sections 3.9 and 3.10 will be used in the definition of one of the main objects of our study in the present book, namely, the Feynman-Kac propagator. In a sense, the Feynman-Kac propagator is a perturbation of a free propagator by a multiplicative functional.

Definition 3.1 A two-parameter family $A = A(\tau, t)$, $0 \le \tau \le t \le T$, of random variables on a filtered probability space $(\Omega, \mathcal{F}, \mathcal{F}_t^\tau, \mathbb{P}_{\tau,x})$ with values in the extended real half-line $\overline{\mathbb{R}}_+$ is called an additive functional provided that the following conditions hold:

(1) For all τ and t with $0 \le \tau \le t \le T$ and all $x \in E$, the random variable $A(\tau, t)$ is finite $\mathbb{P}_{\tau,x}$-almost surely and $\mathcal{F}_t^\tau$-measurable.
(2) For all τ, λ, and t with $0 \le \tau \le \lambda \le t \le T$, and all $x \in E$,

$$A(\tau, t) = A(\tau, \lambda) + A(\lambda, t) \quad \mathbb{P}_{\tau,x}\text{-a.s.}$$

(3) For all $0 \le \tau \le T$ and all $x \in E$,

$$A(\tau, \tau) = 0 \quad \mathbb{P}_{\tau,x}\text{-a.s.}$$

Let P be a transition probability function, and let $(X_t, \mathcal{F}_t^\tau, \mathbb{P}_{\tau,x})$ be a Markov process associated with P. Then a typical example of an additive functional of the process X_t is given by the following. Let $V \geq 0$ be a Borel function on $[0,T] \times E$, and define a family of random variables by

$$A_V(\tau, t) = \begin{cases} \int_\tau^t V(s, X_s)\, ds, & \text{if } \int_\tau^t V(s, X_s)\, ds < \infty \\ 0, & \text{otherwise.} \end{cases} \quad (3.1)$$

It will be shown in the sequel that under certain restrictions on the process X_t and the function V, the family A_V in (3.1) is an additive functional. This functional satisfies several additional conditions. For instance, A_V is non-decreasing and continuous. It will be assumed below that the process X_t is progressively measurable. This condition is needed to guarantee the measurability of the integrand in (3.1).

Definition 3.2 A two-parameter family $M = M(\tau, t)$, $0 \leq \tau \leq t \leq T$, of random variables with values in the extended half-line $\overline{\mathbb{R}}_+$ is called a multiplicative functional provided that

(1) For all τ and t with $0 \leq \tau \leq t \leq T$ and all $x \in E$, the random variable $M(\tau, t)$ is finite $\mathbb{P}_{\tau,x}$-almost surely and $\mathcal{F}_t^\tau$-measurable.
(2) For all τ, λ, and t with $0 \leq \tau \leq \lambda \leq t \leq T$ and all $x \in E$,

$$M(\tau, t) = M(\tau, \lambda) M(\lambda, t) \quad \mathbb{P}_{\tau,x}\text{-a.s.}$$

(3) For all $0 \leq \tau \leq T$ and all $x \in E$,

$$M(\tau, \tau) = 1 \quad \mathbb{P}_{\tau,x}\text{-a.s.}$$

If A is an additive functional, then the functional M defined by $M(\tau, t) = e^{A(\tau,t)}$ is a multiplicative functional. An important example of a multiplicative functional is the Kac functional M_V where V is a Borel function on $[0,T] \times E$. This functional is obtained by exponentiating the functional A_V defined in (3.1). More precisely,

$$M_V(\tau, t) = \begin{cases} \exp\left\{-\int_\tau^t V(s, X_s)\, ds\right\}, & \text{if } \int_\tau^t |V(s, X_s)|\, ds < \infty \\ 1, & \text{otherwise.} \end{cases} \quad (3.2)$$

The functional M_V defined in (3.2) will be used in the definition of the backward Feynman-Kac propagator Y_V associated with the function V (see Section 4.2).

In Section 3.9, we will study the functionals A_μ and M_μ corresponding to a time-dependent measure $\mu = \{\mu(t) : 0 \leq t \leq T\}$.

Definition 3.3 A family $\mu = \{\mu(t) : 0 \leq t \leq T\}$ of signed Borel measures of locally bounded variation on $(E, \mathcal{E})$ is called a time-dependent measure provided that for every set $B \in \mathcal{E}$ the function $s \mapsto \mu(s, B)$ is Borel measurable on $[0, T]$.

Recall that if ν is a signed measure, then at least one of the measures ν^+ and ν^- is finite. In Section 3.9, we will study the functionals in (3.1) and (3.2) in the case of a time-dependent measure μ. The following construction is used in Section 3.9 to define the functional A_μ: the time-dependent measure μ is approximated in a special sense by a sequence of functions V_k on $[0, T] \times E$ so that the sequence A_{V_k} converges in a certain topology. The additive functional A_μ is defined by

$$A_\mu(\tau, t) = \begin{cases} \lim_{k \to \infty} \int_\tau^t V_k(s, X_s)\, ds, & \text{if the limit exists,} \\ 0, & \text{otherwise.} \end{cases} \quad (3.3)$$

In addition, the Kac functional M_μ is given by

$$M_\mu(\tau, t) = \exp\{-A_\mu(\tau, t)\}. \quad (3.4)$$

We conclude the present section by the following examples. Let S be a terminal stopping time (see Definition 2.10). Then

$$A(\tau, t) = \chi_{\{\tau < S \leq t\}}$$

is an additive functional, while

$$M(\tau, t) = 1 - \chi_{\{\tau < S \leq t\}}$$

is a multiplicative functional.

3.2 Potentials of Time-Dependent Measures and Non-Autonomous Kato Classes

Let P be a transition probability function, and let Y be the corresponding free backward propagator. If V is a Borel measurable function on $[0, T] \times E$ such that

$$\int_\tau^T Y(\tau, s)|V(s)|(x)\, ds < \infty$$

for all $(\tau, x) \in [0,T] \times E$, then the function

$$N(V)(\tau, t, x) = \int_\tau^t Y(\tau, s)V(s)(x)ds \qquad (3.5)$$

is defined for all $0 \leq \tau \leq t \leq T$ and $x \in E$. The function $N(V)$ is, in a sense, a potential of V. It is obtained by integrating the free propagator with respect to the time variable.

Now let μ be a time-dependent measure. In order to define the potential of the measure μ, we suppose that the transition function P has a density p with respect to the reference measure m. Let us denote by $|\mu(t)|$ the variation of the measure $\mu(t)$, and assume that the following condition holds:

$$\int_\tau^T Y(\tau, s)|\mu(s)|(x)ds < \infty \qquad (3.6)$$

for all $(\tau, x) \in [0,T] \times E$. In (3.6), $Y(\tau, s)|\mu(s)|(x)$ stands for the integral $\int_E p(\tau, x; s, y)d|\mu(s)|$. If inequality (3.6) holds for a time-dependent measure μ, then the potential $N(\mu)$ of μ is defined by

$$N(\mu)(\tau, t, x) = \int_\tau^t Y(\tau, s)\mu(s)(x)ds = \int_\tau^t ds \int_E p(\tau, x; s, y)d\mu(s) \qquad (3.7)$$

for all $0 \leq \tau \leq t \leq T$ and $x \in E$.

Next, we will introduce various classes of functions and measures. These classes are generalizations of the Kato class of potential functions (see Section 4.1). Note that in the case of classes of functions in Definition 3.4 we do not require the existence of a transition density p, while for the classes of time-dependent measures, the existence of the density p is assumed. In Definition 3.4, the subscripts f and m distinguish the classes of functions from the classes of time-dependent measures.

Definition 3.4 Let P be a transition probability function. Then the class $\widehat{\mathcal{P}}_f^*$ is defined as follows:

$$V \in \widehat{\mathcal{P}}_f^* \iff \sup_{0 \leq \tau \leq t \leq T} \sup_{x \in E} N(|V|)(\tau, t, x) < \infty.$$

Let $V \in \widehat{\mathcal{P}}_f^*$. Then the class $\mathcal{P}_f^*$ is defined by

$$V \in \mathcal{P}_f^* \iff \lim_{t-\tau \downarrow 0} \sup_{x \in E} N(|V|)(\tau, t, x) = 0.$$

Suppose that the transition probability function P has a density p. Then the class $\widehat{\mathcal{P}}_m^*$ is defined as follows:

$$\mu \in \widehat{\mathcal{P}}_m^* \iff \sup_{0 \leq \tau \leq t \leq T} \sup_{x \in E} N(|\mu|)(\tau, t, x) < \infty.$$

If $\mu \in \widehat{\mathcal{P}}_m^*$, then the class $\mathcal{P}_m^*$ is given by

$$\mu \in \mathcal{P}_m^* \iff \lim_{t-\tau \downarrow 0} \sup_{x \in E} N(|\mu|)(\tau, t, x) = 0.$$

We call the class $\widehat{\mathcal{P}}_f^*$ ($\widehat{\mathcal{P}}_m^*$) the extended non-autonomous backward Kato class of functions (measures), while the class $\mathcal{P}_f^*$ ($\mathcal{P}_m^*$) is called the non-autonomous backward Kato class of functions (measures).

Remark 3.1 Suppose that the transition probability function P has a density p. Then the classes $\widehat{\mathcal{P}}_f^*$ and $\mathcal{P}_f^*$ can be identified with subclasses of the classes $\widehat{\mathcal{P}}_m^*$ and $\mathcal{P}_m^*$ as follows: $V(\tau, x) \iff d\mu(\tau, x) = V(\tau, x)dx$.

For a function $V \in \widehat{\mathcal{P}}_f^*$ and a time-dependent measure $\mu \in \widehat{\mathcal{P}}_m^*$, denote

$$||V||_f^* = \sup_{0 \leq \tau \leq t \leq T} \sup_{x \in E} N(|V|)(\tau, t, x),$$

and

$$||\mu||_m^* = \sup_{0 \leq \tau \leq t \leq T} \sup_{x \in E} N(|\mu|)(\tau, t, x).$$

It is clear that

$$||V||_f^* = \sup_{\tau: 0 \leq \tau \leq T} \sup_{x \in E} N(|V|)(\tau, T, x)$$

and

$$||\mu||_m^* = \sup_{\tau: 0 \leq \tau \leq T} \sup_{x \in E} N(|\mu|)(\tau, T, x).$$

Next, we will show that the classes in Definition 3.4 are normed spaces. We will first make several remarks concerning the equivalence relations mod 0 for the elements of these classes. Let l denote the Lebesgue measure on the σ-algebra $\mathcal{B}_{[0,T]}$ of all Borel subsets of $[0, T]$. For every $\tau \in [0, T]$ and $x \in E$, define a measure $\xi_{\tau,x}$ on the σ-algebra $\mathcal{B}_{[\tau,T]} \otimes \mathcal{E}$ by

$$\xi_{\tau,x}(U) = \int \int_U P(\tau, x; u, dy) du, \ U \in \mathcal{B}_{[\tau,T]} \otimes \mathcal{E}.$$

It follows that for $V \in \widehat{\mathcal{P}}_f^*$, the condition $||V||_f^* = 0$ means that for all $\tau \in [0, T)$ and $x \in E$, the equality $V(u, y) = 0$ holds $\xi_{\tau,x}$-a.e. on $[\tau, T] \times E$. If P has a density p such that

$$p(\tau, x; u, y) > 0 \tag{3.8}$$

for all τ, x, u, and y, then we get the following equivalent condition: $V(u, y) = 0$ $l \times m$-a.e. on $[0, T] \times E$. If there exists a density p, and if $\mu \in \widehat{\mathcal{P}}_m^*$, then the condition $||\mu||_m^* = 0$ means that

$$\int_\tau^T \int_E p(\tau, x; u, y) d|\mu(u)|(y) du = 0$$

for all τ and x. If p satisfies (3.8), then we get the following equivalent condition: $\mu(u) = 0$ for l-a.a. $u \in [0, T]$. Taking into account the identifications described above, we see that the spaces $\left(\widehat{\mathcal{P}}_f^*, ||\cdot||_f^*\right)$ and $\left(\widehat{\mathcal{P}}_m^*, ||\cdot||_m^*\right)$ are normed spaces. Next, we will prove that they are Banach spaces.

Lemma 3.1 *Let P be a transition probability function. Then $(\widehat{\mathcal{P}}_f^*, ||\cdot||_f^*)$ is a Banach space, and $(\mathcal{P}_f^*, ||\cdot||_f^*)$ is a closed subspace of $\widehat{\mathcal{P}}_f^*$. Moreover, if P has a strictly positive density p, then $(\widehat{\mathcal{P}}_m^*, ||\cdot||_m^*)$ is a Banach space, and $\mathcal{P}_m^*$ is a closed subspace of $\widehat{\mathcal{P}}_m^*$.*

Proof. We will prove that if p is strictly positive, then the space $\widehat{\mathcal{P}}_m^*$ is complete, and $\mathcal{P}_m^*$ is a closed subspace of $\widehat{\mathcal{P}}_m^*$. The proof of Lemma 3.1 for the spaces $\widehat{\mathcal{P}}_f^*$ and $\mathcal{P}_m^*$ is similar, and we leave it as an exercise for the reader.

Let $\mu_k \in \widehat{\mathcal{P}}_m^*$, $k \geq 1$, be a sequence of time-dependent measures such that

$$\sum_{k=1}^\infty ||\mu_k||_m^* = \sum_k \sup_{\tau: 0 \leq \tau \leq T} \sup_{x \in E} \int_\tau^T du \int_E p(\tau, x; u, y) d|\mu_k(u, y)| < \infty. \tag{3.9}$$

Then for every $x \in E$,

$$\int_0^T du \int_E p(0, x; u, y) d \sum_{k=1}^\infty |\mu_k(u, y)| < \infty.$$

Hence, there exists a Borel set $U_x \in [0, T]$ such that $l(U_x) = T$ and

$$\int_E p(0, x; u, y) d \sum_{k=1}^\infty |\mu_k(u, y)| < \infty \tag{3.10}$$

for all $u \in U_x$. Fix $x \in E$. Then (3.10) implies that for every $j \geq 1$ and $u \in U_x$,

$$\sum_{k=1}^{\infty} |\mu_k(u)|(A_{j,u}) < \infty$$

where $A_{j,u} = \{y \in E : p(0,x;u,y) > j^{-1}\}$. Hence, $\sum \mu_k(u)$ is a finite signed Borel measure on every set $A_{j,u}$ for all $u \in U_x$. Since the strict positivity of p implies $\bigcup_{j=1}^{\infty} A_{j,u} = E$ for all $u \in U_x$, the measure $\mu(u) = \sum \mu_k(u)$ is a signed Borel measure on E for all $u \in U_x$, and hence l-a.e. on $[0,T]$. It follows from (3.9) that $\mu \in \widehat{\mathcal{P}}_m^*$. Moreover, it is not difficult to prove using (3.9) that the series $\sum_{k=1}^{\infty} \mu_k$ converges to μ in the space $\widehat{\mathcal{P}}_m^*$. This establishes the completeness of the space $\widehat{\mathcal{P}}_m^*$.

Now let $\mu_k \in \mathcal{P}_m^*$, $k \geq 1$, be such that $\mu_k \to \mu$ in $\widehat{\mathcal{P}}_m^*$. Then

$$\int_\tau^t Y(\tau,u)|\mu(u)|(x)du \leq \int_\tau^t Y(\tau,u)|\mu(u) - \mu_k(u)|(x)du$$
$$+ \int_\tau^t Y(\tau,u)|\mu_k(u)|(x)du. \quad (3.11)$$

It follows from (3.11) that $\mu \in \mathcal{P}_m^*$. Hence, the class $\mathcal{P}_m^*$ is a closed subspace of the space $\widehat{\mathcal{P}}_m^*$.

This completes the proof of Lemma 3.1. □

The next result provides a description of the classes $\mathcal{P}_f^*$ and $\mathcal{P}_m^*$ in terms of the potential operator N. Note that a function $V \in \widehat{\mathcal{P}}_f^*$ and a time-dependent measure $\mu \in \widehat{\mathcal{P}}_m^*$ are characterized by the uniform boundedness of the functions $N(|V|)$ and $N(|\mu|)$, respectively. For the classes $\mathcal{P}_f^*$ and $\mathcal{P}_m^*$, the following lemma holds.

Lemma 3.2

(a) Let $V \in \widehat{\mathcal{P}}_f^*$. Then $V \in \mathcal{P}_f^*$ if and only if

$$\lim_{t'-t \to 0+} \sup_{\tau: 0 \leq \tau \leq t} \sup_{x \in E} [N(|V|)(\tau, t', x) - N(|V|)(\tau, t, x)] = 0. \quad (3.12)$$

(b) Suppose that P has a density p, and let $\mu \in \widehat{\mathcal{P}}_m^*$. Then $\mu \in \mathcal{P}_m^*$ if and only if (3.12) holds with μ instead of V.

Proof. Let $V \in \mathcal{P}_f^*$. Then we have

$$N(|V|)(\tau, t', x) - N(|V|)(\tau, t, x) = \int_t^{t'} Y(\tau, u)|V(u)|(x)du$$
$$= Y(\tau, t)N(|V|)(t, t')(x),$$

and it follows that

$$\sup_{x \in E}[N(|V|)(\tau, t', x) - N(|V|)(\tau, t, x)] \le \sup_{x \in E} N(|V|)(t, t', x).$$

It is clear that the previous estimate implies (3.12).

Now assume that (3.12) holds. Then we have

$$\lim_{t-\tau \to 0} \sup_{x \in E} N(|V|)(\tau, t, x) = \lim_{t-\tau \to 0} \sup_{x \in E}[N(|V|)(\tau, t, x) - N(|V|)(\tau, \tau, x)] = 0.$$

This implies $V \in \mathcal{P}_f^*$.

The proof of part (b) of Lemma 3.2 is similar. □

Remark 3.2 Let $V \in \mathcal{P}_f^*$. Then, reasoning as in the proof of Lemma 3.2, we see that

$$\lim_{t'-t \to 0+} \sup_{\tau: 0 \le \tau \le t} \sup_{x \in E} |N(V)(\tau, t', x) - N(V)(\tau, t, x)| = 0.$$

Let $V \in \widehat{\mathcal{P}}_f^*$. The following function will be used in Section 3.9:

$$M(V)(\tau, t) = \sup_{r: \tau \le r \le t} \sup_{x \in E} |N(V)(r, t, x)|, \quad 0 \le \tau \le t \le T.$$

Similarly for $\mu \in \widehat{\mathcal{P}}_m^*$, we put

$$M(\mu)(\tau, t) = \sup_{r: \tau \le r \le t} \sup_{x \in E} |N(\mu)(r, t, x)|, \quad 0 \le \tau \le t \le T.$$

3.3 Backward Transition Probability Functions and Non-Autonomous Kato Classes of Functions and Measures

Let $\widetilde{P}(\tau, A; t, y)$ be a backward transition probability function (see Definition 1.3 in Chapter 1). Recall that the free propagator U associated with $\widetilde{P}$ is defined on the space $L_{\mathcal{E}}^\infty$ by

$$\begin{cases} U(t, \tau)g(y) = \int_E g(x)\widetilde{P}(\tau, dx; t, y) \\ U(t, t)g(y) = g(y) \end{cases}$$

for all $0 \leq \tau \leq t \leq T$, $y \in E$, and $g \in L_{\mathcal{E}}^{\infty}$. If $\tilde{P}$ has a density $\tilde{p}$, then

$$\begin{cases} U(t,\tau)g(y) = \int_E g(x)\tilde{p}(\tau,x;t,y)dx \\ U(t,t)g(y) = g(y) \end{cases}$$

for all $0 \leq \tau < t \leq T$, $y \in E$, and $g \in L^{\infty}$.

Suppose that $\tilde{P}$ is a backward transition probability function. If V is a Borel function on $[0,T] \times E$ such that

$$\int_\tau^t U(t,s)|V(s)|(x)ds < \infty$$

for all $0 \leq \tau \leq t \leq T$ and $x \in E$, then we define the potential of the function V by

$$\widetilde{N}(V)(t,\tau,x) = \int_\tau^t U(t,s)V(s)(x)ds.$$

Similarly, if $\tilde{P}$ has a density $\tilde{p}$, then for a time-dependent measure μ satisfying

$$\int_\tau^t U(t,s)|\mu(s)|(x)ds < \infty$$

for all $0 \leq \tau \leq t \leq T$ and $x \in E$, we define the potential of the measure μ by

$$\widetilde{N}(\mu)(t,\tau,x) = \int_\tau^t U(t,s)\mu(s)(x)ds$$

for all $0 \leq \tau \leq t \leq T$ and $x \in E$.

Definition 3.5 Let $\tilde{P}$ be a backward transition probability function. Then the class $\widehat{\mathcal{P}}_f$ is defined as follows:

$$V \in \widehat{\mathcal{P}}_f \iff \sup_{0 \leq \tau \leq t \leq T} \sup_{x \in E} \widetilde{N}(|V|)(t,\tau,x) < \infty.$$

Let $V \in \widehat{\mathcal{P}}_f$. Then the class $\mathcal{P}_f$ is defined by

$$V \in \mathcal{P}_f \iff \lim_{t-\tau \downarrow 0} \sup_{x \in E} \widetilde{N}(|V|)(t,\tau,x) = 0.$$

Suppose that the backward transition probability function $\widetilde{P}$ has a density $\widetilde{p}$. Then the class $\widehat{\mathcal{P}}_m$ is defined as follows:

$$\mu \in \widehat{\mathcal{P}}_m \iff \sup_{0 \leq \tau \leq t \leq T} \sup_{x \in E} \widetilde{N}(|\mu|)(t,\tau,x) < \infty.$$

If $\mu \in \widehat{\mathcal{P}}_m$, then the class $\mathcal{P}_m$ is defined by

$$\mu \in \mathcal{P}_m \iff \lim_{t-\tau \downarrow 0} \sup_{x \in E} \widetilde{N}(|\mu|)(t,\tau,x) = 0.$$

The class $\widehat{\mathcal{P}}_f$ ($\widehat{\mathcal{P}}_m$) is called the extended non-autonomous Kato class of functions (measures). Similarly, the class $\mathcal{P}_f$ ($\mathcal{P}_m$) is called the non-autonomous Kato class of functions (measures).

For the classes in Definition 3.5, the norms are defined by the following:

$$||V||_f = \sup_{0 \leq \tau \leq t \leq T} \sup_{x \in E} \widetilde{N}(|V|)(t,\tau,x)$$

and

$$||\mu||_m = \sup_{0 \leq \tau \leq t \leq T} \sup_{x \in E} \widetilde{N}(|\mu|)(t,\tau,x).$$

It is clear that

$$||V||_f = \sup_{\tau: 0 \leq \tau \leq T} \sup_{x \in E} \widetilde{N}(|V|)(T,\tau,x)$$

and

$$||\mu||_m = \sup_{\tau: 0 \leq \tau \leq T} \sup_{x \in E} \widetilde{N}(|\mu|)(T,\tau,x).$$

As in the case of backward Kato classes of functions and measures in the previous section, we should take into account the equivalence mod 0 for functions and measures. For instance, if $\widetilde{P}$ is a backward transition probability function, we define a measure $\widetilde{\xi}_{t,x}$ by

$$\widetilde{\xi}_{t,y}(B) = \int \int_B \widetilde{P}(t,dx;u,y)du$$

where $B \in \mathcal{B}_{[0,t]} \times \mathcal{E}$. Then for $V \in \widehat{\mathcal{P}}_f$ the condition $||V||_f = 0$ means that for all $t \in [0,T)$ and $y \in E$, we have $V(u,x) = 0$ $\widetilde{\xi}_{t,y}$-a.e. on $[0,t] \times E$. Now it is clear how to define the above-mentioned equivalence relation in the case of functions. The remaining cases are similar. Taking into account the identifications described above, we see that the spaces $\left(\widehat{\mathcal{P}}_f, ||\cdot||_f\right)$ and $\left(\widehat{\mathcal{P}}_m, ||\cdot||_m\right)$ are normed spaces. Moreover, they are Banach spaces.

Lemma 3.3 *Let $\widetilde{P}$ be a backward transition probability function. Then $\left(\widehat{\mathcal{P}}_f, ||\cdot||_f\right)$ is a Banach space, and $(\mathcal{P}_f, ||\cdot||_f)$ is a closed subspace of $\widehat{\mathcal{P}}_f$. Moreover, if $\widetilde{P}$ has a strictly positive density $\widetilde{p}$, then $\left(\widehat{\mathcal{P}}_m, ||\cdot||_m\right)$ is a Banach space, and $\mathcal{P}_m$ is a closed subspace of $\widehat{\mathcal{P}}_m$.*

The proof of Lemma 3.3 is similar to that of Lemma 3.1. The next lemma is analogous to Lemma 3.2.

Lemma 3.4

(a) *Let $V \in \widehat{\mathcal{P}}_f$. Then $V \in \mathcal{P}_f$ if and only if*

$$\lim_{\tau-\tau'\downarrow 0} \sup_{t: 0 \leq \tau \leq t \leq T} \sup_{x \in E} [\widetilde{N}(|V|)(t,\tau,x) - \widetilde{N}(|V|)(t,\tau',x)] = 0. \quad (3.13)$$

(b) *Suppose that $\widetilde{P}$ has a density $\widetilde{p}$, and let $\mu \in \widehat{\mathcal{P}}_m$. Then $\mu \in \mathcal{P}_m$ if and only if (3.13) holds with μ instead of V.*

Remark 3.3 Let $V \in \mathcal{P}_f$. Then it is easy to see that

$$\lim_{\tau-\tau'\downarrow 0} \sup_{t: 0 \leq \tau \leq t \leq T} \sup_{x \in E} \left|\widetilde{N}(V)(t,\tau,x) - \widetilde{N}(V)(t,\tau',x)\right| = 0.$$

3.4 Weighted Non-Autonomous Kato Classes

Let $\varphi(\tau, t)$ be a positive Borel function on $0 \leq \tau < t \leq T$. The classes $\widehat{\mathcal{P}}^*_{f,\varphi}$, $\mathcal{P}^*_{f,\varphi}$, $\widehat{\mathcal{P}}^*_{m,\varphi}$, and $\mathcal{P}^*_{m,\varphi}$ that are introduced in the present section are generalizations of the non-autonomous Kato classes defined in Sections 3.2 and 3.3.

Let P be a transition probability function, and let Y be the corresponding free backward propagator. If V is a Borel measurable function on $[0, T] \times E$ such that

$$\int_\tau^T Y(\tau, s)|V(s)|(x)\varphi(\tau, s)^{-1} ds < \infty$$

for all $(\tau, x) \in [0, T] \times E$, then the function

$$N_\varphi(V)(\tau, t, x) = \int_\tau^T Y(\tau, s) V(s)(x) \varphi(\tau, s)^{-1} ds \qquad (3.14)$$

is defined for $0 \le \tau \le t \le T$ and $x \in E$.

Next assume that the transition function P has a density p. Let μ be a time-dependent measure, and suppose that the following condition holds:

$$\int_\tau^T Y(\tau, s) |\mu(s)|(x) \varphi(\tau, s)^{-1} ds < \infty$$

for all $(\tau, x) \in [0, T] \times E$. Then we put

$$N_\varphi(\mu)(\tau, t, x) = \int_\tau^t Y(\tau, s) \mu(s)(x) \varphi(\tau, s)^{-1} ds \qquad (3.15)$$

for all $0 \le \tau \le t \le T$ and $x \in E$.

Now we are ready to define weighted non-autonomous Kato classes. As in Sections 3.2 and 3.3, we use the subscript f in the case of classes of functions and the subscript m for classes of time-dependent measures.

Definition 3.6 Let P be a transition probability function. Then the class $\widehat{\mathcal{P}}^*_{f,\varphi}$ is defined as follows:

$$V \in \widehat{\mathcal{P}}^*_{f,\varphi} \iff \sup_{0 \le \tau \le t \le T} \sup_{x \in E} N_\varphi(|V|)(\tau, t, x) < \infty.$$

Let $V \in \widehat{\mathcal{P}}^*_{f,\varphi}$. Then the class $\mathcal{P}^*_{f,\varphi}$ is defined by

$$V \in \mathcal{P}^*_{f,\varphi} \iff \lim_{t - \tau \downarrow 0} \sup_{x \in E} N_\varphi(|V|)(\tau, t, x) = 0.$$

Suppose that the transition probability function P has a density p. Then the class $\widehat{\mathcal{P}}^*_{m,\varphi}$ is defined as follows:

$$\mu \in \widehat{\mathcal{P}}^*_{m,\varphi} \iff \sup_{0 \le \tau \le t \le T} \sup_{x \in E} N_\varphi(|\mu|)(\tau, t, x) < \infty.$$

If $\mu \in \widehat{\mathcal{P}}^*_{m,\varphi}$, then the class $\mathcal{P}^*_{m,\varphi}$ is given by

$$\mu \in \mathcal{P}^*_{m,\varphi} \iff \lim_{t - \tau \downarrow 0} \sup_{x \in E} N_\varphi(|\mu|)(\tau, t, x) = 0.$$

Now let $\widetilde{P}$ be a backward transition probability function, and V be a Borel function on $[0,T] \times E$ such that

$$\int_\tau^t U(t,s)|V(s)|(x)\varphi(s,t)^{-1}ds < \infty$$

for all $0 \leq \tau \leq t \leq T$ and $x \in E$. Then we put

$$\widetilde{N}_\varphi(V)(t,\tau,x) = \int_\tau^t U(t,s)V(s)(x)\varphi(s,t)^{-1}ds.$$

Similarly, if $\widetilde{P}$ has a density $\widetilde{p}$, then for a time-dependent measure μ satisfying

$$\int_\tau^t U(t,s)|\mu(s)|(x)\varphi(s,t)^{-1}ds < \infty$$

for all $0 \leq \tau \leq t \leq T$ and $x \in E$, we put

$$\widetilde{N}_\varphi(\mu)(t,\tau,x) = \int_\tau^t U(t,s)\mu(s)(x)\varphi(s,t)^{-1}ds$$

for all $0 \leq \tau \leq t \leq T$ and $x \in E$.

Definition 3.7 Let $\widetilde{P}$ be a backward transition probability function. Then the class $\widehat{\mathcal{P}}_{f,\varphi}$ is defined as follows:

$$V \in \widehat{\mathcal{P}}_{f,\varphi} \iff \sup_{0 \leq \tau \leq t \leq T} \sup_{x \in E} \widetilde{N}_\varphi(|V|)(t,\tau,x) < \infty.$$

Let $V \in \widehat{\mathcal{P}}_{f,\varphi}$. Then the class $\mathcal{P}_{f,\varphi}$ is defined by

$$V \in \mathcal{P}_{f,\varphi} \iff \lim_{t-\tau \downarrow 0} \sup_{x \in E} \widetilde{N}_\varphi(|V|)(t,\tau,x) = 0.$$

Suppose that the backward transition probability function $\widetilde{P}$ has a density $\widetilde{p}$. Then the class $\widehat{\mathcal{P}}_{m,\varphi}$ is defined as follows:

$$\widehat{\mathcal{P}}_{m,\varphi} \iff \sup_{0 \leq \tau \leq t \leq T} \sup_{x \in E} \widetilde{N}_\varphi(|\mu|)(t,\tau,x) < \infty.$$

If $\mu \in \widehat{\mathcal{P}}_m$, then the class $\mathcal{P}_{m,\varphi}$ is defined by

$$\mu \in \mathcal{P}_{m,\varphi} \iff \lim_{t-\tau \downarrow 0} \sup_{x \in E} \widetilde{N}_\varphi(|\mu|)(t,\tau,x) = 0.$$

For a function $V \in \widehat{\mathcal{P}}^*_{f,\varphi}$ and a time-dependent measure $\mu \in \widehat{\mathcal{P}}^*_{m,\varphi}$, we put

$$||V||^*_{f,\varphi} = \sup_{0 \leq \tau \leq t \leq T} \sup_{x \in E} N_\varphi(|V|)(\tau, t, x)$$

and

$$||\mu||^*_{m,\varphi} = \sup_{0 \leq \tau \leq t \leq T} \sup_{x \in E} N_\varphi(|\mu|)(\tau, t, x).$$

Similarly, for a function $V \in \widehat{\mathcal{P}}_{f,\varphi}$ and a time-dependent measure $\mu \in \widehat{\mathcal{P}}_{m,\varphi}$, we put

$$||V||_{f,\varphi} = \sup_{0 \leq \tau \leq t \leq T} \sup_{x \in E} \widetilde{N}_\varphi(|V|)(t, \tau, x)$$

and

$$||\mu||_{m,\varphi} = \sup_{0 \leq \tau \leq t \leq T} \sup_{x \in E} \widetilde{N}_\varphi(|\mu|)(t, \tau, x).$$

As in Sections 3.2 and 3.3, the symbol l stands for the Lebesgue measure on the σ-algebra $\mathcal{B}_{[0,T]}$. For every $\tau \in [0, T]$ and $x \in E$, we define a measure $\xi_{\tau,x,\varphi}$ on the σ-algebra $\mathcal{B}_{[\tau,T]} \otimes \mathcal{E}$ as follows. For $U \in \mathcal{B}_{[\tau,T]} \otimes \mathcal{E}$,

$$\xi_{\tau,x,\varphi}(U) = \int_\tau^T \int_U P(\tau, x; u, dy) \varphi(\tau, u)^{-1} du.$$

Then for $V \in \widehat{\mathcal{P}}^*_{f,\varphi}$ the condition $||V||^*_{f,\varphi} = 0$ means that for all $\tau \in [0, T)$ and $x \in E$, we have $V(u, y) = 0$ $\xi_{\tau,x,\varphi}$-a.e. on $[\tau, T] \times E$. If P has a density p such that

$$p(\tau, x; u, y) > 0 \qquad (3.16)$$

for all τ, x, u, and y, then we can formulate an equivalent condition: $V(u, y) = 0$ $l \times m$-a.e. on $[0, T] \times E$. If the density p exists and $\mu \in \widehat{\mathcal{P}}^*_{m,\varphi}$, then the condition $||\mu||^*_{m,\varphi} = 0$ means that

$$\int_\tau^T \int_E p(\tau, x; u, y) d|\mu(u)|(y) \varphi(\tau, u)^{-1} du = 0$$

for all τ and x. If p satisfies condition (3.16), then we get the following equivalent condition: $\mu(u) = 0$ for l-a.a. $u \in [0, T]$. If we take into account the identifications described above, then the spaces $\left(\widehat{\mathcal{P}}^*_{f,\varphi}, ||\cdot||^*_{f,\varphi}\right)$ and $\left(\widehat{\mathcal{P}}^*_{m,\varphi}, ||\cdot||^*_{m,\varphi}\right)$ become normed spaces.

Lemma 3.5 *Let P be a transition probability function. Then $\left(\widehat{\mathcal{P}}^*_{f,\varphi}, ||\cdot||^*_{f,\varphi}\right)$ is a Banach space, and $\left(\mathcal{P}^*_{f,\varphi}, ||\cdot||^*_{f,\varphi}\right)$ is a closed subspace of the space $\widehat{\mathcal{P}}^*_{f,\varphi}$. Moreover, if P has a strictly positive density p, then $\left(\widehat{\mathcal{P}}^*_{m,\varphi}, ||\cdot||^*_{m,\varphi}\right)$ is a Banach space, and $\mathcal{P}^*_{m,\varphi}$ is a closed subspace of the space $\widehat{\mathcal{P}}^*_{m,\varphi}$.*

Lemma 3.6 *Let $\widetilde{P}$ be a backward transition probability function. Then $\left(\widehat{\mathcal{P}}_{f,\varphi}, ||\cdot||_{f,\varphi}\right)$ is a Banach space, and $(\mathcal{P}_{f,\varphi}, ||\cdot||_{f,\varphi})$ is a closed subspace of the space $\widehat{\mathcal{P}}_{f,\varphi}$. Moreover, if $\widetilde{P}$ has a strictly positive density $\widetilde{p}$, then $\left(\widehat{\mathcal{P}}_{m,\varphi}, ||\cdot||_{m,\varphi}\right)$ is a Banach space, and $\mathcal{P}_{m,\varphi}$ is a closed subspace of the space $\widehat{\mathcal{P}}_{m,\varphi}$.*

The proof of Lemmas 3.5 and 3.6 is similar to that of Lemma 3.1.

3.5 Examples of Functions and Measures in Non-Autonomous Kato Classes

In this section we give examples of functions and time-dependent measures from the classes $\mathcal{P}^*_f$, $\mathcal{P}^*_m$, $\mathcal{P}_f$, and $\mathcal{P}_m$. Recall that E denotes a locally compact Hausdorff topological space satisfying the second axiom of countability. It is equipped with a metric $\rho: E \times E \to [0,\infty)$. The symbol $B_r(x)$ stands for the open ball of radius $r > 0$ centered at $x \in E$. It is defined by $B_r(x) = \{y \in E : \rho(x,y) < r\}$.

Let p and $\widetilde{p}$ be a transition probability density and a backward transition probability density, respectively. The next two lemmas provide sufficient conditions for a time-dependent measure to belong to the backward Kato class $\mathcal{P}^*_m$ associated with p, or to the Kato class $\mathcal{P}_m$ associated with p^*.

Lemma 3.7 *Let p be a transition probability density, and suppose that there exist positive Borel functions γ_1 and γ_2 on $0 \leq \tau \leq s \leq T$ and a positive function Φ on $[0,\infty)$ such that*

(1) The function Φ is bounded, strictly decreasing, and such that $\Phi(\lambda) \to 0$ as $\lambda \to \infty$.

(2) The following estimate holds:

$$p(\tau, x; s, y) \leq \gamma_1(\tau, s)\Phi\left(\gamma_2(\tau, s)\rho(x,y)\right) \qquad (3.17)$$

for all $0 \leq \tau < s \leq T$, $x \in E$, and $y \in E$.

Let μ be a time-dependent measure, and assume that there exist a number $\delta \geq 0$ and a positive function C on $[0,T]$ such that

$$|\mu(s)|\,(B_r(x)) \leq C(s)r^\delta \tag{3.18}$$

for all $0 \leq s \leq T$, $r > 0$, and $x \in E$. Moreover, suppose that

$$\int_0^\infty \lambda^\delta d\,(-\Phi(\lambda)) < \infty \tag{3.19}$$

and

$$\lim_{t-\tau \downarrow 0} \int_\tau^t C(s) \frac{\gamma_1(\tau,s)}{\gamma_2^\delta(\tau,s)} ds = 0. \tag{3.20}$$

Then the time-dependent measure μ belongs to the class $\mathcal{P}_m^*$.

Lemma 3.8 Let $\widetilde{p}$ be a backward transition probability density, and suppose that there exist positive Borel functions $\widetilde{\gamma}_1$ and $\widetilde{\gamma}_2$ on $0 \leq \tau \leq s \leq T$, and a positive function $\widetilde{\Phi}$ on $[0,\infty)$ such that

(1) The function $\widetilde{\Phi}$ is bounded, strictly decreasing, and such that $\widetilde{\Phi}(\lambda) \to 0$ as $\lambda \to \infty$.
(2) The following estimate holds:

$$\widetilde{p}(s,x;t,y) \leq \widetilde{\gamma}_1(s,t)\widetilde{\Phi}\left(\widetilde{\gamma}_2(s,t)\rho(x,y)\right) \tag{3.21}$$

for all $0 \leq s \leq t \leq T$, $x \in E$, and $y \in E$.

Let $\widetilde{\mu}$ be a time-dependent measure, and assume that there exist a number $\delta \geq 0$ and a positive function $\widetilde{C}$ on $[0,T]$ such that

$$|\widetilde{\mu}(s)|\,(B_r(y)) \leq \widetilde{C}(s)r^\delta \tag{3.22}$$

for all $0 \leq s \leq T$, $r > 0$, and $y \in E$. Moreover, suppose that

$$\int_0^\infty \lambda^\delta d\left(-\widetilde{\Phi}(\lambda)\right) < \infty \tag{3.23}$$

and

$$\lim_{t-\tau \downarrow 0} \int_\tau^t \widetilde{C}(s) \frac{\widetilde{\gamma}_1(s,t)}{\widetilde{\gamma}_2^\delta(s,t)} ds = 0. \tag{3.24}$$

Then the time-dependent measure $\widetilde{\mu}$ belongs to the class $\mathcal{P}_m$.

Proof. We will only prove Lemma 3.7. The proof of Lemma 3.8 is similar. It follows from (3.17) and (3.18) that

$$\int_E p(\tau,x;s,y) d\,|\mu(s)|\,(y) = \int_0^\infty |\mu(s)|\,(y : p(\tau,x;s,y) > \lambda)\,d\lambda$$

$$\leq \int_0^{\Phi(0)\gamma_1(\tau,s)} |\mu(s)|\,(y : \gamma_1(\tau,s)\Phi\,(\gamma_2(\tau,s)\rho(x,y)) > \lambda)\,d\lambda$$

$$= \int_0^{\Phi(0)\gamma_1(\tau,s)} |\mu(s)|\left(y : \rho(x,y) < \frac{1}{\gamma_2(\tau,s)}\Phi^{-1}\left(\frac{\lambda}{\gamma_1(\tau,s)}\right)\right)d\lambda$$

$$\leq C(s)\frac{1}{\gamma_2^\delta(\tau,s)}\int_0^{\Phi(0)\gamma_1(\tau,s)} \Phi^{-1}\left(\frac{\lambda}{\gamma_1(\tau,s)}\right)^\delta d\lambda$$

$$= C(s)\frac{\gamma_1(\tau,s)}{\gamma_2^\delta(\tau,s)}\int_0^{\Phi(0)} \Phi^{-1}(\eta)^\delta\,d\eta. \tag{3.25}$$

By (3.25),

$$\sup_{x \in E}\int_\tau^t \int_E p(\tau,x;s,y) d\,|\mu(s)|\,(y)$$

$$\leq \int_0^{\Phi(0)} \Phi^{-1}(\eta)^\delta\,d\eta \int_\tau^t C(s)\frac{\gamma_1(\tau,s)}{\gamma_2^\delta(\tau,s)}\,ds. \tag{3.26}$$

Now we see that (3.19), (3.20), and (3.26) imply Lemma 3.7. □

The next two lemmas allow us to construct examples of time-dependent measures not belonging to the backward Kato class $\mathcal{P}_m^*$ associated with p, or to the forward Kato class $\mathcal{P}_m$ associated with p^*.

Lemma 3.9 *Let p be a transition probability density, and suppose that there exist positive Borel functions γ_1 and γ_2 on $0 \leq \tau \leq s \leq T$ and a positive function Φ on $[0,\infty)$ satisfying the following conditions:*

(1) The function Φ is bounded, strictly decreasing, and such that $\Phi(\lambda) \to 0$ as $\lambda \to \infty$.
(2) The estimate

$$p(\tau,x;s,y) \geq \gamma_1(\tau,s)\Phi\,(\gamma_2(\tau,s)\rho(x,y)) \tag{3.27}$$

holds for all $0 \leq \tau < s \leq T$, $x \in E$, and $y \in E$.

Let μ be a time-dependent measure, and assume that there exist a number $\delta \geq 0$, a positive Borel function $u \mapsto r(u)$ on $[0,T]$ with values in the extended half-line $\bar{\mathbb{R}}_+$, a positive Borel function $u \mapsto C(u)$ on $[0,T]$, and a family D_s, $0 \leq s \leq T$, of Borel subsets of E such that the function $(s,x) \mapsto \chi_{D_s}(x)$ is Borel measurable and

$$|\mu(s)|(B_r(x)) \geq C(s)r^\delta \tag{3.28}$$

for all $0 \leq s \leq T$, $0 < r < r(s)$, and $x \in D_s$. Moreover, suppose that

$$\limsup_{t-\tau \downarrow 0} \left\{ \sup_x \int_\tau^t \chi_{D_s}(x) C(s) \frac{\gamma_1(\tau,s)}{\gamma_2^\delta(\tau,s)} ds \right. \\ \left. \int_0^{\gamma_2(\tau,s)r(s)} \lambda^\delta d(-\Phi(\lambda)) \right\} > 0. \tag{3.29}$$

Then the time-dependent measure μ does not belong to the class $\mathcal{P}_m^*$.

Lemma 3.10 *Let $\widetilde{p}$ be a backward transition probability density, and suppose that there exist positive Borel functions $\widetilde{\gamma}_1$ and $\widetilde{\gamma}_2$ on $0 \leq \tau \leq s \leq T$ and a positive function $\widetilde{\Phi}$ satisfying the following conditions:*

(1) The function $\widetilde{\Phi}$ is bounded, strictly decreasing, and such that $\widetilde{\Phi}(\lambda) \to 0$ as $\lambda \to \infty$.

(2) The estimate

$$\widetilde{p}(s,x;t,y) \geq \widetilde{\gamma}_1(s,t)\widetilde{\Phi}(\widetilde{\gamma}_2(s,t)\rho(x,y)) \tag{3.30}$$

holds for all $0 \leq s \leq t \leq T$, $x \in E$, and $y \in E$.

Let $\widetilde{\mu}$ be a time-dependent measure, and assume that there exist a number $\delta \geq 0$, a positive Borel function $u \mapsto r(u)$ on $[0,T]$ with values in $\mathbb{R}_+$, a positive Borel function $u \mapsto \widetilde{C}(u)$ on $[0,T]$, and a family $\widetilde{D}_s$, $0 \leq s \leq T$, of Borel subsets of E such that the function $(s,x) \mapsto \chi_{\widetilde{D}_s}(x)$ is Borel measurable and

$$|\widetilde{\mu}(s)|(B_r(y)) \geq \widetilde{C}(s)r^\delta \tag{3.31}$$

for all $0 \leq s \leq T$, $0 < r < r(s)$, and $y \in \widetilde{D}_s$. Moreover, suppose that

$$\limsup_{t-\tau \downarrow 0} \left\{ \sup_x \int_\tau^t \chi_{\widetilde{D}_s}(x) \widetilde{C}(s) \frac{\widetilde{\gamma}_1(s,t)}{\widetilde{\gamma}_2^\delta(s,t)} ds \right. \\ \left. \int_0^{\gamma_2(s,t)r(s)} \lambda^\delta d\left(-\widetilde{\Phi}(\lambda)\right) \right\} > 0. \tag{3.32}$$

Then the time-dependent measure $\widetilde{\mu}$ does not belong to the class $\mathcal{P}_m$.

Proof. We will only prove Lemma 3.9. The proof of Lemma 3.10 is similar.

Let $s \in [0, T]$. Then, assuming that $x \in D_s$ and using (3.27), we obtain

$$\int_E p(\tau, x; s, y) d\,|\mu(s)|\,(y) = \int_0^\infty |\mu(s)|\,(y : p(\tau, x; s, y) > \lambda)\,d\lambda$$

$$\geq \int_0^{\Phi(0)\gamma_1(\tau,s)} |\mu(s)|\,(y : \gamma_1(\tau, s)\Phi\,(\gamma_2(\tau, s)\rho(x, y)) > \lambda)\,d\lambda$$

$$= \int_0^{\Phi(0)\gamma_1(\tau,s)} |\mu(s)|\left(y : \rho(x, y) < \frac{1}{\gamma_2(\tau, s)}\Phi^{-1}\left(\frac{\lambda}{\gamma_1(\tau, s)}\right)\right) d\lambda$$

$$= \gamma_1(\tau, s) \int_0^{\Phi(0)\gamma_1(\tau,s)} |\mu(s)|\left(y : \rho(x, y) < \frac{1}{\gamma_2(\tau, s)}\Phi^{-1}(\eta)\right) d\eta. \quad (3.33)$$

Our next goal is to use estimate (3.28) in (3.33). However, the lower estimate in (3.28) holds only if

$$\frac{1}{\gamma_2(\tau, s)}\Phi^{-1}(\eta) \leq r(s),$$

or equivalently,

$$\eta > \Phi\left(\gamma_2(\tau, s)r(s)\right).$$

Taking this into account, we see that (3.33) implies the estimate

$$\int_E p(\tau, x; s, y) d\,|\mu(s)|\,(y) \geq C(s)\frac{\gamma_1(\tau, s)}{\gamma_2^\delta(\tau, s)} \int_{\Phi(\gamma_2(\tau,s)r(s))}^{\Phi(0)} \Phi^{-1}(\eta)^\delta\,d\eta \quad (3.34)$$

for all $x \in D_s$. Now it is clear that Lemma 3.9 follows from (3.29) and (3.34). □

Assume that $\gamma_2(\tau, s)r(s) > \epsilon_1$ for all τ and s such that $s - \tau < \epsilon_2$ where ϵ_1 and ϵ_2 are strictly positive constants. Then (3.29) is equivalent to the following condition:

$$\limsup_{t-\tau \downarrow 0}\left\{\sup_x \int_\tau^t \chi_{D_s}(x)C(s)\frac{\gamma_1(\tau, s)}{\gamma_2^\delta(\tau, s)}\,ds\right\} > 0. \quad (3.35)$$

Similarly, (3.32) is equivalent to the condition

$$\limsup_{t-\tau \downarrow 0}\left\{\sup_x \int_\tau^t \chi_{\widetilde{D}_s}(x)\widetilde{C}(s)\frac{\widetilde{\gamma}_1(s, t)}{\widetilde{\gamma}_2^\delta(s, t)}\,ds\right\} > 0. \quad (3.36)$$

Remark 3.4 Conditions (3.17), (3.21), (3.27), and (3.30) are modelled after the two-sided Gaussian estimates for the heat kernels. Conditions

(3.18), (3.22), (3.28), and (3.31) are often used in Geometric Measure Theory. In a sense, they mean that the measure is δ-dimensional. In the case of a function V defined on $[0, T] \times E$, condition (3.18) becomes

$$\int_{B_r(x)} |V(s,y)| \, dy \leq C(s) r^\delta.$$

Example 3.1 Let E be d-dimensional Euclidean space $\mathbb{R}^d$ equipped with the metric $\rho(x, y) = |x - y|$, where $|\cdot|$ stands for the Euclidean norm on $\mathbb{R}^d$. The reference measure m in this case is the Lebesgue measure m_d. Let G_t be the Gaussian density on $\mathbb{R}^d$ given by

$$G_t(x) = \frac{1}{(2\pi t)^{\frac{d}{2}}} \exp\left\{-\frac{|x|^2}{2t}\right\}$$

where $t > 0$ and $x \in \mathbb{R}^d$. The function G_t generates a transition probability density p and a backward transition probability density $\widetilde{p}$ as follows:

$$p(\tau, x; t, y) = \widetilde{p}(\tau, x; t, y) = G_{t-\tau}(x - y).$$

Next we choose

$$\gamma_1(\tau, s) = \widetilde{\gamma}_1(\tau, s) = \frac{1}{(2\pi(s-\tau))^{\frac{d}{2}}},$$

$$\gamma_2(\tau, s) = \widetilde{\gamma}_2(\tau, s) = \frac{1}{\sqrt{s-\tau}},$$

and

$$\Phi(\lambda) = \widetilde{\Phi}(\lambda) = \exp\left\{-\frac{\lambda^2}{2}\right\}.$$

Here we use the notation in Lemmas 3.7–3.10. It follows that conditions (3.19) and (3.23) hold for the functions Φ and $\widetilde{\Phi}$. Indeed,

$$\int_0^\infty \lambda^\delta d(-\Phi(\lambda)) \, d\lambda = \int_0^\infty \lambda^\delta d\left(-\widetilde{\Phi}(\lambda)\right) d\lambda$$

$$= \int_0^\infty \lambda^{\delta+1} \exp\left\{-\frac{\lambda^2}{2}\right\} d\lambda < \infty.$$

Let us consider the following family of functions on $[0, T] \times \mathbb{R}^d$:

$$\Lambda_{\alpha,\beta,\nu}(t, x) = \frac{1}{t^\alpha \ln^\beta \frac{Te}{t}} \frac{1}{|x|^\nu} \tag{3.37}$$

where $\alpha > 0$, $\beta \geq 0$, and $0 < \nu < d$. Note that for all $\alpha > 0$ and $\beta \geq 0$, there exists a number $a(\alpha, \beta) > 0$ such that

$$t_1^\alpha \ln^\beta \frac{Te}{t_1} \leq a(\alpha, \beta) t_2^\alpha \ln^\beta \frac{Te}{t_2} \quad (3.38)$$

if $0 \leq t_1 \leq t_2 \leq T$. Indeed, we have

$$\frac{d}{dt}\left(t^\alpha \ln^\beta \frac{Te}{t}\right) = t^{\alpha-1} \ln^{\beta-1} \frac{Te}{t}\left(\alpha \ln \frac{Te}{t} - \beta\right).$$

Therefore, if $\alpha \geq \beta$, then the function $t \mapsto t^\alpha \ln^\beta \frac{Te}{t}$ is increasing on $[0, T]$. If $\alpha < \beta$, then the maximum of this function is attained at $t = T \exp\left\{\frac{\alpha - \beta}{\alpha}\right\}$, and (3.38) easily follows. Property (3.38) will be used in the estimates below.

We have

$$\int_{B_r(x)} \frac{dy}{|y|^\nu} \leq \int_{y:|y|\leq r} \frac{dy}{|y|^\nu} = c_{\nu,d} r^{d-\nu}$$

for all $x \in \mathbb{R}^d$ and $r > 0$. Hence, if $\delta = d - \nu$, $r(u) = \infty$ for $0 \leq u \leq T$, $D_s = \widetilde{D}_s = \{0\}$ for all $0 \leq s \leq T$, and

$$C(s) = \widetilde{C}(s) = c_{\nu,d} \frac{1}{s^\alpha \ln^\beta \frac{Te}{s}},$$

then conditions (3.18), (3.22), (3.28), and (3.31) hold. Next we will determine for what values of the parameters α, β, and ν the remaining conditions in Lemmas 3.7–3.10 also hold.

Let us start with condition (3.20). By property (3.38),

$$\limsup_{t-\tau\downarrow 0} \int_\tau^t C(s) \frac{\gamma_1(\tau, s)}{\gamma_2^\delta(\tau, s)} ds = \frac{c_{\nu,d}}{(2\pi)^{\frac{d}{2}}} \limsup_{t-\tau\downarrow 0} \int_\tau^t \frac{ds}{(s-\tau)^{\frac{\nu}{2}} s^\alpha \ln^\beta \frac{Te}{s}}$$

$$= \frac{c_{\nu,d}}{(2\pi)^{\frac{d}{2}}} \limsup_{t-\tau\downarrow 0} \int_0^{t-\tau} \frac{d\eta}{\eta^{\frac{\nu}{2}}(\eta + \tau)^\alpha \ln^\beta \frac{Te}{\eta+\tau}}$$

$$\leq c_1(\alpha, \beta, \nu, d) \limsup_{\epsilon\downarrow 0} \int_0^\epsilon \frac{d\eta}{\eta^{\alpha+\frac{\nu}{2}} \ln^\beta \frac{Te}{\eta}}. \quad (3.39)$$

It follows from (3.39) that if $\alpha + \frac{\nu}{2} < 1$ and $0 \leq \beta < \infty$, then condition (3.20) holds. This condition also holds for $\alpha + \frac{\nu}{2} = 1$ and $\beta > 1$.

Next, we turn our attention to condition (3.24). Reasoning as in the proof of (3.39) and using (3.38), we see that

$$\limsup_{t-\tau\downarrow 0}\int_\tau^t \widetilde{C}(s)\frac{\widetilde{\gamma}_1(s,t)}{\gamma_2^\delta(s,t)}ds = \frac{c_{\nu,d}}{(2\pi)^{\frac{d}{2}}}\limsup_{t-\tau\downarrow 0}\int_\tau^t \frac{ds}{(t-s)^{\frac{\nu}{2}}s^\alpha \ln^\beta \frac{Te}{s}}$$

$$= \frac{c_{\nu,d}}{(2\pi)^{\frac{d}{2}}}\limsup_{t-\tau\downarrow 0}\int_0^{t-\tau}\frac{d\eta}{\eta^{\frac{\nu}{2}}(t-\eta)^\alpha \ln^\beta \frac{Te}{t-\eta}}$$

$$= \frac{c_{\nu,d}}{(2\pi)^{\frac{d}{2}}}\limsup_{\epsilon\downarrow 0}\sup_{t:\epsilon\le t\le T}\int_0^\epsilon \frac{d\eta}{\eta^{\frac{\nu}{2}}(t-\eta)^\alpha \ln^\beta \frac{Te}{t-\eta}}$$

$$\le c_2(\alpha,\beta,\nu,d)\limsup_{\epsilon\downarrow 0}\int_0^\epsilon \frac{d\eta}{\eta^{\frac{\nu}{2}}(\epsilon-\eta)^\alpha \ln^\beta \frac{Te}{\epsilon-\eta}}. \qquad (3.40)$$

Let us assume that $0 < \nu < 2$. Then we have

$$\limsup_{\epsilon\downarrow 0}\int_0^{\frac{\epsilon}{2}}\frac{d\eta}{\eta^{\frac{\nu}{2}}(\epsilon-\eta)^\alpha \ln^\beta \frac{Te}{\epsilon-\eta}} \le 2^\alpha \limsup_{\epsilon\downarrow 0}\frac{1}{\epsilon^\alpha \ln^\beta \frac{Te}{\epsilon}}\int_0^{\frac{\epsilon}{2}}\frac{d\eta}{\eta^{\frac{\nu}{2}}}$$

$$= c_3(\alpha,\nu)\limsup_{\epsilon\downarrow 0}\frac{\epsilon^{1-\alpha-\frac{\nu}{2}}}{\ln^\beta \frac{Te}{\epsilon}}. \qquad (3.41)$$

Moreover,

$$\limsup_{\epsilon\downarrow 0}\int_{\frac{\epsilon}{2}}^\epsilon \frac{d\eta}{\eta^{\frac{\nu}{2}}(\epsilon-\eta)^\alpha \ln^\beta \frac{Te}{\epsilon-\eta}} \le 2^{\frac{\nu}{2}}\limsup_{\epsilon\downarrow 0}\epsilon^{-\frac{\nu}{2}}\int_{\frac{\epsilon}{2}}^\epsilon \frac{d\eta}{(\epsilon-\eta)^\alpha \ln^\beta \frac{Te}{\epsilon-\eta}}$$

$$\le c_4(\alpha,\nu)\limsup_{\epsilon\downarrow 0}\frac{\epsilon^{1-\alpha-\frac{\nu}{2}}}{\ln^\beta \frac{Te}{\epsilon}}. \qquad (3.42)$$

It follows from (3.40), (3.41), and (3.42) that condition (3.24) holds provided that $\alpha + \frac{\nu}{2} < 1$ and $0 \le \beta < \infty$, or $0 < \nu < 2$, $\alpha + \frac{\nu}{2} = 1$, and $\beta > 0$.

Next, we will analyze the range of applicability of condition (3.35) in the case of the functions $\Lambda_{\alpha,\beta,\nu}$. Reasoning as in the proof of (3.39), we get

$$\limsup_{t-\tau\downarrow 0}\left\{\sup_x \int_\tau^t \chi_{D_s}(x)C(s)\frac{\gamma_1(\tau,s)}{\gamma_2^\delta(\tau,s)}ds\right\} = \limsup_{t-\tau\downarrow 0}\int_\tau^t C(s)\frac{\gamma_1(\tau,s)}{\gamma_2^\delta(\tau,s)}ds$$

$$= \frac{c_{\nu,d}}{(2\pi)^{\frac{d}{2}}}\limsup_{t-\tau\downarrow 0}\int_\tau^t \frac{ds}{(s-\tau)^{\frac{\nu}{2}}s^\alpha \ln^\beta \frac{Te}{s}}$$

$$\ge \frac{c_{\nu,d}}{(2\pi)^{\frac{d}{2}}}\limsup_{t\downarrow 0}\int_0^t \frac{ds}{s^{\alpha+\frac{\nu}{2}}\ln^\beta \frac{Te}{s}}.$$

Therefore, condition (3.35) holds provided that $\alpha + \frac{\nu}{2} > 1$ and $0 \leq \beta < \infty$, or $\alpha + \frac{\nu}{2} = 1$ and $0 \leq \beta \leq 1$.

Our next goal is to utilize condition (3.36). Arguing as in the proof of (3.40), we obtain

$$\limsup_{t-\tau \downarrow 0} \left\{ \sup_x \int_\tau^t \chi_{\widetilde{D}_s}(x)\widetilde{C}(s) \frac{\widetilde{\gamma}_1(s,t)}{\widetilde{\gamma}_2^\delta(s,t)} ds \right\} = \limsup_{t-\tau \downarrow 0} \int_\tau^t \widetilde{C}(s) \frac{\widetilde{\gamma}_1(s,t)}{\widetilde{\gamma}_2^\delta(s,t)} ds$$

$$= \frac{c_{\nu,d}}{(2\pi)^{\frac{d}{2}}} \limsup_{\epsilon \downarrow 0} \sup_{t:\epsilon \leq t \leq T} \int_0^\epsilon \frac{d\eta}{\eta^{\frac{\nu}{2}}(t-\eta)^\alpha \ln^\beta \frac{Te}{t-\eta}}$$

$$\geq \frac{c_{\nu,d}}{(2\pi)^{\frac{d}{2}}} \limsup_{\epsilon \downarrow 0} \int_0^\epsilon \frac{d\eta}{\eta^{\frac{\nu}{2}}(\epsilon-\eta)^\alpha \ln^\beta \frac{Te}{\epsilon-\eta}}$$

$$\geq \frac{c_{\nu,d}}{(2\pi)^{\frac{d}{2}}} \limsup_{\epsilon \downarrow 0} \int_0^{\frac{\epsilon}{2}} \frac{d\eta}{\eta^{\frac{\nu}{2}}(\epsilon-\eta)^\alpha \ln^\beta \frac{Te}{\epsilon-\eta}}$$

$$\geq \frac{c_{\nu,d}}{(2\pi)^{\frac{d}{2}}} \limsup_{\epsilon \downarrow 0} \frac{1}{\epsilon^\alpha \ln^\beta \frac{2Te}{\epsilon}} \int_0^{\frac{\epsilon}{2}} \frac{d\eta}{\eta^{\frac{\nu}{2}}}. \qquad (3.43)$$

It follows from (3.43) that condition (3.36) holds provided that either $\nu \geq 2$, or $0 < \nu < 2$, $\alpha + \frac{\nu}{2} > 1$, and $0 \leq \beta < \infty$, or $0 < \nu < 2$, $\alpha + \frac{\nu}{2} = 1$, and $\beta = 0$. Therefore, Lemmas 3.7–3.10 imply that

(1) The function $\Lambda_{\alpha,\beta,\nu}$ belongs to the class $\mathcal{P}_f^*$ if and only if either $\alpha + \frac{\nu}{2} < 1$ and $0 \leq \beta < \infty$, or $\alpha + \frac{\nu}{2} = 1$ and $\beta > 1$.
(2) The function $\Lambda_{\alpha,\beta,\nu}$ belongs to the class $\mathcal{P}_f$ if and only if either $\alpha + \frac{\nu}{2} < 1$ and $0 \leq \beta < \infty$; or $0 < \nu < 2$, $\alpha + \frac{\nu}{2} = 1$, and $\beta > 0$.

The previous characterizations show that in the case of the Gaussian transition density the classes $\mathcal{P}_f^*$ and $\mathcal{P}_f$ do not coincide. Indeed, the function $\Lambda_{\frac{1}{2},1,1}$ belongs to the class $\mathcal{P}_f$, but does not belong to the class $\mathcal{P}_f^*$. This result was first obtained in [Gulisashvili (2002a)].

Example 3.2 This example is a continuation of Example 3.1. Here we consider a family $\Lambda_\gamma(\alpha, \beta)$ of time-dependent measures defined by

$$\Lambda_\gamma(\alpha, \beta) = \frac{1}{t^\alpha \ln^\beta \frac{Te}{t}} \gamma \qquad (3.44)$$

where $\alpha > 0$, $\beta \geq 0$, and γ is a signed Radon measure of locally bounded variation on $\mathbb{R}^d$. Although the family of time-dependent measures in (3.44) is more general than the family of functions in (3.37), the methods used in example (3.37) can also be employed in the case of the time-dependent measures defined by (3.44).

Let us suppose that there exists a number $\delta \geq 0$ such that

$$|\gamma|(B_r(x)) \leq c_1 r^\delta \qquad (3.45)$$

for all $x \in \mathbb{R}^d$ and $r > 0$, where c_1 is a positive constant. It is also assumed that there exist numbers $r_0 > 0$, $c_2 > 0$, and a point $x_0 \in \mathbb{R}^d$ such that

$$|\gamma|(B_r(x_0)) \geq c_2 r^\delta \qquad (3.46)$$

for all $0 < r < r_0$. It is not hard to see, reasoning as in Example 3.1, that the following assertions hold

(1) The time-dependent measure $\Lambda_\gamma(\alpha, \beta)$ belongs to the class $\mathcal{P}_m^*$ if and only if either $\alpha + \frac{d-\delta}{2} < 1$ and $0 \leq \beta < \infty$, or $\alpha + \frac{d-\delta}{2} = 1$ and $\beta > 1$.
(2) The time-dependent measure $\Lambda_\gamma(\alpha, \beta)$ belongs to the class $\mathcal{P}_m$ if and only if either $\alpha + \frac{d-\delta}{2} < 1$ and $0 \leq \beta < \infty$, or $0 < d-\delta < 2$, $\alpha + \frac{d-\delta}{2} = 1$, and $\beta > 0$.

Let us suppose that $\gamma = \delta_{x_0}$, where δ_{x_0} is the Dirac measure concentrated at $x_0 \in \mathbb{R}^d$. Then we can take $\delta = 0$ in conditions (3.45) and (3.46). It is not difficult to show using assertions (1) and (2), that for such a measure γ, $\Lambda_\gamma(\alpha, \beta) \in \mathcal{P}_m^*$ or $\Lambda_\gamma(\alpha, \beta) \in \mathcal{P}_m$ only if $d = 1$. Moreover, $\Lambda_\gamma(\alpha, \beta) \in \mathcal{P}_m^*$ if and only if either $0 < \alpha < \frac{1}{2}$ and $0 \leq \beta < \infty$, or $\alpha = \frac{1}{2}$ and $\beta > 1$. Similarly, $\Lambda_\gamma(\alpha, \beta) \in \mathcal{P}_m$ if and only if either $0 < \alpha < \frac{1}{2}$ and $0 \leq \beta < \infty$, or $\alpha = \frac{1}{2}$ and $\beta > 0$.

In the case where $d > 1$, examples of time-dependent singular measures in non-homogeneous Kato classes can be constructed utilizing the measures satisfying conditions (3.45) and (3.46) with $0 < \delta < d$. For instance, if δ is an integer, we can take any δ-dimensional subspace of $\mathbb{R}^d$ and restrict the δ-dimensional Hausdorff measure $\mathcal{H}_\delta$ to it. If δ is fractional, then the Hausdorff measure $\mathcal{H}_\delta$ restricted to the standard Cantor set of Hausdorff dimension δ in the unit cube of $\mathbb{R}^d$ is an example of a measure γ satisfying conditions (3.45) and (3.46). It follows that for such a measure γ assertions (1) and (2) provide necessary and sufficient conditions for $\Lambda_\gamma(\alpha, \beta) \in \mathcal{P}_m^*$ or $\Lambda_\gamma(\alpha, \beta) \in \mathcal{P}_m$.

3.6 Transition Probability Densities and Fundamental Solutions to Parabolic Equations in Non-Divergence Form

Parabolic initial and final value problems are rich sources of transition densities. For instance, under rather general conditions, a fundamental solution $\widetilde{p}$ to a second order parabolic conservative initial value problem is also a backward transition probability density. Similarly, if p is a fundamental solution to a second order parabolic conservative final value problem, then p is a transition probability density.

Let us consider the following partial differential equations on d-dimensional Euclidean space $\mathbb{R}^d$:

$$\frac{\partial u}{\partial \tau} + Lu = 0 \tag{3.47}$$

and

$$\frac{\partial \widetilde{u}}{\partial t} - L\widetilde{u} = 0. \tag{3.48}$$

In equations (3.47) and (3.48), the symbol L stands for the differential operator given by

$$L = \sum_{i,j=1}^{d} a_{ij}(\tau, x) \frac{\partial^2}{\partial x_i \partial x_j} + \sum_{i=1}^{d} b_i(\tau, x) \frac{\partial}{\partial x_i}. \tag{3.49}$$

It is said that the operator L in formula (3.49) is in non-divergence form. Let us assume that the coefficients a_{ij} and b_i are Borel measurable on $[0, T] \times \mathbb{R}^d$, and that the uniform parabolicity condition holds for the second order part of the operator L. This means that there exist constants $\gamma_1 > 0$ and $\gamma_2 > 0$ such that for all $(\tau, x) \in [0, T] \times \mathbb{R}^d$ and all collections of real numbers $\lambda_1, \cdots, \lambda_d$, the following inequalities hold:

$$\gamma_1 \sum_{i=1}^{d} \lambda_i^2 \leq \sum_{i,j=1}^{d} a_{ij}(\tau, x) \lambda_i \lambda_j \leq \gamma_2 \sum_{i=1}^{d} \lambda_i^2. \tag{3.50}$$

Inequalities (3.50) imply the boundedness of the coefficients a_{ij}.

In the present section, we will study final value problems of the following form:

$$\begin{cases} \dfrac{\partial u}{\partial \tau} + Lu = 0, \ 0 \leq \tau < t \leq T \\ u(t) = f, \end{cases} \tag{3.51}$$

where f is a function on $\mathbb{R}^d$. The function f in (3.51) is called the final condition. We will also consider initial value problems given by

$$\begin{cases} \dfrac{\partial \widetilde{u}}{\partial t} - L\widetilde{u} = 0, \ 0 \le \tau < t \le T \\ \widetilde{u}(\tau) = g. \end{cases} \quad (3.52)$$

The function g in (3.52) is called the initial condition.

Definition 3.8 Suppose that L is a differential operator given by (3.49). A function $p(\tau, x; t, y)$ where $0 \le \tau < t \le T$, $x \in \mathbb{R}^d$, and $y \in \mathbb{R}^d$, is called a fundamental solution to equation (3.47) provided that the following conditions hold:

(1) For all $(t, y) \in (0, T] \times \mathbb{R}^d$, the function $(\tau, x) \mapsto p(\tau, x; t, y)$ is continuously differentiable in τ on the interval $(0, t)$ and twice continuously differentiable in x on $\mathbb{R}^d$. Moreover, p satisfies equation (3.47).

(2) For any bounded and continuous function f on $\mathbb{R}^d$,

$$f(x) = \lim_{\tau \uparrow t} \int_{\mathbb{R}^d} f(y) p(\tau, x; t, y) dy$$

uniformly on compact subsets of $\mathbb{R}^d$.

Definition 3.9 Suppose that L is a differential operator given by (3.49). A function $\widetilde{p}(\tau, x; t, y)$ where $0 \le \tau < t \le T$, $x \in \mathbb{R}^d$, and $y \in \mathbb{R}^d$, is called a fundamental solution to equation (3.48) provided that the following conditions hold:

(1) For all $(\tau, x) \in [0, T) \times \mathbb{R}^d$, the function $(t, y) \mapsto \widetilde{p}(\tau, x; t, y)$ is continuously differentiable in t on the interval (τ, T) and twice continuously differentiable in y on $\mathbb{R}^d$. Moreover, $\widetilde{p}$ satisfies equation (3.48).

(2) For any bounded and continuous function g on $\mathbb{R}^d$,

$$g(y) = \lim_{t \downarrow \tau} \int_{\mathbb{R}^d} g(x) p(\tau, x; t, y) dx$$

uniformly on compact subsets of $\mathbb{R}^d$.

The next result is known (see the references in Section 3.12). It concerns the existence of fundamental solutions and the Gaussian estimates.

Theorem 3.1 *Let L be as in formula (3.49), and assume that*

(a) The functions a_{ij} and b_i are bounded and measurable on $[0, T] \times \mathbb{R}^d$, and the matrix (a_{ij}) is symmetric, i.e. $a_{ij}(t, y) = a_{ji}(t, y)$.

(b) The uniform parabolicity condition (3.50) holds.
(c) There exists a constant δ with $0 < \delta \le 1$ such that

$$\sum_{i,j=1}^{d} |a_{ij}(\tau_1, x_1) - a_{ij}(\tau_2, x_2)| + \sum_{i=1}^{d} |b_i(\tau_1, x_1) - b_i(\tau_2, x_2)|$$
$$\le C\left(|x_1 - x_2|^{\delta} + |\tau_1 - \tau_2|^{\delta}\right)$$

for all $(\tau_1, x_1), (\tau_2, x_2) \in [0, T] \times \mathbb{R}^d$ (the Hölder continuity condition).

Then there exists a fundamental solution $\widetilde{p}(\tau, x; t, y)$ to equation (3.48). The function $\widetilde{p}$ is jointly continuous, strictly positive, and the following estimates hold:

$$\widetilde{p}(\tau, x; t, y) \le \frac{M}{(t-\tau)^{\frac{d}{2}}} \exp\left\{-\frac{|x-y|^2}{\alpha(t-\tau)}\right\}, \qquad (3.53)$$

(the upper Gaussian estimate for $\widetilde{p}$);

$$\widetilde{p}(\tau, x; t, y) \ge \frac{1}{M(t-\tau)^{\frac{d}{2}}} \exp\left\{-\frac{\alpha|x-y|^2}{t-\tau}\right\} \qquad (3.54)$$

(the lower Gaussian estimate for $\widetilde{p}$);

$$\frac{\partial \widetilde{p}(\tau, x; t, y)}{\partial t} \le \frac{M}{(t-\tau)^{\frac{d}{2}+1}} \exp\left\{-\frac{|x-y|^2}{\alpha(t-\tau)}\right\} \qquad (3.55)$$

(the upper Gaussian estimate for the time-derivative of $\widetilde{p}$);

$$\frac{\partial \widetilde{p}(\tau, x; t, y)}{\partial y_i} \le \frac{M}{(t-\tau)^{\frac{d+1}{2}}} \exp\left\{-\frac{|x-y|^2}{\alpha(t-\tau)}\right\} \qquad (3.56)$$

for all $1 \le i \le d$ (the upper Gaussian estimate for the first order space-derivatives of $\widetilde{p}$); and

$$\frac{\partial^2 \widetilde{p}(\tau, x; t, y)}{\partial y_i \partial y_j} \le \frac{M}{(t-\tau)^{\frac{d}{2}+1}} \exp\left\{-\frac{|x-y|^2}{\alpha(t-\tau)}\right\} \qquad (3.57)$$

for all $1 \le i \le d$ and $1 \le j \le d$ (the upper Gaussian estimate for the second order space-derivatives of $\widetilde{p}$). In estimates (3.53)-(3.57), M and α are positive constants.

Remark 3.5 The fundamental solution $\widetilde{p}$ in Theorem 3.1 is unique.

Let $\widetilde{p}$ be the fundamental solution to equation (3.48) in Theorem 3.1, and define a function p by

$$p(\tau, x; t, y) = \widetilde{p}(T - t, y; T - \tau, x). \tag{3.58}$$

Then

$$\frac{\partial p(\tau, x; t, y)}{\partial \tau} = -D_3 \widetilde{p}(T - t, y; T - \tau, x) = -L\widetilde{p}(T - t, y; T - \tau, x)$$
$$= -Lp(\tau, x; t, y)$$

where D_3 denotes the partial derivative with respect to the third variable, and the operator L acts on the variable x. It follows that the function p given by (3.58) is the unique fundamental solution to equation (3.47). Moreover, formulas (3.58), (3.53), and (3.54) imply the validity of the two-sided Gaussian estimates

$$p(\tau, x; t, y) \leq \frac{M}{(t-\tau)^{\frac{d}{2}}} \exp\left\{-\frac{|x-y|^2}{\alpha(t-\tau)}\right\} \tag{3.59}$$

and

$$p(\tau, x; t, y) \geq \frac{1}{M(t-\tau)^{\frac{d}{2}}} \exp\left\{-\frac{\alpha|x-y|^2}{t-\tau}\right\} \tag{3.60}$$

for the function p. The Gaussian estimates (3.55)-(3.57) for the derivatives of p are also satisfied.

The following assertion holds:

Theorem 3.2 *Let L be the operator given by (3.49), and suppose that L satisfies conditions (a)-(c) in Theorem 3.1. Then the unique fundamental solution p to equation (3.47) is a transition probability density, while the unique fundamental solution $\widetilde{p}$ to equation (3.48) is a backward transition probability density.*

Proof. We will prove the first part of Theorem 3.2. The proof of the second part is similar.

It is known that for every final condition $f \in BC$, where BC is the space of bounded continuous functions on $\mathbb{R}^d$, the final value problem in (3.51) has a unique solution given by

$$u(\tau, x) = \int_{\mathbb{R}^d} f(y) p(\tau, x; t, y) dy. \tag{3.61}$$

We will first prove that the function p satisfies the condition

$$\int_{\mathbb{R}^d} p(\tau, x; t, y) dy = 1 \tag{3.62}$$

for all $0 \leq \tau < t \leq T$ and $x \in \mathbb{R}^d$. Fix $0 < t \leq T$, and consider the following functions: $(\tau, x) \mapsto 1$ and $(\tau, x) \mapsto \int_{\mathbb{R}^d} p(\tau, x; t, y) dy$. It is not hard to see that these functions are solutions to final value problem (3.51). In order to prove that the second function is a solution to final value problem (3.51), we differentiate under the integral sign. This is possible, since the upper Gaussian estimates (3.55), (3.56), and (3.57) hold. Now the uniqueness result implies that (3.62) is true.

Next, we will establish the Chapman-Kolmogorov equation for p; that is, the equation

$$p(\tau, x; v, y) = \int_{\mathbb{R}^d} p(\tau, x; s, z) p(s, z; v, y) dz,$$

$$0 \leq \tau < s < v \leq T,\ x \in \mathbb{R}^d,\text{ and } y \in \mathbb{R}^d. \tag{3.63}$$

Consider the following two functions on the open set $(0, s) \times \mathbb{R}^d$:

$$(\tau, x) \mapsto p(\tau, x; v, y) \quad \text{and} \quad (\tau, x) \mapsto \int_{\mathbb{R}^d} p(\tau, x; s, z) p(s, z; v, y) dz.$$

These functions are solutions to the final value problem in (3.51) with $t = s$ and $f(x) = p(s, x; v, y)$. Here we use the upper Gaussian bounds for the function p in order to justify the differentiation under the integral sign. By the uniqueness of solutions to final value problem (3.51), the Chapman-Kolmogorov equation holds. $\square$

Let $\mathcal{P}_m^*$ be the non-autonomous Kato class of time-dependent measures associated with p, and $\mathcal{P}_m$ be the non-autonomous Kato class associated with $\widetilde{p}$. Next, we will continue our discussion of the properties of the family of time-dependent measures defined by

$$\Lambda_\gamma(\alpha, \beta) = \frac{1}{t^\alpha \ln^\beta \frac{Te}{t}} \gamma \tag{3.65}$$

where $\alpha > 0$, $\beta \geq 0$, and γ is a signed Borel measure of locally bounded variation on $\mathbb{R}^d$. We assume that there exists a number $\delta \geq 0$ such that

$$|\gamma|(B_r(x)) \leq c_1 r^\delta$$

for all $x \in \mathbb{R}^d$ and $r > 0$, where c_1 is a positive constant. Moreover, we also assume that there exist numbers $r_0 > 0$, $c_2 > 0$, and a point $x_0 \in \mathbb{R}^d$ such

that
$$|\gamma|(B_r(x_0)) \geq c_2 r^\delta$$

for all $0 < r < r_0$. It follows from the two-sided Gaussian estimates for p and $\widetilde{p}$ and from the results obtained in Examples (3.1) and (3.2) that

$$\Lambda_\gamma(\alpha,\beta) \in \mathcal{P}_m^* \iff \alpha + \frac{d-\delta}{2} < 1,\ 0 \leq \beta < \infty,$$

$$\text{or } \alpha + \frac{d-\delta}{2} = 1,\ \beta > 1,$$

and

$$\Lambda_\gamma(\alpha,\beta) \in \mathcal{P}_m \iff \alpha + \frac{d-\delta}{2} < 1,\ 0 \leq \beta < \infty,$$

$$\text{or } 0 < d-\delta < 2,\ \alpha + \frac{d-\delta}{2} = 1,\ \beta > 0.$$

3.7 Transition Probability Densities and Fundamental Solutions to Parabolic Equations in Divergence Form

In this section we discuss equations (3.47) and (3.48) with L given in divergence form. This means that the operator L can be represented by the following formula:

$$L = \sum_{i,j=1}^d \frac{\partial}{\partial x_i}\left[a_{ij}(\tau,x)\frac{\partial}{\partial x_j}\right] + \sum_{i=1}^d b_i(\tau,x)\frac{\partial}{\partial x_i}. \tag{3.66}$$

The fundamental solutions to equations (3.47) and (3.48) with L in divergence form are defined in the weak sense. Let $0 \leq a < b \leq T$, and let $C_0^\infty\left((a,b) \times \mathbb{R}^d\right)$ denote the space of infinitely differentiable functions with compact support contained in the set $(a,b) \times \mathbb{R}^d$. We denote by $C_b^\infty\left(\mathbb{R}^d\right)$ the space of all functions $f \in C_b^\infty\left(\mathbb{R}^d\right)$ such that f is infinitely differentiable on $\mathbb{R}^d$ and all partial derivatives of f are bounded on $\mathbb{R}^d$. The symbol $C\left([0,t]; L^2\left(\mathbb{R}^d\right)\right)$ will stand for the space of all L^2-valued continuous functions on the interval $[0,t]$, and $H^1\left(\mathbb{R}^d\right)$ will denote the Sobolev space of all functions $g \in L^2\left(\mathbb{R}^d\right)$ such that $\dfrac{\partial g}{\partial x_i} \in L^2\left(\mathbb{R}^d\right)$ for all $1 \leq i \leq d$. Here the partial derivatives are understood in the sense of distributions. The space of all locally square integrable functions on $\mathbb{R}^d$ is denoted by $L^2_{\text{loc}}\left(\mathbb{R}^d\right)$, and $H^1_{\text{loc}}\left(\mathbb{R}^d\right)$ denotes the local Sobolev space, consisting of all functions from the space $L^2_{\text{loc}}\left(\mathbb{R}^d\right)$ with all first order generalized partial derivatives in the space $L^2_{\text{loc}}(\mathbb{R}^d)$. The symbol $L^2((0,t);H^1(\mathbb{R}^d))$ stands for the space of all strongly measurable functions u on the interval $(0,t)$ with values in the space $H^1(\mathbb{R}^d)$ for which

$$\int_0^t \|u(s)\|_{H^1}^2 \, ds < \infty.$$

We denote by $L^2_{\text{loc}}((0,t); H^1_{\text{loc}}(\mathbb{R}^d))$ the space of all strongly measurable functions u on the interval $(0,t)$ with values in the space $H^1_{\text{loc}}(\mathbb{R}^d)$ such that

$$\int_\epsilon^{t-\epsilon} ds \int_C \left(|u(s,x)|^2 + |\nabla u(s,x)|^2\right) dx < \infty$$

for every ϵ with $0 < \epsilon < t$ and every compact subset C of $\mathbb{R}^d$. We will also need the space $L^q((0,T); L^p(B(0,r)))$ where $q \geq 1$, $p \geq 1$, and $B(0,r)$ stands for the ball of radius r in $\mathbb{R}^d$ centered at the origin. This space consists of all strongly measurable functions u defined on the interval $(0,T)$ with values in the space $L^p(B(0,r))$, and such that

$$\int_0^T \|u(s)\|_{L^p(B(0,r))}^q \, ds < \infty.$$

Now we are ready to introduce the notion of a weak solution.

Definition 3.10 Suppose that L is a differential operator given by (3.66). Let $0 < t \leq T$ and $f \in L^2_{\text{loc}}(\mathbb{R}^d)$. A function $u(\tau,x)$ where $0 \leq \tau < t \leq T$ and $x \in \mathbb{R}^d$, is called a weak solution to final value problem (3.51) provided that the following conditions hold:

(1) The function u belongs to the space $L^2_{\text{loc}}((0,t); H^1_{\text{loc}}(\mathbb{R}^d))$.
(2) The function

$$(\tau,x) \mapsto \sum_{i=1}^d b_i(\tau,x) \frac{\partial u}{\partial x_i}(\tau,x),$$

belongs to the space $L^1_{\text{loc}}((0,t) \times \mathbb{R}^d)$.
(3) For all test functions $\varphi \in C_0^\infty((0,t) \times \mathbb{R}^d)$,

$$\int_0^t \int_{\mathbb{R}^d} \left(u \frac{\partial \varphi}{\partial \tau} + \sum_{i,j=1}^d a_{i,j} \frac{\partial u}{\partial x_j} \frac{\partial \varphi}{\partial x_i} - \sum_{i=1}^d b_i \varphi \frac{\partial u}{\partial x_i} \right) dx d\tau = 0.$$

(4) The following equality holds:

$$\lim_{\tau \uparrow t} \int_{\mathbb{R}^d} u(\tau, x)\psi(x)dx = \int_{\mathbb{R}^d} f(x)\psi(x)dx$$

for all L^2-functions ψ on $\mathbb{R}^d$ with compact support.

Definition 3.11 Suppose that L is a differential operator given by (3.66). Let $0 \leq \tau < T$ and $g \in L^2_{\text{loc}}(\mathbb{R}^d)$. A function $\widetilde{u}(t, y)$ where $\tau < t \leq T$ and $y \in \mathbb{R}^d$ is called a weak solution to initial value problem (3.52) provided that the following conditions hold:

(1) The function $\widetilde{u}$ belongs to the space $L^2_{\text{loc}}((\tau, T); H^1_{\text{loc}}(\mathbb{R}^d))$.
(2) The function

$$(t, y) \mapsto \sum_{i=1}^d b_i(t, y)\frac{\partial \widetilde{u}}{\partial y_i}(t, y),$$

belongs to the space $L^1_{\text{loc}}((\tau, T) \times \mathbb{R}^d)$.
(3) For all test functions $\varphi \in C_0^\infty((\tau, T) \times \mathbb{R}^d)$,

$$\int_\tau^T \int_{\mathbb{R}^d} \left(\widetilde{u}\frac{\partial \varphi}{\partial t} + \sum_{i,j=1}^d a_{i,j}\frac{\partial \widetilde{u}}{\partial y_j}\frac{\partial \varphi}{\partial y_i} - \sum_{i=1}^d b_i\varphi\frac{\partial \widetilde{u}}{\partial y_i} \right) dy dt = 0.$$

(4) The following equality holds:

$$\lim_{t \downarrow \tau} \int_{\mathbb{R}^d} \widetilde{u}(t, y)\psi(y)dy = \int_{\mathbb{R}^d} g(y)\psi(y)dy$$

for all L^2 functions ψ on $\mathbb{R}^d$ with compact support.

Definition 3.12 Suppose that L is a differential operator given by (3.66). A Borel function $p(\tau, x; t, y)$ where $0 \leq \tau < t \leq T$, $x \in \mathbb{R}^d$, and $y \in \mathbb{R}^d$, is called a weak fundamental solution to equation (3.47) if for every t with $0 < t \leq T$ and every bounded measurable function f with compact support, the function u defined by

$$u(\tau, x) = \int_{\mathbb{R}^d} f(y)p(\tau, x; t, y)dy \tag{3.67}$$

is a bounded weak solution to final value problem (3.51).

Definition 3.13 Suppose that L is a differential operator given by (3.66). A Borel function $\widetilde{p}(\tau, x; t, y)$ where $0 \leq \tau < t \leq T$, $x \in \mathbb{R}^d$, and $y \in \mathbb{R}^d$, is called a weak fundamental solution to equation (3.48) if for every τ with

$0 \leq \tau < T$ and every bounded measurable function g of compact support, the function $\widetilde{u}$ defined by

$$\widetilde{u}(t, y) = \int_{\mathbb{R}^d} g(x) \widetilde{p}(\tau, x; t, y) dx \qquad (3.68)$$

is a bounded weak solution to initial value problem (3.52).

Definitions 3.10, 3.11, 3.12, and 3.13 can be found in [Porper and Eidel'man (1984)]. In [Liskevich and Semenov (2000)], the authors studied weak solutions and weak fundamental solutions in the global case. The definitions used in [Liskevich and Semenov (2000)] are as follows:

Definition 3.14 Suppose that L is a differential operator given by (3.66). Let $0 < t \leq T$ and $f \in L^1(\mathbb{R}^d) \cap L^\infty(\mathbb{R}^d)$. A function $u(\tau, x)$ where $0 \leq \tau < t \leq T$ and $x \in \mathbb{R}^d$, is called a weak solution to final value problem (3.51) provided that the following conditions hold:

(1) The function u belongs to the space

$$C\left([0, t]; L^2(\mathbb{R}^d)\right) \cap L^2\left((0, t); H^1(\mathbb{R}^d)\right).$$

(2) The function

$$(\tau, x) \mapsto \sum_{i=1}^{d} b_i(\tau, x) \frac{\partial u}{\partial x_i}(\tau, x),$$

belongs to the space $L^1\left((0, t) \times \mathbb{R}^d\right)$.

(3) For all $\varphi \in C_0^\infty\left((0, t) \times \mathbb{R}^d\right)$,

$$\int_0^t \int_{\mathbb{R}^d} \left(u \frac{\partial \varphi}{\partial \tau} + \sum_{i,j=1}^{d} a_{i,j} \frac{\partial u}{\partial x_j} \frac{\partial \varphi}{\partial x_i} - \sum_{i=1}^{d} b_i \varphi \frac{\partial u}{\partial x_i} \right) dx d\tau = 0.$$

(4) $u(t, x) = f(x)$ m-almost everywhere.

By condition (1), the function $x \mapsto u(t, x)$ belongs to the space $L^2(\mathbb{R}^d)$. Therefore, the equality in condition (4) makes sense.

Definition 3.15 Suppose that L is a differential operator given by (3.66). A Borel function $\widetilde{p}(\tau, x; t, y)$ where $0 \leq \tau < t \leq T$, $x \in \mathbb{R}^d$, and $y \in \mathbb{R}^d$, is called a weak fundamental solution to equation (3.48) provided that the following conditions hold:

(1) For every $0 \leq \tau < T$ and $g \in L^1(\mathbb{R}^d) \cap L^\infty(\mathbb{R}^d)$, the function $\widetilde{u}$ defined by (3.68) belongs to the space
$$C([\tau, T]; L^2(\mathbb{R}^d)) \cap L^2((\tau, T); H^1(\mathbb{R}^d)).$$

(2) The function
$$(t, y) \mapsto \sum_{i=1}^{d} b_i(t, y) \frac{\partial \widetilde{u}}{\partial y_i}(t, y)$$
belongs to the space $L^1((\tau, T) \times \mathbb{R}^d)$.

(3) For all $\varphi \in C_0^\infty([\tau, T] \times \mathbb{R}^d)$,
$$\int_\tau^T \int_{\mathbb{R}^d} \left(\widetilde{u} \frac{\partial \varphi}{\partial t} - \sum_{i,j=1}^d a_{i,j} \frac{\partial \widetilde{u}}{\partial y_j} \frac{\partial \varphi}{\partial y_i} + \sum_{i=1}^d b_i \varphi \frac{\partial \widetilde{u}}{\partial y_i} \right) dy\, dt = 0.$$

(4) $\widetilde{u}(\tau, y) = g(y)$ m-almost everywhere.

Definition 3.16 Suppose that L is a differential operator given by (3.66). A Borel function $p(\tau, x; t, y)$ where $0 \leq \tau < t \leq T$, $x \in \mathbb{R}^d$, and $y \in \mathbb{R}^d$, is called a weak fundamental solution to equation (3.47) if for every t with $0 < t \leq T$ and $f \in L^1(\mathbb{R}^d) \cap L^\infty(\mathbb{R}^d)$, the function u defined by
$$u(\tau, x) = \int_{\mathbb{R}^d} f(y) p(\tau, x; t, y) dy \tag{3.69}$$
is a weak solution to final value problem (3.51).

Definition 3.17 Suppose that L is a differential operator given by (3.66). A Borel function $\widetilde{p}(\tau, x; t, y)$ where $0 \leq \tau < t \leq T$, $x \in \mathbb{R}^d$, and $y \in \mathbb{R}^d$, is called a weak fundamental solution to equation (3.48) if for every τ with $0 \leq \tau < T$ and every $g \in L^1(\mathbb{R}^d) \cap L^\infty(\mathbb{R}^d)$, the function $\widetilde{u}$ defined by
$$\widetilde{u}(t, y) = \int_{\mathbb{R}^d} g(x) \widetilde{p}(\tau, x; t, y) dx \tag{3.70}$$
is a weak solution to initial value problem (3.52).

The following result concerns the existence and uniqueness of weak fundamental solutions (see [Porper and Eidel'man (1984)], Theorem 6.1).

Theorem 3.3 Let L be given by (3.66), and suppose that the following assumptions hold:

(1) The coefficients a_{ij} satisfy the uniform parabolicity condition (3.50).

(2) For some $\vartheta_0 \in (0,1)$, $r_0 > 0$, $s \geq \dfrac{2}{1-\vartheta_1}$, and $\vartheta_1 = \dfrac{1}{2} - \dfrac{d}{2s} - \dfrac{1}{q} \geq \dfrac{\vartheta_0}{2}$, the coefficients $\{b_i\}$ belong to the space $L^q((0,T); L^s(B(0,r_0)))$.

(3) For almost all $x \in \mathbb{R}^d$ with $|x| > r_0$ and all $t \in [0,T]$, the coefficients $\{b_i\}$ satisfy $|b_i(t,x)| \leq M$, $1 \leq i \leq d$, where M is a positive constant.

Then there exists the unique weak fundamental solution $\widetilde{p}$ to the equation $\dfrac{\partial \widetilde{u}}{\partial t} - L\widetilde{u} = 0$ in the sense of Definition 3.13 such that the two-sided Gaussian estimates (3.53) and (3.54) hold for $\widetilde{p}$ with the constants M and α depending only on T, d, the uniform parabolicity constants γ_1 and γ_2, and the constants q, s, r_0, M, and ϑ_0.

Fundamental solutions in Theorem 3.3 are unique almost everywhere with respect to the measure $m_d \times m_d$. A theorem, similar to Theorem 3.3, holds for weak fundamental solutions in the case of equation (3.48). Here we take into account formula (3.58). It is also true that bounded solutions to initial value problem (3.52) are unique.

Next we formulate a recent result from [Liskevich and Semenov (2000)] concerning the existence of the unique weak fundamental solution $\widetilde{p}$ to the equation $\dfrac{\partial \widetilde{u}}{\partial t} - L\widetilde{u} = 0$ and the two-sided Gaussian estimates for $\widetilde{p}$ under more general conditions on the coefficients $\{b_i\}$ than those in Theorem 3.3. The existence theorem in [Liskevich and Semenov (2000)] was obtained in the case where the coefficients a_{ij} are Borel measurable on $[0,T] \times \mathbb{R}^d$ and satisfy the uniform parabolicity condition (3.50), while the restrictions on the coefficients b_i are expressed in terms of the function

$$w(t,x) = \sum_{i,j=1}^{d} \left(\dfrac{a(t,x) + a^*(t,x)}{2}\right)^{-1}_{i,j}(t,x) b_i(t,x) b_j(t,x), \qquad (3.71)$$

where $a^*(t,x)$ denotes the transpose of the matrix $a(t,x)$. More precisely, it is assumed that the function w belongs to the intersection of the extended weighted non-autonomous Kato classes $\widehat{\mathcal{P}}_{f,\varphi}$ and $\widehat{\mathcal{P}}^*_{f,\varphi}$. These classes are associated with a scaled Gaussian transition density

$$\widetilde{p}_\lambda(\tau,x;t,y) = \dfrac{1}{(2\pi\lambda(t-\tau))^{\frac{d}{2}}} \exp\left\{-\dfrac{|x-y|^2}{2\lambda(t-\tau)}\right\}, \qquad (3.72)$$

and the function φ is such that $\varphi(\tau,t)$ depends only on the difference $t-\tau$

and satisfies the conditions

$$\varphi(0) = 0, \int_0^T \frac{\varphi(s)}{s} ds < \infty, \int_0^T [\varphi'(s)]_+ ds < \infty, \text{ and } \int_0^T \frac{ds}{\varphi(s)} < \infty. \quad (3.73)$$

In (3.73), we use the notation $\varphi(t-\tau)$ instead of $\varphi(\tau,t)$.

Theorem 3.4 *Let L be given by (3.66), and suppose that the following assumptions hold:*

(1) The coefficients a_{ij} satisfy condition (3.50).
(2) There exist a constant $\lambda > 0$ and a function φ such that the conditions in (3.73) are satisfied and

$$w \in \widehat{\mathcal{P}}_{f,\varphi} \cap \widehat{\mathcal{P}}^*_{f,\varphi}. \quad (3.74)$$

In (3.74), the function w is defined by formula (3.71), and the classes $\widehat{\mathcal{P}}_{f,\varphi}$ and $\widehat{\mathcal{P}}^*_{f,\varphi}$ are associated with the function φ and the scaled Gaussian density p_λ given by (3.72).

Then there exists the unique weak fundamental solution $\widetilde{p}$ to the equation $\dfrac{\partial \widetilde{u}}{\partial t} - L\widetilde{u} = 0$ in the sense of Definition 3.17. Moreover, the two-sided Gaussian estimates in (3.53) and (3.54) hold for $\widetilde{p}$ with the constants M and α depending only on T, d, the uniform parabolicity constants γ_1 and γ_2, the function φ, and the norm of the function w in the space $\widehat{\mathcal{P}}_{f,\varphi} \cap \widehat{\mathcal{P}}^*_{f,\varphi}$.

It was shown in [Liskevich and Semenov (2000)] that the uniqueness theorem holds for weak solutions to initial value problem (3.51) (see [Liskevich and Semenov (2000)], Lemma 4.7). It is also clear that the function $\widetilde{p}(\tau, x; t, y)$ is defined for all $0 \leq \tau < t \leq T$ $m \times m$-almost everywhere on $\mathbb{R}^d \times \mathbb{R}^d$. In the proof of Theorem 1 in [Liskevich and Semenov (2000)], the weak fundamental solution $\widetilde{p}$ is constructed as the $L^1_{\text{loc}}(\mathbb{R}^d \times \mathbb{R}^d, dxdy)$-limit of a double sequence $r_{n,m}(\tau, x; t, y)$ of classical weak fundamental solutions. The limit in this construction is uniform with respect to the variables τ and t with $0 \leq \tau < t \leq T$ (see the remark after the proof of Lemma 4.2 in [Liskevich and Semenov (2000)]). A theorem similar to Theorem 3.4 also holds for the weak fundamental solution to equation (3.48). Here we take into account formula (3.58).

The next lemma states that the restrictions on the coefficients $\{b_i\}$ in Theorem 3.3 are stronger than those in Theorem 3.4.

Lemma 3.11 *Let L be an operator given by (3.66), and suppose that the uniform parabolicity condition (3.50) holds for the coefficients $\{a_{ij}\}$, and* conditions (2) and (3) in Theorem 3.3 hold for the coefficients $\{b_i\}$. Then there exist a number $\lambda > 0$ and a function φ such that the conditions in (3.73) are satisfied and condition (2) in Theorem 3.4 holds.

Proof. Since the matrix $\dfrac{a(t,x) + a^*(t,x)}{2}$ is symmetric and satisfies the uniform parabolicity condition with the same constants γ_1 and γ_2 as the matrix $a(t,x)$, its inverse matrix $\left(\dfrac{a(t,x) + a^*(t,x)}{2}\right)^{-1}$ is bounded by γ_1^{-1} uniformly with respect to t and x. It follows that

$$\left|\left(\frac{a(t,x) + a^*(t,x)}{2}\right)^{-1}_{ij}\right| \leq \frac{1}{\gamma_1}$$

for all $1 \leq i \leq d$, $1 \leq j \leq d$, $t \in [0,T]$, and $x \in \mathbb{R}^d$. By formula (3.71),

$$|w(t,x)| \leq M_d \sum_{i=1}^{d} b_i^2(t,x).$$

It suffices to prove that there exist a number $\lambda > 0$ and a function φ such that the conditions in (3.73) are satisfied and

$$b_i^2 \in \widehat{\mathcal{P}}_{f,\varphi} \cap \widehat{\mathcal{P}}^*_{f,\varphi}, \quad 1 \leq i \leq d. \tag{3.75}$$

Take $\lambda = 1$ and $\varphi(\tau,t) = (t-\tau)^\epsilon$, where $\epsilon > 0$ is a small number that will be chosen later. By Definition 3.6, we see that in order to establish (3.75), it suffices to show that

$$\sup_{0 \leq \tau \leq T} \sup_{x \in \mathbb{R}^d} \int_\tau^T \frac{dv}{(v-\tau)^{\epsilon+\frac{d}{2}}} \int_{\mathbb{R}^d} b_i^2(v,y) \exp\left\{-\frac{|x-y|^2}{2(v-\tau)}\right\} dy < \infty \tag{3.76}$$

and

$$\sup_{0 \leq t \leq T} \sup_{x \in \mathbb{R}^d} \int_0^t \frac{dv}{(t-v)^{\epsilon+\frac{d}{2}}} \int_{\mathbb{R}^d} b_i^2(v,y) \exp\left\{-\frac{|x-y|^2}{2(t-v)}\right\} dy < \infty \tag{3.77}$$

for all $1 \leq i \leq d$. We will prove only inequality (3.76). Inequality (3.77) can be obtained similarly.

It follows from condition (2) in Theorem 3.3 that $q > 2$ and $s > 2$. It is also clear from (2) and (3) in Theorem 3.3 that every function b_i^2 with $1 \leq i \leq d$ can be represented as the sum $b_i^2 = \eta_i + \zeta_i$ of two functions such that $\eta_i \in L^{\frac{q}{2}}\left((0,T); L^{\frac{s}{2}}(\mathbb{R}^d)\right)$ and $\zeta_i \in L^\infty\left([0,T] \times \mathbb{R}^d\right)$. Now, using

Hölder's inequality twice, we get

$$\int_\tau^T \frac{dv}{(v-\tau)^{\epsilon+\frac{d}{2}}} \int_{\mathbb{R}^d} \eta_i(v,y) \exp\left\{-\frac{|x-y|^2}{2(v-\tau)}\right\} dy$$

$$\leq \int_\tau^T \frac{dv}{(v-\tau)^{\epsilon+\frac{d}{2}}} \left\{\int_{\mathbb{R}^d} \eta_i(v,y)^{\frac{s}{2}} dy\right\}^{\frac{2}{s}}$$

$$\cdot \left\{\int_{\mathbb{R}^d} \exp\left\{-\frac{s|x-y|^2}{2(s-2)(v-\tau)}\right\} dy\right\}^{\frac{s-2}{s}}$$

$$\leq c_s \int_\tau^T \frac{dv}{(v-\tau)^{\epsilon+\frac{d}{s}}} \left\{\int_{\mathbb{R}^d} \eta_i(v,y)^{\frac{s}{2}} dy\right\}^{\frac{2}{s}}$$

$$\leq c_s \|\eta_i\|_{L^{\frac{q}{2}}\left((0,T);L^{\frac{s}{2}}(\mathbb{R}^d)\right)} \left\{\int_\tau^T \frac{dv}{(v-\tau)^\lambda}\right\}^{\frac{q-2}{q}} \quad (3.78)$$

where $\lambda = \left(\epsilon + \frac{d}{s}\right)\frac{q}{q-2}$. By condition (2) in Theorem 3.3, we have $\frac{1}{2} - \frac{d}{2s} - \frac{1}{q} > 0$. Therefore, $\frac{dq}{s(q-2)} < 1$, and for small values of ϵ we have $\lambda < 1$. It follows that the last integral in (3.78) is finite. Hence, inequality (3.76) holds with η_i instead of b_i^2. We also have

$$\int_\tau^T \frac{dv}{(v-\tau)^{\epsilon+\frac{d}{2}}} \int_{\mathbb{R}^d} \zeta_i(v,y) \exp\left\{-\frac{|x-y|^2}{2(v-\tau)}\right\} dy$$

$$\leq c_d \|\zeta_i\|_\infty \int_\tau^T \frac{dv}{(v-\tau)^\epsilon} < \infty. \quad (3.79)$$

Now it is clear that (3.78) and (3.79) imply (3.76).

This completes the proof of Lemma 3.11. □

The next theorem is similar to Theorem 3.2.

Theorem 3.5 *Let L be the operator given by (3.66), and suppose that the conditions in Theorem 3.4 hold. Then the unique weak fundamental solution p to equation (3.47) is a transition probability density, while the unique weak fundamental solution $\tilde{p}$ to equation (3.48) is a backward transition probability density.*

Proof. The normality condition can be established using the results obtained in [Liskevich and Semenov (2000)]. We have already mentioned that in the proof of Theorem 1 in [Liskevich and Semenov (2000)], the unique fundamental solution p is approximated in the space $L^1_{\text{loc}}\left(\mathbb{R}^d \times \mathbb{R}^d, dxdy\right)$

by a double sequence $r_{n,m}(\tau, x; t, y)$ of classical fundamental solutions. The limit in this construction is uniform with respect to the variables τ and t with $0 \leq \tau < t \leq T$. This was established in the remark after the proof of Lemma 4.2 in [Liskevich and Semenov (2000)]. Moreover, the upper Gaussian estimate holds for $r_{n,m}$ uniformly with respect to n and m. By Theorem 3.2, $\int r_{n,m}(\tau, x; t, y) dy = 1$. Therefore, the dominated convergence theorem implies that $\int p(\tau, x; t, y) dy = 1$.

Our next goal is to prove that the Chapman-Kolmogorov equation holds for p. Let $0 \leq \tau < s < v \leq T$ and $\psi \in L^1(\mathbb{R}^d) \cap L^\infty(\mathbb{R}^d)$. Then the function

$$u_1(\tau, x) = \int_{\mathbb{R}^d} \psi(y) p(\tau, x; v, y) dy$$

is a weak solution to the final value problem (3.51) with $t = v$ and with the final condition $u_1(v) = \psi$. It follows from the properties of weak solutions that the function u_1 is a weak solution to final value problem (3.51) with $t = s$ and the final condition given by

$$u_1(s, x) = \int_{\mathbb{R}^d} \psi(y) p(s, x; v, y) dy.$$

Moreover, the function

$$u_2(\tau, x) = \int_{\mathbb{R}^d} \psi(y) dy \int_{\mathbb{R}^d} p(\tau, x; s, z) p(s, z; v, y) dz$$
$$= \int_{\mathbb{R}^d} p(\tau, x; s, z) dz \int_{\mathbb{R}^d} \psi(y) p(s, z; v, y) dy$$

is a weak solution to final value problem (3.51) with $t = s$ and the final condition given by

$$u_2(s, x) = \int_{\mathbb{R}^d} \psi(y) p(s, x; v, y) dy.$$

By the upper Gaussian estimate, $u_2(s) \in L^1(\mathbb{R}^d) \cap L^\infty(\mathbb{R}^d)$. The uniqueness of weak solutions established by Liskevich and Semenov yields that for all τ with $0 \leq \tau < s$, the equality $u_1(\tau, x) = u_2(\tau, x)$ holds m-almost everywhere on $\mathbb{R}^d$. This implies that for all $\psi \in L^1(\mathbb{R}^d) \cap L^\infty(\mathbb{R}^d)$ and $0 \leq \tau < s < v \leq T$,

$$\int_{\mathbb{R}} \psi(y) p(\tau, x; v, y) dy = \int_{\mathbb{R}} \psi(y) dy \int_{\mathbb{R}} p(\tau, x; s, z) p(s, z; v, y) dz.$$

Therefore, p satisfies the Chapman-Kolmogorov equation.

This completes the proof of Theorem 3.5. □

Let $\Lambda_\gamma(\alpha, \beta)$ be the family of functions defined in Example 3.2. Since the two-sided Gaussian estimates hold for all transition probability densities discussed in this section, assertions (1) and (2) in Example 3.2 hold for these densities. Similar results are true for the classes $\mathcal{P}^*_{m,\varphi}$ and $\widetilde{\mathcal{P}}_{m,\varphi}$ where the function φ is defined by $\varphi(\tau, t) = (t - \tau)^a$ with $0 < a < 1$. Let us assume that the class $\mathcal{P}^*_{m,\varphi}$ is associated with one of the transition probability functions P discussed in this section, while the class $\widetilde{\mathcal{P}}_{m,\varphi}$ corresponds to one of the backward transition probability functions $\widetilde{P}$ appearing in this section. Then the following assertions hold:

(1) For any a with $0 < a < 1$, the time-dependent measure $\Lambda_\gamma(\alpha, \beta)$ belongs to the class $\mathcal{P}^*_{m,\varphi}$ associated with the function $\varphi(\tau, t) = (t - \tau)^a$ if and only if either $\alpha + a + \frac{d-\delta}{2} < 1$ and $0 \leq \beta < \infty$, or $\alpha + a + \frac{d-\delta}{2} = 1$ and $\beta > 1$.

(2) For any a with $0 < a < 1$, the time-dependent measure $\Lambda_\gamma(\alpha, \beta)$ belongs to the class $\widetilde{\mathcal{P}}_{m,\varphi}$ associated with the function $\varphi(\tau, t) = (t - \tau)^a$ if and only if either $\alpha + a + \frac{d-\delta}{2} < 1$, and $0 \leq \beta < \infty$, or $0 < d + 2a - \delta < 2$, $\alpha + a + \frac{d-\delta}{2} = 1$, and $\beta > 0$.

3.8 Diffusion Processes and Stochastic Differential Equations

This section is devoted to stochastic processes generated by second order partial differential operators. Such processes are called diffusions (see Definition 3.18 below). Let P be a transition probability function on $\mathbb{R}^d$, and consider the following second order partial differential operator with time-dependent coefficients:

$$L_\tau = \frac{1}{2} \sum_{i,j=1}^{d} a_{i,j}(\tau, x) \frac{\partial^2}{\partial x_i \partial x_j} + \sum_{i=1}^{d} b_i(\tau, x) \frac{\partial}{\partial x_i} \quad (3.80)$$

where $\tau \in [0, T]$ and $x \in \mathbb{R}^d$. We use the subscript τ in (3.80) to emphasize the time-dependence. It is assumed that the following conditions hold:

(1) The coefficients $a_{i,j}$ and b_i are Borel measurable and locally bounded functions on $[0, T] \times \mathbb{R}^d$.
(2) The matrix field $a_{i,j}(\tau, x)$ is symmetric; that is, $a_{i,j}(\tau, x) = a_{j,i}(\tau, x)$.

(3) The matrix field $a_{i,j}(\tau, x)$ is nonnegative; that is,

$$\sum_{i,j=1}^{d} a_{i,j}(\tau, x)\xi_i \xi_j \geq 0$$

for all $(\tau, x) \in [0,T] \times \mathbb{R}^d$ and all $\xi = (\xi_1, \ldots, \xi_d) \in \mathbb{R}^d$.

In the next definition, we introduce non-homogeneous diffusion processes.

Definition 3.18 A Markov process $(X_t, \mathcal{F}_t^\tau, \mathbb{P}_{\tau,x})$ with state space $\mathbb{R}^d$ is called a diffusion process with generator L_τ provided that

(1) The process X_t has continuous paths.
(2) For any function $f \in C_0^\infty$, the equality

$$\mathbb{E}_{\tau,x}\left[f(X_t)\right] = f(x) + \mathbb{E}_{\tau,x}\left[\int_\tau^t L_s f(X_s)\, ds\right] \quad (3.81)$$

holds for all $0 \leq \tau \leq t \leq T$ and $x \in \mathbb{R}^d$.

The matrix field $a_{i,j}$ in Definition 3.18 is called the covariance or the diffusion coefficient of the process X_t, while the vector field b_i is called the drift of X_t.

The next assertion provides an equivalent martingale characterization of a diffusion process.

Lemma 3.12 *The following are equivalent:*

(1) Condition (2) in Definition 3.18 holds for the process X_t.
(2) For any $\tau \in [0,T]$, $x \in \mathbb{R}^d$, and $f \in C_0^\infty$, the process

$$M_t^{\tau,f} = f(X_t) - f(X_\tau) - \int_\tau^t L_s f(X_s)\, ds,\ t \in [\tau, T], \quad (3.82)$$

is a $\mathcal{F}_t^\tau$-martingale with respect to the measure $\mathbb{P}_{\tau,x}$.

Proof. Suppose that condition (2) in Lemma 3.12 holds. Then the expression $\mathbb{E}_{\tau,x}\left[M_u^{\tau,f}\right]$ does not depend on $u \in [\tau, T]$. By substituting $u = t$ and $u = \tau$ into this expression, we see that condition (1) in Lemma 3.12 holds.

Conversely, suppose that condition (1) holds and let $\tau \leq t_1 < t_2 \leq T$. By the Markov property, we get

$$\mathbb{E}_{\tau,x}\left[M_{t_2}^{\tau,f} \mid \mathcal{F}_{t_1}^\tau\right] - M_{t_1}^{\tau,f} = \mathbb{E}_{\tau,x}\left[M_{t_2}^{\tau,f} - M_{t_1}^{\tau,f} \mid \mathcal{F}_{t_1}^\tau\right]$$

$$= \mathbb{E}_{\tau,x}\left[f(X_{t_2}) - f(X_{t_1}) - \int_{t_1}^{t_2} L_s f(X_s)\, ds \mid \mathcal{F}_{t_1}^{\tau}\right]$$

$$= \mathbb{E}_{t_1, X_{t_1}}\left[f(X_{t_2}) - f(X_{t_1}) - \int_{t_1}^{t_2} L_s f(X_s)\, ds\right]. \tag{3.83}$$

By substituting $\tau = t_1$, $x = X_{t_1}$, and $t = t_2$ into (3.81), we see that (3.83) yields condition (2) in Lemma 3.12.

This completes the proof of Lemma 3.12. $\square$

Let Y be the free backward propagator on the space $L_{\mathcal{E}}^{\infty}$ defined by

$$Y(\tau, t)f(x) = \int_{\mathbb{R}} f(y) P(\tau, x; t, dy).$$

It is clear that formula (3.81) in Definition 3.18 is equivalent to the following formula:

$$Y(\tau, t)f(x) = f(x) + \int_{\tau}^{t} Y(\tau, s) L_s f(x)\, ds \tag{3.84}$$

for all $0 \leq \tau \leq t \leq T$, $x \in \mathbb{R}^d$, and $f \in C_0^{\infty}$.

In Section 2.3, we defined the right $\sigma(L_{\mathcal{E}}^{\infty}, M)$-generators $A_+^M(\tau)$, $0 \leq \tau < T$, of the backward propagator Y. They are given by the formula

$$A_+^M(\tau)f = \sigma(L_{\mathcal{E}}^{\infty}, M)\text{-}\lim_{h \downarrow 0} \frac{Y(\tau, \tau + h)f - f}{h} \tag{3.85}$$

where f belongs to the subspace $D(A_+^M(\tau))$ of the space $L_{\mathcal{E}}^{\infty}$, consisting of all functions for which the limit on the right-hand side of formula (3.85) exists. The next lemma states that under certain restrictions, the generator of the diffusion process X_t coincides with the right generator of the backward propagator Y.

Lemma 3.13 *Suppose that the free backward propagator Y in (3.84) is a strongly continuous backward Feller-Dynkin propagator, and let X_t be a diffusion process with generator L_{τ}. Suppose also that the coefficients $a_{i,j}$ and b_i in (3.80) are continuous functions. Then $L_{\tau} f(x) = A_+^M(\tau) f(x)$ for all $\tau \in [0, T)$, $x \in \mathbb{R}^d$, and $f \in C_0^2$.*

Proof. It follows from equality (3.84) with $t = \tau + h$ that

$$\frac{Y(\tau, \tau + h)f(x) - f(x)}{h} = \frac{1}{h}\int_{\tau}^{\tau + h} Y(\tau, s) L_s f(x)\, ds. \tag{3.86}$$

Passing to the limit as $h \downarrow 0$ in equality (3.86) and using the assumptions in Lemma 3.13, we see that

$$\lim_{h \downarrow 0} \frac{Y(\tau, \tau + h) f(x) - f(x)}{h} = L_\tau f(x) \qquad (3.87)$$

for all $x \in \mathbb{R}^d$. Moreover, equality (3.86) and the fact that the backward propagator Y is a family of contractions on the space $L_\mathcal{E}^\infty$, imply

$$\sup_{h: 0 < h < T - \tau} \sup_{x \in \mathbb{R}^d} \left| \frac{Y(\tau, \tau + h) f(x) - f(x)}{h} \right| \leq \sup_{(s,x) \in [\tau, T] \times \mathbb{R}^d} |L_s f(x)| < \infty \qquad (3.88)$$

for all $f \in C_0^2$. It follows from (3.87) and (3.88) that

$$\sigma\left(L_\mathcal{E}^\infty, M\right)\text{-}\lim_{h \downarrow 0} \frac{Y(\tau, \tau + h) f - f}{h} = L_\tau f$$

for all $f \in C_0^2$. Hence $C_0^2 \subset D\left(A_+^M(\tau)\right)$ for all $\tau \in [0, T)$, and $A_+^M(\tau) f = L_\tau f$ for all $\tau \in [0, T)$ and $f \in C_0^2$. □

What second order partial differential operators on $\mathbb{R}^d$ can be generators of diffusion processes? This question will be partially answered in the sequel. We will need the basic facts from the theory of stochastic integration and stochastic differential equations. Some of the results gathered below are formulated without proofs. We refer the reader to Section 3.12 for the bibliography concerning stochastic integrals and stochastic differential equations.

Let us consider the following stochastic differential equation:

$$dX_t = b(t, X_t) \, dt + \sigma(t, X_t) \, dB_t, \qquad (3.89)$$

where B_t is a standard Brownian motion. Since Brownian paths are not of finite variation, the equation in (3.89) cannot be integrated pathwise using Riemann-Stieltjes integral. If we integrate the stochastic differential equation in (3.89) informally, the resulting integral equation is as follows:

$$X_t = X_\tau + \int_\tau^t b(s, X_s) \, ds + \int_\tau^t \sigma(s, X_s) \, dB_s. \qquad (3.90)$$

Hence, in order to understand the meaning of the stochastic differential equation in (3.89), one should know how to integrate stochastic processes with respect to Brownian motion. The theory of stochastic integration and stochastic differential equations was developed by Itô in 1950's. Next we will discuss the elements of this theory. Most of the proofs of the results formulated below are omitted, and we refer the reader to Section 3.12 for the bibliographical information on stochastic integration and stochastic differential equations.

Let $(B_t, \mathcal{F}_t, \mathbb{P})$ be a standard one-dimensional Brownian motion (see Subsection 1.12.1). A stochastic process H_s, $s \in [\tau, T]$, is called a simple process if there exists a partition $\tau = s_0 < s_1 < \cdots < s_m = T$ of the interval $[\tau, T]$ and a finite sequence of random variables h_i, $0 \leq i \leq m-1$, such that h_i is $\mathcal{F}_{s_i}$-measurable and $H_s = h_i$ for all $s_i < s \leq s_{i+1}$. The stochastic integral or the Itô integral of the process H_s is the stochastic process defined as follows:

$$\int_\tau^t H_s dB_s = \sum_{i=0}^{j-1} h_i \left(B_{s_{i+1}} - B_{s_i}\right) + h_j \left(B_t - B_{s_j}\right) \tag{3.91}$$

for $s_j < t \leq s_{j+1}$, $0 \leq j \leq m-1$.

The next lemma concerns Itô integrals of simple processes.

Lemma 3.14 *Let H_s, $\tau \leq s \leq T$, be a simple process. Then the following assertions hold:*

(1) For every $\tau \in [0, T)$, the process $I(H)_t = \int_\tau^t H_s dB_s$ is a continuous $\mathcal{F}_t^\tau$-martingale on the interval $[\tau, T]$.
(2) For all t with $\tau \leq t \leq T$,

$$\mathbb{E}\left[(I(H)_t)^2\right] = \mathbb{E}\left[\int_\tau^t (H_s)^2 ds\right]. \tag{3.92}$$

(3) The following inequality holds:

$$\mathbb{E}\left[\sup_{t: \tau \leq t \leq T} (I(H)_t)^2\right] \leq 4\mathbb{E}\left[\int_\tau^T (H_s)^2 ds\right]. \tag{3.93}$$

Next we will explain how to integrate more general stochastic processes. Let $\tau \in [0, T)$, and denote by $\mathcal{H}_\tau$ the class of all stochastic processes H_t, $t \in [\tau, T]$, such that H_t is an $\mathcal{F}_t^\tau$-adapted process, and

$$\mathbb{E}\left[\int_\tau^T (H_s)^2 ds\right] < \infty.$$

The next assertion concerns the existence of Itô integrals of stochastic processes from the class $\mathcal{H}_\tau$.

Theorem 3.6 *There exists a unique linear mapping J from the space $\mathcal{H}_\tau$ into the space of all continuous $\mathcal{F}_t^\tau$-martingales on the interval $[\tau, T]$ such that*

(1) For any simple process H_t, $t \in [\tau, T]$,

$$J(H)_t = \int_\tau^t H_s dB_s.$$

(2) For all t with $\tau \leq t \leq T$,

$$\mathbb{E}\left[(J(H)_t)^2\right] = \mathbb{E}\left[\int_\tau^t (H_s)^2 ds\right].$$

The mapping J in Theorem 3.6 is unique. More precisely, if J_1 and J_2 are two such mappings, and if $H \in \mathcal{H}_\tau$, then $J_1(H)_t = J_2(H)_t$ $\mathbb{P}$-almost surely for all $t \in [\tau, T]$ (see, e.g., [Lamberton and Lapeyre (1996)]). For all $H \in \mathcal{H}_\tau$ and $t \in [\tau, T]$, the Itô integral $J(H)_t$ is denoted by the symbol $\int_\tau^t H_s dB_s$. The equality in (2) of Theorem 3.6 can be rewritten as follows:

$$\mathbb{E}\left[\left|\int_\tau^t H_s dB_s\right|^2\right] = \mathbb{E}\left[\int_\tau^t (H_s)^2 ds\right]. \tag{3.94}$$

The next inequality for Itô integrals is a generalization of the inequality in (3.93):

$$\mathbb{E}\left[\sup_{t:\tau \leq t \leq T}\left(\int_\tau^t H_s dB_s\right)^2\right] \leq 4\mathbb{E}\left[\int_\tau^T (H_s)^2 ds\right]. \tag{3.95}$$

It is also known that for all $\tau \in [0, T]$, $q > 0$, and $H \in \mathcal{H}_\tau$, there exist positive constants a_q and A_q depending only on q and such that

$$a_q \mathbb{E}\left[\left(\int_\tau^T |H_s|^2 ds\right)^{\frac{q}{2}}\right] \leq \mathbb{E}\left[\sup_{t:\tau \leq t \leq T}\left|\int_\tau^t H_s dB_s\right|^q\right]$$

$$\leq A_q \mathbb{E}\left[\left(\int_\tau^T |H_s|^2 ds\right)^{\frac{q}{2}}\right]. \tag{3.96}$$

This result is a special case of the Burkholder–Davis–Gundy inequalities (see [Ikeda and Watanabe (1989); Revuz and Yor (1991)]).

Our next goal is to explain what is the precise definition of a solution to the stochastic differential equation in (3.90). Let d and n be positive integers, and let $\sigma_{ij}(\tau, x)$, $1 \leq i \leq d$, $1 \leq j \leq n$, be a matrix field on

$[0, T] \times \mathbb{R}^d$ with values in the space of $d \times n$ matrices. In this section we use column vectors in the spaces $\mathbb{R}^n$ and $\mathbb{R}^d$. For a $d \times n$ matrix σ, the symbol $|\sigma|$ stands for the norm of σ defined by

$$|\sigma|^2 = \sum_{1 \le i \le d, 1 \le j \le n} \sigma_{ij}^2.$$

Let $b(\tau, x) = (b_1(\tau, x), \ldots, b_d(\tau, x))^*$ be a vector field on $[0, T] \times \mathbb{R}^d$, and let $B_t = (B_t^1, \ldots, B_t^n)^*$ be a standard Brownian motion in $\mathbb{R}^n$, where v^* denotes the transpose of the row vector v. Next we explain when a stochastic process $X_t = (X_t^1, \ldots, X_t^d)^*$ on Ω with values in $\mathbb{R}^d$ is a solution to the stochastic differential equation in (3.89). The process X_t should be adapted to the filtration $\mathcal{F}_t$ generated by B_t, satisfy the conditions

$$\mathbb{E}\left[\int_\tau^T |b_i(s, X_s)| \, ds\right] < \infty, \ 1 \le i \le d,$$

and

$$\mathbb{E}\left[\int_\tau^T \sigma_{ij}^2(s, X_s) \, ds\right] < \infty, \ 1 \le i \le d, \ 1 \le j \le n,$$

and be such that

$$X_t^i = X_\tau^i + \int_\tau^t b_i(s, X_s) \, ds + \sum_{j=1}^n \int_\tau^t \sigma_{ij}(s, X_s) \, dB_s^j \qquad (3.97)$$

$\mathbb{P}$-almost surely for all $1 \le i \le d$ and $0 \le \tau \le t \le T$. Next we formulate and prove an existence and uniqueness theorem for stochastic differential equations. Note that by $\mathcal{F}_t^\tau$ is denoted the two-parameter filtration generated by Brownian motion.

Theorem 3.7 *Suppose that σ is a continuous matrix field and b is a continuous vector field on the space $[0, T] \times \mathbb{R}^d$. Suppose also that the following conditions hold:*

(1) There exists a constant $c_1 > 0$ such that

$$|b(t, y) - b(t, z)| + |\sigma(t, y) - \sigma(t, z)| \le c_1 |y - z|$$

for all $t \in [0, T]$, $y \in \mathbb{R}^d$, and $z \in \mathbb{R}^d$.
(2) There exists a constant $c_2 > 0$ such that

$$|b(t, y)| + |\sigma(t, y)| \le c_2(1 + |y|)$$

for all $t \in [0,T]$ and $y \in \mathbb{R}^d$.

Let $\tau \in [0,T]$, $q \geq 1$, and let Z be an $\mathbb{R}^d$-valued $\mathcal{F}_\tau^c$-measurable random variable on Ω, where c is such that $0 \leq c \leq \tau$. Assume that

$$\mathbb{E}\left[|Z|^q\right] < \infty.$$

Then there exists a solution $X_t = X_t^{\tau,Z}$, $t \in [\tau,T]$, to the equation

$$X_t = Z + \int_\tau^t b(s, X_s)\, ds + \int_\tau^t \sigma(s, X_s)\, dB_s. \tag{3.98}$$

The process X_t, $\tau \leq t \leq T$, is continuous, and for every $t \in [\tau, T]$, the random variable X_t is measurable with respect to the σ-algebra $\sigma(Z, B_{s_2} - B_{s_1} : \tau \leq s_1 \leq s_2 \leq t)$ augmented by the events A such that $\mathbb{P}(A) = 0$. Moreover, the following inequality holds:

$$\mathbb{E}\left[\sup_{t:\tau \leq t \leq T} |X_t|^q\right] < \infty. \tag{3.99}$$

If X_t and Y_t are two solutions to equation (3.98) satisfying the conditions in Theorem 3.7, then X_t and Y_t are indistinguishable.

Proof. Given $\tau \in [0,T]$ and $c \in [0,\tau]$, consider the vector space $\mathcal{S}_{\tau,c}^q$ consisting of all $\mathcal{F}_t^c$-adapted continuous processes X_t on $[\tau, T]$ with state space $\mathbb{R}^d$ such that

$$\|X\|_q = \left\{\mathbb{E}\left[\sup_{t:\tau \leq t \leq T} |X_t|^q\right]\right\}^{\frac{1}{q}} < \infty. \tag{3.100}$$

Then $|\cdot|_q$ is a norm on the space $\mathcal{S}_{\tau,c}^q$. It is not hard to prove that the space $\mathcal{S}_{\tau,c}^q$ equipped with the norm defined in (3.100) is a Banach space. Next we will show that the function

$$\Phi(X)_t = Z + \int_\tau^t b(s, X_s)\, ds + \int_\tau^t \sigma(s, X_s)\, dB_s, \tag{3.101}$$

where Z is the initial condition in the formulation of Theorem 3.7, maps the space $\mathcal{S}_{\tau,c}^q$ into itself. Note that the process $\Phi(X)$ is continuous. This follows from the continuity of the coefficients b and σ and from the fact that stochastic integrals are continuous processes (see Theorem 3.6). Moreover, the process $\Phi(X)$ is $\mathcal{F}_t^c$-adapted, since Z is $\mathcal{F}_\tau^c$-measurable, X is $\mathcal{F}_t^c$-adapted, and the stochastic integral in (3.101) is $\mathcal{F}_t^c$-adapted (see Theorem

3.6). Using the linear growth condition for b, we get

$$\mathbb{E}\left[\sup_{t:\tau\leq t\leq T}\left|\int_\tau^t b(s,X_s)\,ds\right|^q\right] \leq (T-\tau)^q \mathbb{E}\left[\sup_{s:\tau\leq s\leq T}|b(s,X_s)|^q\right]$$

$$\leq c_2^q(T-\tau)^q \mathbb{E}\left[\sup_{s:\tau\leq s\leq T}(1+|X_s|)^q\right] < \infty. \tag{3.102}$$

In addition, using the linear growth condition for σ and inequality (3.96), we obtain

$$\mathbb{E}\left[\sup_{t:\tau\leq t\leq T}\left|\int_\tau^t \sigma(s,X_s)\,dB_s\right|^q\right] \leq A_q \mathbb{E}\left[\left(\int_\tau^T |\sigma(s,X_s)|^2\,ds\right)^{\frac{q}{2}}\right]$$

$$\leq A_q c_2^q (T-\tau)^{\frac{q}{2}} \mathbb{E}\left[\sup_{s:\tau\leq s\leq T}(1+|X_s|)^q\right] < \infty. \tag{3.103}$$

Now it is not hard to see that the condition $\Phi(X) \in \mathcal{S}_{\tau,c}^q$ follows from (3.102), (3.103), and the estimate $\mathbb{E}[|Z|^q] < \infty$.

Our next goal is to establish that for small values of the difference $T-\tau$, the mapping $\Phi: \mathcal{S}_{\tau,c}^q \to \mathcal{S}_{\tau,c}^q$ is a contraction. Let $X \in \mathcal{S}_{\tau,c}^q$ and $Y \in \mathcal{S}_{\tau,c}^2$. Then

$$\|\Phi(X) - \Phi(Y)\|_q^q$$

$$\leq 2^{q-1} \left\{ \mathbb{E}\left[\sup_{t:\tau\leq t\leq T}\left(\int_\tau^t |b(s,X_s) - b(s,Y_s)|\,ds\right)^q\right] \right.$$

$$\left. + \mathbb{E}\left[\sup_{t:\tau\leq t\leq T}\left|\int_\tau^t (\sigma(s,X_s) - \sigma(s,Y_s))\,dB_s\right|^q\right] \right\}. \tag{3.104}$$

Next, using the same ideas as in the proof of (3.102) and (3.103), we see that condition (1) in Theorem 3.7 implies

$$\mathbb{E}\left[\sup_{t:\tau\leq t\leq T}\left(\int_\tau^t |b(s,X_s) - b(s,Y_s)|\,ds\right)^q\right]$$

$$\leq (T-\tau)^q \mathbb{E}\left[\sup_{s:\tau\leq s\leq T}|b(s,X_s) - b(s,Y_s)|^q\right]$$

$$\leq c_1^q(T-\tau)^q \mathbb{E}\left[\sup_{s:\tau\leq s\leq T}|X_s - Y_s|^q\right]$$

$$= c_1^q(T-\tau)^q \|X-Y\|_q^q \tag{3.105}$$

and

$$\mathbb{E}\left[\sup_{t:\tau\le t\le T}\left|\int_\tau^t (\sigma(s,X_s)-\sigma(s,Y_s))\,dB_s\right|^q\right]$$
$$\le A_q(T-\tau)^{\frac{q}{2}}\mathbb{E}\left[\sup_{s:\tau\le s\le T}|\sigma(s,X_s)-\sigma(s,Y_s)|^q\right]$$
$$\le A_q c_1^q(T-\tau)^{\frac{q}{2}}\mathbb{E}\left[\sup_{s:\tau\le s\le T}|X_s-Y_s|^q\right]$$
$$= A_q c_1^q(T-\tau)^{\frac{q}{2}}\|X-Y\|_q^q. \tag{3.106}$$

Next, combining (3.104), (3.105), and (3.106), we see that

$$\|\Phi(X)-\Phi(Y)\|_q \le 2^{\frac{q-1}{q}}\left(c_1^q(T-\tau)^q + A_q c_1^q(T-\tau)^{\frac{q}{2}}\right)^{\frac{1}{q}}\|X-Y\|_q. \tag{3.107}$$

Estimate (3.107) shows that if

$$2^{q-1}\left(c_1^q(T-\tau)^q + A_q c_1^q(T-\tau)^{\frac{q}{2}}\right) < 1,$$

then the mapping $\Phi : \mathcal{S}_{\tau,c}^q \to \mathcal{S}_{\tau,c}^q$ is a contraction. Let us denote by $\bar{Z}$ the stochastic process defined by $\bar{Z}_t = Z$, $\tau \le t \le T$. Then it is clear that $\bar{Z} \in \mathcal{S}_{\tau,c}^q$. It follows from the fixed point theorem that the sequence $\Phi^n(\bar{Z})$ converges in the space $\mathcal{S}_{\tau,c}^q$ to a stochastic process X_t, $\tau \le t \le T$, which is a solution of equation (3.98). Moreover, the random variables $\int_\tau^t b(s, \Phi^n(\bar{Z})_s)\,ds$ and $\int_\tau^t \sigma(s, \Phi^n(\bar{Z})_s)\,dB_s$ are measurable with respect to the σ-algebra $\sigma(Z, B_{s_2}-B_{s_1} : \tau \le s_1 \le s_2 \le t)$ augmented by the events A such that $\mathbb{P}(A) = 0$. Therefore, for every $t \in [\tau, T]$, the random variable X_t is measurable with respect to this σ-algebra, and hence, the process X_t is adapted to the filtration $\mathcal{F}_t^c$. The continuity of the process X_t follows from the definition of the iterates $\Phi^n(\bar{Z})_t$ and the definition of the norm in the space $\mathcal{S}_{\tau,c}^q$. Here we need the fact that the stochastic integral is a continuous process. It is also clear that the convergence of the sequence $\Phi^n(\bar{Z})$ to X in the space $\mathcal{S}_{\tau,c}^q$ implies inequality (3.99). This proves Theorem 3.7 in the case where $T - \tau$ is a small number. The general case can be established by subdividing the interval $[\tau, T]$ into small subintervals and arguing as in the proof above. This proves the existence result in Theorem 3.7.

We will next prove the uniqueness of solutions in Theorem 3.7. We will need the Gronwall lemma in the proof.

Lemma 3.15 *Let f be a continuous function on the interval $[\tau, t]$, and suppose that there exist α and β such that $-\infty < \alpha < \infty$, $\beta \geq 0$, and*

$$f(s) \leq \alpha + \beta \int_\tau^s f(u) du$$

for all $s \in [\tau, t]$. Then $f(t) \leq \alpha e^{\beta(t-\tau)}$.

Proof. The case $\beta = 0$ is trivial. Next suppose that $\beta > 0$. Set

$$g(s) = e^{-\beta(s-\tau)} \int_\tau^s f(u) du.$$

Then

$$g'(s) = e^{-\beta(s-\tau)} \left[-\beta \int_\tau^s f(u) du + f(s) \right] \leq \alpha e^{-\beta(s-\tau)}$$

for all $\tau \leq s \leq t$. Integrating the previous inequality from τ to t, we get

$$g(t) \leq \alpha \int_0^{t-\tau} e^{-\beta v} dv = \frac{\alpha}{\beta} \left(1 - e^{-\beta(t-\tau)} \right).$$

Therefore,

$$\int_\tau^t f(u) du \leq \frac{\alpha}{\beta} \left(e^{\beta(t-\tau)} - 1 \right),$$

and it follows that

$$f(t) \leq \alpha + \beta \int_\tau^t f(u) du \leq \alpha + \alpha \left(e^{\beta(t-\tau)} - 1 \right) = \alpha e^{\beta(t-\tau)}.$$

This completes the proof of the Gronwall lemma. □

Let X_t and Y_t be two solutions to equation (3.97). Then, using the Lipschitz estimates for b and σ, Hölder's inequality, and equality (3.94), we see that for every $t \in [\tau, T]$,

$$\mathbb{E}\left[|X_t - Y_t|^2\right] \leq 2\mathbb{E}\left[\left|\int_\tau^t (b(s, X_s) - b(s, Y_s)) ds\right|^2\right]$$

$$+ 2\mathbb{E}\left[\left|\int_\tau^t (\sigma(s, X_s) - \sigma(s, Y_s)) dB_s\right|^2\right]$$

$$\leq 2(t-\tau) \int_\tau^t \mathbb{E}\left[|b(s, X_s) - b(s, Y_s)|^2\right] ds$$

$$+ 2 \int_\tau^t \mathbb{E}\left[|\sigma(s, X_s) - \sigma(s, Y_s)|^2\right] ds$$

$$\le 2c_1^2(T-\tau+1)\int_\tau^t \mathbb{E}\left[|X_s-Y_s|^2\right]ds. \qquad (3.108)$$

It is not hard to see that the function

$$f(t) = \mathbb{E}\left[|X_t - Y_t|^2\right]$$

is continuous on the interval $[\tau, T]$. Therefore, (3.108) implies that

$$f(t) \le \beta \int_\tau^t f(s)ds$$

for all $t \in [\tau, T]$ where $\beta = 2c^2(T-\tau+1)$. By the Gronwall lemma (Lemma 3.15), $f(t) = 0$ for all $t \in [\tau, T]$. It follows from the previous assertion that for all $t \in [\tau, T]$, $X_t = Y_t$ $\mathbb{P}$-almost surely. Since X_t and Y_t are continuous processes, they are indistinguishable (see Subsection 1.1.4). This proves the uniqueness of solutions in Theorem 3.7. $\square$

Let $x \in \mathbb{R}^d$, and assume that $X_\tau(\omega) = Z(\omega) = x$ for $\mathbb{P}$-almost all $\omega \in \Omega$. In this case, we will use the symbol $X_t^{\tau,x}$ for the unique solution of equation (3.97) in Theorem 3.7. Note that for fixed τ and x, the process $X_t^{\tau,x}$ is defined $\mathbb{P}$-almost surely with the exceptional set depending on τ and x.

Theorem 3.8 *There exists a modification of the process $(\tau, t, x) \mapsto X_t^{\tau,x}$ such that its sample paths are locally Hölder continuous of order $\alpha < \frac{1}{2}$.*

Proof. Fix a pair $(\tau, x) \in [0, T] \times \mathbb{R}^d$ and put $X_t^{\tau,x} = x$ for all $0 \le t < \tau$. Then the function $t \mapsto X_t^{\tau,x}$ is continuous on the interval $[0, T]$ $\mathbb{P}$-almost surely. Here the exceptional set depends on τ and x. For $\tau \in [0, T]$, $x \in \mathbb{R}^d$, and $y \in \mathbb{R}^d$, consider the processes $X_t^{\tau,x}$ and $X_t^{\tau,y}$ where $\tau \le t \le T$. Since these processes are solutions to the stochastic differential equation in Theorem 3.7, we have

$$X_t^{\tau,x} - X_t^{\tau,y} = x - y + \int_\tau^t [b(s, X_s^{\tau,x}) - b(s, X_s^{\tau,y})]ds$$
$$+ \int_\tau^t [\sigma(s, X_s^{\tau,x}) - \sigma(s, X_s^{\tau,y})]dB_s \qquad (3.109)$$

$\mathbb{R}^d$-almost surely. The exceptional set in (3.109) depends on τ, x, and y. For $q \ge 2$, set

$$f(t) = f(t; \tau, x, y, q) = \mathbb{E}\left[\sup_{u: \tau \le u \le t} |X_u^{\tau,x} - X_u^{\tau,y}|^q\right], \quad \tau \le t \le T. \qquad (3.110)$$

Then f is a continuous function on $[\tau, T]$. Indeed, for $0 \leq t \leq s \leq T$, we have

$$f(s) - f(t) \leq \mathbb{E}\left[\sup_{u:t \leq u \leq s} |X_u^{\tau,x} - X_u^{\tau,y}|^q\right].$$

Now the continuity of the process X_t, estimate (3.99), and the monotone convergence theorem imply the continuity of f. It follows from (3.109), (3.110), the Lipschitz estimate in (1) of Theorem (3.7), Hölder's inequality, and (3.96) that

$$f(t) \leq 3^{q-1} |x-y|^q + 3^{q-1} c_1^q \mathbb{E}\left[\left(\int_\tau^t |X_s^{\tau,x} - X_s^{\tau,y}| ds\right)^q\right]$$

$$+ 3^{q-1} c_1^q A_q \mathbb{E}\left[\left(\int_\tau^t |X_s^{\tau,x} - X_s^{\tau,y}|^2 ds\right)^{\frac{q}{2}}\right]$$

$$\leq 3^{q-1} |x-y|^q + 3^{q-1} c_1^q \left((t-\tau)^{q-1} + A_q (t-\tau)^{\frac{q-2}{2}}\right)$$

$$\mathbb{E}\left[\int_\tau^t |X_s^{\tau,x} - X_s^{\tau,y}|^q ds\right]$$

$$\leq 3^{q-1} |x-y|^q + 3^{q-1} c_1^q \left((t-\tau)^{q-1} + A_q (t-\tau)^{\frac{q-2}{2}}\right) \int_\tau^t f(s) ds.$$
(3.111)

By (3.111) and the Gronwall lemma,

$$\mathbb{E}\left[\sup_{u:\tau \leq u \leq t} |X_u^{\tau,x} - X_u^{\tau,y}|^q\right]$$

$$\leq 3^{q-1} |x-y|^q \exp\left\{3^{q-1} c_1^q \left((t-\tau)^q + A_q (t-\tau)^{\frac{q}{2}}\right)\right\}$$

for all $t \in [\tau, T]$, $x \in \mathbb{R}^d$, and $y \in \mathbb{R}^d$. Hence,

$$\mathbb{E}\left[\sup_{u:\tau \leq u \leq t} |X_u^{\tau,x} - X_u^{\tau,y}|^q\right] \leq \alpha_1 |x-y|^q \quad (3.112)$$

for all $0 \leq \tau \leq t \leq T$, $x \in \mathbb{R}^d$, and $y \in \mathbb{R}^d$, where the constant α_1 depends only on q, T, and c_1. Fix $x \in \mathbb{R}^d$ and $t \in [0, T]$. Then for all τ_1 and τ_2 with $0 \leq \tau_1 \leq \tau_2 \leq t$,

$$X_t^{\tau_1,x} - X_t^{\tau_2,x}$$

$$= \int_{\tau_2}^t [b(s, X_s^{\tau_1,x}) - b(s, X_s^{\tau_2,x})] ds + \int_{\tau_1}^{\tau_2} b(s, X_s^{\tau_1,x}) ds$$

$$+ \int_{\tau_2}^{t} [\sigma(s, X_s^{\tau_1,x}) - \sigma(s, X_s^{\tau_2,x})] \, dB_s + \int_{\tau_1}^{\tau_2} \sigma(s, X_s^{\tau_1,x}) \, dB_s \quad (3.113)$$

$\mathbb{P}$-almost surely. The exceptional set in (3.113) depends on τ_1, τ_2, and x. For every $q \geq 2$, put

$$g(t) = g(t; \tau_1, \tau_2, x) = \mathbb{E}\left[\sup_{u:\tau_2 \leq u \leq t} |X_u^{\tau_1,x} - X_u^{\tau_2,x}|^q\right], \quad \tau_2 \leq t \leq T. \quad (3.114)$$

Then, using (3.113), (3.114), the Lipschitz and the linear growth estimates in conditions (1) and (2) of Theorem (3.7), Hölder's inequality, and (3.96), we obtain

$$g(t) \leq 4^{q-1} c_1^q (t - \tau_2)^{q-1} \mathbb{E}\left[\int_{\tau_2}^{t} |X_s^{\tau_1,x} - X_s^{\tau_2,x}|^q \, ds\right]$$

$$+ 4^{q-1} c_2^q (\tau_2 - \tau_1)^{q-1} \mathbb{E}\left[\int_{\tau_1}^{\tau_2} (1 + |X_s^{\tau_1,x}|)^q \, ds\right]$$

$$+ 4^{q-1} A_q c_1^q (t - \tau_2)^{\frac{q-2}{2}} \mathbb{E}\left[\int_{\tau_2}^{t} |X_s^{\tau_1,x} - X_s^{\tau_2,x}|^q \, ds\right]$$

$$+ 4^{q-1} A_q c_2^q (\tau_2 - \tau_1)^{\frac{q-2}{2}} \mathbb{E}\left[\int_{\tau_1}^{\tau_2} (1 + |X_s^{\tau_1,x}|)^q \, ds\right]$$

$$\leq 4^{q-1} c_2^q \left((\tau_2 - \tau_1)^{q-1} + A_q (\tau_2 - \tau_1)^{\frac{q-2}{2}}\right) \mathbb{E}\left[\int_{\tau_1}^{\tau_2} (1 + |X_s^{\tau_1,x}|)^q \, ds\right]$$

$$+ 4^{q-1} c_1^q \left((t - \tau_2)^{q-1} + A_q (t - \tau_2)^{\frac{q-2}{2}}\right) \int_{\tau_2}^{t} g(s) \, ds. \quad (3.115)$$

It follows from (3.115) and the Gronwall lemma that

$$\mathbb{E}\left[\sup_{u:\tau_2 \leq u \leq t} |X_u^{\tau_1,x} - X_u^{\tau_2,x}|^q\right]$$

$$\leq 4^{q-1} c_2^q \left((\tau_2 - \tau_1)^{q-1} + A_q (\tau_2 - \tau_1)^{\frac{q-2}{2}}\right) \mathbb{E}\left[\int_{\tau_1}^{\tau_2} (1 + |X_s^{\tau_1,x}|)^q \, ds\right]$$

$$\exp\left\{4^{q-1} c_1^q \left((t - \tau_2)^q + A_q (t - \tau_2)^{\frac{q}{2}}\right)\right\} \quad (3.116)$$

for all $t \in [\tau_2, T]$.

Next let $\tau_1 \leq u \leq \tau_2$. Then

$$X_u^{\tau_1,x} - X_u^{\tau_2,x} = X_u^{\tau_1,x} - x$$

$$= \int_{\tau_1}^{u} b(s, X_s^{\tau_1,x}) \, ds + \int_{\tau_1}^{u} \sigma(s, X_s^{\tau_1,x}) \, dB_s \quad (3.117)$$

P-almost surely. Using (3.117) and reasoning as in the proof of (3.115), we see that

$$\mathbb{E}\left[\sup_{u:\tau_1\leq u\leq \tau_2}|X_u^{\tau_1,x}-X_u^{\tau_2,x}|^q\right]$$
$$\leq 2^{q-1}c_2^q\left((\tau_2-\tau_1)^{q-1}+A_q\,(\tau_2-\tau_1)^{\frac{q-2}{2}}\right)\mathbb{E}\left[\int_{\tau_1}^{\tau_2}(1+|X_s^{\tau_1,x}|)^q\,ds\right]. \tag{3.118}$$

It follows from (3.116) and (3.118) that

$$\mathbb{E}\left[\sup_{u:\tau_1\leq u\leq t}|X_u^{\tau_1,x}-X_u^{\tau_2,x}|^q\right]$$
$$\leq 2^{q-1}c_2^q\left((\tau_2-\tau_1)^{q-1}+A_q\,(\tau_2-\tau_1)^{\frac{q-2}{2}}\right)\mathbb{E}\left[\int_{\tau_1}^{\tau_2}(1+|X_s^{\tau_1,x}|)^q\,ds\right]$$
$$\times\left(2^{q-1}\exp\left\{4^{q-1}c_1^q\left((t-\tau_2)^q+A_q\,(t-\tau_2)^{\frac{q}{2}}\right)\right\}+1\right) \tag{3.119}$$

for all $t\in[\tau_1,T]$. Moreover, for all $0\leq\tau\leq t_1\leq t_2\leq T$ and $x\in\mathbb{R}^d$, the following estimate holds:

$$\mathbb{E}\left[|X_{t_1}^{\tau,x}-X_{t_2}^{\tau,x}|^q\right]$$
$$\leq 2^{q-1}c_2^q\left((t_2-t_1)^{q-1}+A_q\,(t_2-t_1)^{\frac{q-2}{2}}\right)\mathbb{E}\left[\int_{t_1}^{t_2}(1+|X_s^{\tau,x}|)^q\,ds\right]. \tag{3.120}$$

We also have

$$\mathbb{E}\left[\sup_{u:\tau\leq u\leq t}|X_u^{\tau,x}|^q\right]$$
$$\leq 3^{q-1}|x|^q+3^{q-1}c_2^q\left((t-\tau)^{q-1}+A_q\,(t-\tau)^{\frac{q-2}{2}}\right)\mathbb{E}\left[\int_\tau^t(1+|X_s^{\tau,x}|)^q\,ds\right]$$
$$\leq 3^{q-1}|x|^q+6^{q-1}c_2^q\left((t-\tau)^q+A_q\,(t-\tau)^{\frac{q}{2}}\right)$$
$$+6^{q-1}c_2^q\left((t-\tau)^{q-1}+A_q\,(t-\tau)^{\frac{q-2}{2}}\right)\mathbb{E}\left[\int_\tau^t\sup_{u:\tau\leq u\leq s}|X_u^{\tau,x}|^q\,ds\right].$$

Therefore, by the Gronwall lemma,

$$\mathbb{E}\left[\sup_{u:\tau\leq u\leq t}|X_u^{\tau,x}|^q\right]\leq\left[3^{q-1}|x|^q+6^{q-1}c_2^q\left((t-\tau)^q+A_q(t-\tau)^{\frac{q}{2}}\right)\right]$$
$$\times\exp\left\{6^{q-1}c_2^q\left((t-\tau)^q+A_q(t-\tau)^{\frac{q}{2}}\right)\right\}. \tag{3.121}$$

It follows from (3.119) and (3.121) that

$$\mathbb{E}\left[\sup_{u:\tau_1 \leq u \leq t} |X_u^{\tau_1,x} - X_u^{\tau_2,x}|^q\right] \leq \alpha_2 \left((\tau_2 - \tau_1)^q + (\tau_2 - \tau_1)^{\frac{q}{2}}\right)(\alpha_3 + |x|^q)$$
(3.122)

for all $0 \leq \tau_1 \leq \tau_2 \leq T$, $\tau_1 \leq t \leq T$, and $x \in \mathbb{R}^d$. The constant $\alpha_2 > 0$ in (3.122) depends only on q, T, c_1 and c_2, while the constant $\alpha_3 > 0$ depends only on q, T, and c_2. It is not hard to see that (3.120) and (3.121) imply

$$\mathbb{E}\left[|X_{t_1}^{\tau,x} - X_{t_2}^{\tau,x}|^q\right] \leq \alpha_4 \left((t_2 - t_1)^q + (t_2 - t_1)^{\frac{q}{2}}\right)(\alpha_5 + |x|^q) \quad (3.123)$$

for all $0 \leq \tau \leq t_1 \leq t_2 \leq T$ and $x \in \mathbb{R}^d$. The constants $\alpha_4 > 0$ and $\alpha_5 > 0$ in (3.123) depend only on q, T, and c_2.

Let $0 \leq \tau_1 \leq \tau_2 \leq T$, $\tau_1 \leq t_1$, $\tau_1 \leq t_2$, $x \in \mathbb{R}^d$, $y \in \mathbb{R}^d$, and suppose that $|\tau_2 - \tau_1| < 1$ and $|t_2 - t_1| < 1$. Since

$$\left|X_{t_2}^{\tau_2,x_2} - X_{t_1}^{\tau_1,x_1}\right| \leq \left|X_{t_2}^{\tau_2,x_2} - X_{t_2}^{\tau_2,x_1}\right| + \left|X_{t_2}^{\tau_2,x_1} - X_{t_2}^{\tau_1,x_1}\right|$$
$$+ \left|X_{t_2}^{\tau_1,x_1} - X_{t_1}^{\tau_1,x_1}\right|,$$

the estimates in (3.112), (3.122), and (3.123) imply

$$\mathbb{E}\left[|X_{t_2}^{\tau_2,x_2} - X_{t_1}^{\tau_1,x_1}|^q\right] \leq 3^{q-1}\alpha_1 |x_2 - x_1|^q$$
$$+ 2 \times 3^{q-1}\alpha_2 |\tau_2 - \tau_1|^{\frac{q}{2}} (\alpha_3 + |x_1|^q) + 2 \times 3^{q-1}\alpha_4 |t_2 - t_1|^{\frac{q}{2}} (\alpha_5 + |x_1|^q).$$

Hence, there exists a constant $\alpha > 0$, depending only on q, T, c_1, and c_2, for which

$$\mathbb{E}\left[|X_{t_2}^{\tau_2,x_2} - X_{t_1}^{\tau_1,x_1}|^q\right]$$
$$\leq \alpha \left\{|x_2 - x_1|^q + \left(|\tau_2 - \tau_1|^{\frac{q}{2}} + |t_2 - t_1|^{\frac{q}{2}}\right)(1 + |x_1|^q)\right\}. \quad (3.124)$$

In order to complete the proof of Theorem 3.8, we need a multi-dimensional version of Kolmogorov's criterion for the existence of continuous modifications of a stochastic process. This criterion concerns stochastic processes indexed by subsets of Euclidean space $\mathbb{R}^n$. Let $C = [a_1, b_1] \times \cdots \times [a_n, b_n]$ be a rectangular box in $\mathbb{R}^n$, and consider a stochastic process X_s, $s \in C$, on a probability space $(\Omega, \mathcal{F}, \mathbb{P})$ with state space $\mathbb{R}^d$. The following assertion is a multi-dimensional version of Kolmogorov's criterion (see, e.g., [Revuz and Yor (1991)]).

Theorem 3.9 *Suppose that there exist three strictly positive constants q, c, and ε such that*

$$\mathbb{E}\left[|X_s - X_r|_d^q\right] \leq c|s-r|_n^{n+\varepsilon} \tag{3.125}$$

for all $s \in C$ and $r \in C$. Then there is a modification $\widetilde{X}_s$ of the process X_s such that the paths of $\widetilde{X}_s$ are Hölder continuous of order α with $0 \leq \alpha < \dfrac{\varepsilon}{q}$.

The symbols $|\cdot|_d$ and $|\cdot|_n$ in the formulation of Theorem 3.9 stand for the Euclidean norms in $\mathbb{R}^d$ and $\mathbb{R}^n$, respectively. Next we will rewrite inequality (3.124) using the notation in the formulation of Theorem 3.9. Let $k \geq 1$, and denote by C_k a rectangular box in $\mathbb{R}^{d+2}$ given by $C_k = [0,T]^2 \times D_k$, where D_k is the cube in $\mathbb{R}^d$ defined by $D_k = \{x \in \mathbb{R}^d : |x_i| \leq k,\ 1 \leq i \leq d\}$. Put $s = (\tau_2, t_2, x_2)$ and $r = (\tau_1, t_1, x_1)$. Then estimate (3.124) implies that

$$\mathbb{E}\left[|X_s - X_r|_d^q\right] \leq \beta_k |s-r|_{d+2}^{\frac{q}{2}} \tag{3.126}$$

for all $q \geq 2$, $s \in C_k$ and $r \in C_k$. The constant β_k in (3.126) depends only on k, q, T, c_1 and c_2. It follows that if $n = d+2$, $q > 2(d+2)$, and $\varepsilon = \frac{q}{2} - d - 2$, then Kolmogorov's criterion can be applied. Therefore, there exists a modification $\widetilde{X}_s^k$ of the process X_s such that the paths of $\widetilde{X}_s^k$ are Hölder continuous of order α with $\alpha < \frac{1}{2}$. Here we assume that $s \in C_k$. Since any two continuous modifications of a stochastic process are indistinguishable, it is not hard to construct a required modification of the process $(\tau, t, x) \mapsto X_t^{\tau,x}$ in Theorem 3.8 using the processes $\widetilde{X}_s^k$.

This completes the proof of Theorem 3.8. □

The next assertion concerns the Markov property of the process $X_t^{\tau,x}$ with respect to the Brownian filtration. The uniqueness of solutions of equation (3.97) will be used in the proof.

Theorem 3.10 *Suppose that b and σ satisfy the conditions in Theorem 3.7, and let $X_t^{\tau,x}$ be the unique solution to equation (3.97). Then for all $\tau \leq s < t \leq T$ and all bounded Borel functions f on $\mathbb{R}^d$,*

$$\mathbb{E}\left[f\left(X_t^{\tau,x}\right) \mid \mathcal{F}_s^{\tau}\right] = \mathbb{E}\left[f\left(X_t^{\tau,x}\right) \mid \sigma\left(X_s^{\tau,x}\right)\right]$$
$$= \mathbb{E}\left[f\left(X_t^{s,y}\right)\right]\Big|_{y=X_s^{\tau,x}}. \tag{3.127}$$

Proof. We will first establish the flow property of the process $X_t^{\tau,x}$; that is,

$$X_t^{\tau,x} = X_t^{s,X_s^{\tau,x}} \quad \mathbb{P}\text{-almost surely.} \tag{3.128}$$

Indeed, using the fact that $X_t^{\tau,x}$ is the unique solution to equation (3.97), we get

$$\begin{aligned} X_t^{\tau,x} &= x + \int_\tau^t b\left(\rho, X_\rho^{\tau,x}\right) d\rho + \int_\tau^t \sigma\left(\rho, X_\rho^{\tau,x}\right) dB_\rho \\ &= x + \int_\tau^s b\left(\rho, X_\rho^{\tau,x}\right) d\rho + \int_\tau^s \sigma\left(\rho, X_\rho^{\tau,x}\right) dB_\rho \\ &\quad + \int_s^t b\left(\rho, X_\rho^{\tau,x}\right) d\rho + \int_s^t \sigma\left(\rho, X_\rho^{\tau,x}\right) dB_\rho \\ &= X_s^{\tau,x} + \int_s^t b\left(\rho, X_\rho^{\tau,x}\right) d\rho + \int_s^t \sigma\left(\rho, X_\rho^{\tau,x}\right) dB_\rho. \end{aligned} \quad (3.129)$$

Therefore, (3.97) with the initial value equal to $X_s^{\tau,x}$ implies that

$$X_t^{s,X_s^{\tau,x}} = X_s^{\tau,x} + \int_s^t b\left(\rho, X_\rho^{s,X_s^{\tau,x}}\right) d\rho + \int_s^t \sigma\left(\rho, X_\rho^{s,X_s^{\tau,x}}\right) dB_\rho. \quad (3.130)$$

Now (3.129) and (3.130) imply that the processes $X_t^{\tau,x}$ and $X_t^{s,X_s^{\tau,x}}$ satisfy the same stochastic differential equation

$$X_t = X_s^{\tau,x} + \int_s^t b\left(\rho, X_\rho\right) d\rho + \int_s^t \sigma\left(\rho, X_\rho\right) dB_\rho.$$

By the uniqueness of solutions in Theorem 3.6, equality (3.128) holds.

It follows from the flow property in (3.128) and Theorem 3.7 that for $\tau \leq s \leq t \leq T$, the random variable $X_t^{\tau,x}$ is measurable with respect to the σ-algebra $\sigma\left(X_s^{\tau,x}, B_{s_2} - B_{s_1} : s \leq s_1 \leq s_2 \leq t\right)$ augmented by the family of $\mathbb{P}$-negligible sets. Therefore, for every Borel function f on $\mathbb{R}^d$, the random variable $f\left(X_t^{\tau,x}\right)$ is measurable with respect to the same σ-algebra. Since the Brownian increments are independent, and the process $X_t^{\tau,x}$ is adapted to the Brownian filtration $\mathcal{F}_t^\tau$, the random variable $X_s^{\tau,x}$ is independent of the Brownian increments after time s. Next let $A_1 \in \sigma\left(X_s^{\tau,x}\right)$ and $A_2 \in \sigma\left(B_{s_2} - B_{s_1} : s \leq s_1 \leq s_2 \leq t\right)$. By the properties of conditional expectations, we get

$$\mathbb{E}\left[\chi_{A_1}\chi_{A_2} \mid \mathcal{F}_s^\tau\right] = \chi_{A_1}\mathbb{E}\left[\chi_{A_2} \mid \mathcal{F}_s^\tau\right] = \chi_{A_1}\mathbb{P}\left[A_2\right] \quad (3.131)$$

and

$$\mathbb{E}\left[\chi_{A_1}\chi_{A_2} \mid \sigma\left(X_s^{\tau,x}\right)\right] = \chi_{A_1}\mathbb{E}\left[\chi_{A_2} \mid \sigma\left(X_s^{\tau,x}\right)\right] = \chi_{A_1}\mathbb{P}\left[A_2\right]. \quad (3.132)$$

By (3.131) and (3.132), the Markov property holds for the functions $\chi_{A_1}\chi_{A_2}$. It follows from the monotone class theorem that the first equality in (3.127) holds. The proof of the second equality in (3.127) is similar.

This completes the proof of Theorem 3.10. □

The next result is the celebrated Itô's formula. It is known that Itô's formula holds for more general stochastic processes X_t and functions f than those in the formulation of Theorem 3.11 below (for more information, see the references in Section 3.12).

Theorem 3.11 *Let $(\tau, x) \in [0, T] \times \mathbb{R}^d$, and suppose that b and σ satisfy the conditions in Theorem 3.7. Let $X_t = X_t^{\tau,x}$ be the unique solution to the stochastic differential equation in Theorem 3.7. Then for every function $f \in C^{1,2}\left((\tau, T) \times \mathbb{R}^d\right)$ such that the partial derivatives $\dfrac{\partial f}{\partial x_i}$ are bounded for all $1 \leq i \leq d$, the following formula holds:*

$$f(t, X_t) = f(\tau, X_\tau) + \int_\tau^t \frac{\partial f}{\partial s}(s, X_s)\, ds$$
$$+ \sum_{i=1}^d \int_\tau^t b_i(s, X_s) \frac{\partial f}{\partial x_i}(s, X_s)\, ds$$
$$+ \sum_{i=1}^d \sum_{j=1}^n \sigma_{ij}(s, X_s) \frac{\partial f}{\partial x_i}(s, X_s)\, dB_s^j$$
$$+ \frac{1}{2} \sum_{i,j=1}^d \int_\tau^t a_{ij}(s, X_s) \frac{\partial^2 f}{\partial x_i \partial x_j}(s, X_s)\, ds, \qquad (3.133)$$

where

$$a_{ij}(s, y) = (\sigma\sigma^*)_{ij}(s, y) = \sum_{k=1}^n \sigma_{ik}(s, y) \sigma_{jk}(s, y). \qquad (3.134)$$

The last term on the right-hand side of formula (3.133) is called the Itô correction. The next lemma follows from the Itô formula and from the fact that under the conditions in Theorem 3.6, stochastic integrals are continuous martingales.

Lemma 3.16 *Let $(\tau, x) \in [0, T] \times \mathbb{R}^d$, and suppose that b and σ satisfy the conditions in Theorem 3.7. Let $X_t = X_t^{\tau,x}$ be the unique solution to the stochastic differential equation in Theorem 3.7. Then for every function $f \in C^{1,2}\left((\tau, T) \times \mathbb{R}^d\right)$ such that the partial derivatives $\dfrac{\partial f}{\partial x_i}$ are bounded for* all $1 \le i \le d$, the process

$$f(t, X_t) - f(\tau, X_\tau) - \int_\tau^t L_s f(s, X_s)\, ds - \int_\tau^t \frac{\partial f}{\partial s}(s, X_s)\, ds \quad (3.135)$$

is a continuous $\mathcal{F}_t^\tau$-martingale on $[\tau, T]$. In (3.135), the operator L_s is defined by (3.80) with a_{ij} given by formula (3.134).

Our next goal is to show that the operator L_τ in (3.80) generates a diffusion process (see Definition 3.18).

Theorem 3.12 *Let b and σ satisfy the conditions in Theorem 3.7, and let $a = \sigma\sigma^*$. Then the family of operators L_τ defined by (3.80) is the generator of a diffusion process X_t such that its sample paths are Hölder continuous of order $\alpha < \frac{1}{2}$.*

Proof. We will first show that it suffices to construct a continuous Markov process $(X_t, \mathcal{F}_t^\tau, \mathbb{P}_{\tau,x})$, $0 \le \tau \le t \le T$, with state space $\mathbb{R}^d$ such that

$$\mathbb{E}_{\tau,x}[f(X_t)] = \mathbb{E}[f(X_t^{\tau,x})] \quad (3.136)$$

for all $0 \le \tau \le t \le T$, $x \in \mathbb{R}^d$, and all bounded Borel functions f on $\mathbb{R}^d$. Indeed, if f is a C_0^2-function on $\mathbb{R}^d$, then Lemma 3.16 implies that the process

$$f(X_t^{\tau,x}) - f(X_\tau^{\tau,x}) - \int_\tau^t L_s f(X_s^{\tau,x})\, ds$$

is a martingale with respect to the measure $\mathbb{P}$. It follows from (3.136) that formula (3.81) holds.

Our next goal is to prove equality (3.136). Put

$$P(\tau, x; t, B) = \mathbb{P}(X_t^{\tau,x} \in B), \quad (3.137)$$

where $0 \le \tau < t \le T$, $x \in \mathbb{R}^d$, and $B \in \mathcal{B}_{\mathbb{R}^d}$, and define a family of operators by

$$Y(\tau, t) f(x) = \int_{\mathbb{R}^d} f(y) P(\tau, x; t, dy) = \mathbb{E}[f(X_t^{\tau,x})], \quad (3.138)$$

where $0 \le \tau < t \le T$, $x \in \mathbb{R}^d$, and f is a Borel function on $\mathbb{R}^d$. We will need the following lemma:

Lemma 3.17 *Suppose that b and σ satisfy the conditions in Theorem 3.7, and let $X_t^{\tau,x}$ be the unique solution to the stochastic differential equation* in Theorem 3.7. Then the function P defined by (3.137) is a transition probability function, and the family of operators $Y(\tau,t)$ given by (3.138) is a backward Feller propagator.

Proof. It follows from Theorem 3.8 that without loss of generality we may assume that the sample paths of the process $(\tau,t,x) \mapsto X_t^{\tau,x}$ are continuous. The family of operators defined by (3.138) is a backward propagator on the space $L_\mathcal{E}^\infty$. Here the symbol $\mathcal{E}$ stands for the Borel σ-algebra of the space $\mathbb{R}^d$. This assertion can be obtained from Theorem 3.10. Moreover, it is not hard to see that the function P defined by (3.137) is a transition probability function. It only remains to prove that Y is a Feller propagator. Let f be a bounded continuous function on $\mathbb{R}^d$. Then for all $0 \leq \tau \leq t \leq T$, the function $Y(\tau,t)f(x)$ is continuous on $\mathbb{R}^d$. This fact can be obtained using the continuity of the process $(\tau,t,x) \mapsto X_t^{\tau,x}$ and equality (3.138).

This completes the proof of Lemma 3.17. □

Let us go back to the proof of Theorem 3.12. Since the function P defined by (3.137) is a transition probability function (see Lemma 3.17), there exists a Markov process $(Y_t, \mathcal{F}_t^\tau, \mathbb{P}_{\tau,x})$ with state space $\mathbb{R}^d$ such that P is its transition function. Here $\mathcal{F}_t^\tau = \sigma(Y_s : \tau \leq s \leq t)$. It follows from (3.137) and the Markov property in Theorem 3.10 that

$$\mathbb{E}_{\tau,x}\left[F\left(Y_{t_1},\ldots,Y_{t_m}\right)\right] = \mathbb{E}\left[F\left(X_{t_1}^{\tau,x},\ldots,X_{t_m}^{\tau,x}\right)\right] \tag{3.139}$$

for all $\tau \leq t_1 \leq \cdots \leq t_m \leq T$, $x \in \mathbb{R}^d$, and all positive Borel functions F on $\left[\mathbb{R}^d\right]^m$. Now it is clear that (3.123) and (3.139) imply

$$\mathbb{E}_{\tau,x}\left[|X_{t_1} - X_{t_2}|^q\right] = \mathbb{E}\left[\left|X_{t_1}^{\tau,x} - X_{t_2}^{\tau,x}\right|^q\right]$$
$$\leq \alpha_4 \left((t_2 - t_1)^q + (t_2 - t_1)^{\frac{q}{2}}\right)(\alpha_5 + |x|^q) \tag{3.140}$$

for all $\tau \leq t_1 \leq t_2 \leq T$ and $x \in \mathbb{R}^d$ and $q \geq 2$. Reasoning as in the proof of Theorem 3.8, we see that there exists a modification X_t of the process Y_t such that its sample paths are Hölder continuous of order $\alpha < \frac{1}{2}$. It follows from (3.139) that equality (3.136) holds. Therefore, the process X_t is a diffusion process satisfying the conditions in Theorem 3.12.

The proof of Theorem 3.12 is thus completed. □

The next lemma concerns the Feller-Dynkin property of the backward propagator Y. In our opinion, this lemma has an independent interest.

Lemma 3.18 *Suppose that b and σ are bounded on $\mathbb{R}^d$ and satisfy the conditions in Theorem 3.7. Let $X_t^{\tau,x}$ be the unique solution to the stochastic* differential equation in Theorem 3.7. Then the family of operators $Y(\tau,t)$ defined by (3.138) is a backward Feller-Dynkin propagator.

Proof. By Lemma 3.17, it suffices to prove that for all $0 \le \tau \le t \le T$, $f \in C_0 \Longrightarrow Y(\tau,t)f \in C_0$. Let $f \in C_0$. Then for all $M > 0$, we have

$$|Y(\tau,t)f(x)| \le |Y(\tau,t)f\chi_{B(x,M)}(x)| + |Y(\tau,t)f\chi_{\mathbb{R}^d \setminus B(x,M)}(x)|$$
$$\le \sup_{y:|x-y|\le M} |f(y)| + \|f\|_\infty \mathbb{P}\left[|X_t^{\tau,x} - x| > M\right]$$
$$\le \sup_{y:|x-y|\le M} |f(y)| + \frac{\|f\|_\infty}{M^2} \mathbb{E}\left[|X_t^{\tau,x} - x|^2\right]. \tag{3.141}$$

Since

$$X_t^{\tau,x} - x = \int_\tau^t b(s, X_s^{\tau,x})\,ds + \int_\tau^t \sigma(s, X_s^{\tau,x})\,dB_s, \tag{3.142}$$

we have

$$\sup_{x \in \mathbb{R}^d} \mathbb{E}\left[|X_t^{\tau,x} - x|^2\right] \le 2\left(T^2\gamma_1^2 + T\gamma_2^2\right), \tag{3.143}$$

where

$$\gamma_1 = \sup_{(s,x)\in[0,T]\times\mathbb{R}^d} |b(s,x)| \quad \text{and} \quad \gamma_2 = \sup_{(s,x)\in[0,T]\times\mathbb{R}^d} |\sigma(s,x)|.$$

In the proof of estimate (3.143), we used (3.142) and (3.94). It follows from (3.141) and (3.143) that

$$|Y(\tau,t)f(x)| \le \sup_{y:|x-y|\le M} |f(y)| + \gamma \frac{\|f\|_\infty}{M^2}, \tag{3.144}$$

where the constant $\gamma > 0$ does not depend on x. Now it is clear that (3.144) implies

$$\lim_{M\to\infty} \sup_{x:|x|>M} |Y(\tau,t)f(x)| = 0.$$

This completes the proof of Lemma 3.18. □

In the proof of Theorem 3.12, we implicitly solved the so-called martingale problem. This problem was formulated and studied by Stroock and Varadhan (see [Stroock and Varadhan (1979)]). Let $\Omega = C\left([0,T],\mathbb{R}^d\right)$, where $C\left([0,T],\mathbb{R}^d\right)$ is the space of all $\mathbb{R}^d$-valued continuous functions on the interval $[0,T]$, and denote by X_t the coordinate process on Ω. Put $\mathcal{F}_t^\tau = \sigma\left(X_s : \tau \le s \le t\right)$.

Definition 3.19 For $(\tau, x) \in [0, T] \times \mathbb{R}^d$, a probability measure $\mathbb{P}_{\tau,x}$ on Ω is a solution to the martingale problem $\pi(\tau, x)$, provided that

(1) $\mathbb{P}_{\tau,x}[X_\tau = x] = 1$.
(2) For any function $f \in C_0^\infty$, the process

$$t \mapsto f(X_t) - f(X_\tau) - \int_\tau^t L_s f(X_s)\, ds$$

is an $\mathcal{F}_t^\tau$-martingale on $[\tau, T]$ with respect to the measure $\mathbb{P}_{\tau,x}$.

By Theorem 3.12, if $(X_t, \Omega, \mathcal{F}_t^\tau, \mathbb{P}_{\tau,x})$ is a diffusion process generated by L_τ, then for every pair $(\tau, x) \in [0, T] \times \mathbb{R}^d$, the measure $\mathbb{P}_{\tau,x}$ is a solution to the martingale problem $\pi(\tau, x)$. On the other hand, suppose that the operator L_τ is given, and the problem is to determine whether there exists a diffusion process X_t generated by the operator L_τ. In order to find such a process, one can start with the martingale problem $\pi(\tau, x)$. If the solution $\mathbb{P}_{\tau,x}$ exists for all pairs $(\tau, x) \in [0, T] \times \mathbb{R}^d$, then one can try to prove that the coordinate process X_t is a Markov process with respect to the family $\{\mathbb{P}_{\tau,x}\}$. It is known that if for every pair $(\tau, x) \in [0, T] \times \mathbb{R}^d$, the martingale problem $\pi(\tau, x)$ is uniquely solvable, then the Markov property holds (see, e.g., [Stroock and Varadhan (1979)]). We refer the reader to [Stroock and Varadhan (1979)] for more information on the martingale problem.

Next, we will show that the diffusion process in Theorem 3.12 possesses the strong Markov property. Let us first recall what was established in Corollary 2.1. Let P be a transition probability function, and let $(X_t, \mathcal{G}_t^\tau, \mathbb{P}_{\tau,x})$ be an adapted right-continuous Markov process on Ω associated with P. Suppose that the following conditions hold:

(1) For every $B \in \mathcal{E}$, the function $(\tau, x, t) \mapsto P(\tau, x; t, B)$ is $\mathcal{B}_{[0,T]} \otimes \mathcal{E} \otimes \mathcal{B}_{[0,T]}$-measurable.
(2) For any $f \in C_0$ and $t \in (0, T]$, the function $(\tau, x) \mapsto Y(\tau, t) f(x)$ is right-continuous in τ on the interval $[0, t)$ and continuous in $x \in E$.

Then X_t is a strong Markov process with respect to the family of measures $\{\mathbb{P}_{\tau,x}\}$ (see Definition 2.20). This means that for any admissible family $\mathcal{M}$ of stopping times, $\tau \in [0, T]$, $x \in E$, $S_1 \in \mathcal{M}(\tau)$, $S_2 \in \mathcal{M}(\tau)$ with $\tau \leq S_1 \leq S_2 \leq T$, and any bounded Borel function f defined on $[\tau, T] \times E$, the equality

$$\mathbb{E}_{\tau,x}\left[f(S_2, X_{S_2}) \mid \mathcal{G}_{S_1}^\tau\right] = \mathbb{E}_{S_1, X_{S_1}}[f(S_2, X_{S_2})]$$

holds $\mathbb{P}_{\tau,x}$-almost surely. The next assertion concerns the diffusion process X_t in Theorem 3.12.

Theorem 3.13 *Suppose that the conditions in Theorem 3.12 hold, and let X_t be the diffusion process whose existence was established in Theorem 3.12. Then X_t is a strong Markov process with respect to the family of measures $\{\mathbb{P}_{\tau,x}\}$.*

Proof. By Theorem 3.12, the process X_t is continuous. Moreover, condition (3) in Corollary 2.1 holds. Indeed, let $x_0 \in \mathbb{R}^d$, $\tau \le s_0 < T$, and $f \in C_0$. By equality (3.135), condition (3) in Corollary 2.1 is equivalent to the equality

$$\lim_{s\downarrow s_0, x\to x_0} \mathbb{E}\left[f\left(X_t^{s,x}\right)\right] = \mathbb{E}\left[f\left(X_t^{s_0,x_0}\right)\right]. \qquad (3.145)$$

Since the process $X_t^{\tau,x}$ is continuous with respect to the variables τ, t, and x, the equality in (3.145) can be established using the dominated convergence theorem. Next, applying Corollary 2.1, we see that the process X_t is a strong Markov process with respect to the family $\{\mathbb{P}_{\tau,x}\}$.

This completes the proof of Theorem 3.13. □

3.9 Additive Functionals Associated with Time-Dependent Measures

This section is devoted to additive functionals of non-homogeneous progressively measurable Markov processes. The existence of additive functionals associated with time-dependent measures has already been mentioned (see Section 3.1). The next definition introduces an important class of transition probability functions.

Definition 3.20 The class $\mathcal{PM}$ of transition probability functions is defined as follows: $P \in \mathcal{PM}$ if and only if there exists a progressively measurable Markov process X_t with P as its transition function.

Given a transition probability function $P \in \mathcal{PM}$, we will always choose a progressively measurable process X_t to represent P. Suppose that V is a nonnegative function from the non-autonomous Kato class $\mathcal{P}_f^*$. Then for all τ and t with $0 \le \tau \le t \le T$, the random variable $(s,\omega) \mapsto V(s, X_s(\omega))$, $\tau \le s \le t$, is $\mathcal{B}_{[\tau,t]} \otimes \mathcal{F}_t^\tau$-measurable. Moreover, the random variable

$$\omega \mapsto \int_\tau^t V(s, X_s(\omega))\,ds$$

is $\mathcal{F}_t^\tau$-measurable and finite $\mathbb{P}_{\tau,x}$-almost surely. It follows that the functional A_V defined by

$$A_V(\tau,t) = \begin{cases} \int_\tau^t V(s, X_s)\, ds, & \text{if } \int_\tau^t V(s, X_s)\, ds < \infty \\ 0, & \text{otherwise,} \end{cases} \quad (3.146)$$

satisfies the following conditions:

(1) For all τ and t with $0 \leq \tau \leq t \leq T$, the random variable $A_V(\tau,t)$ is finite everywhere on Ω and $\mathcal{F}_t^\tau$-measurable.
(2) For all $0 \leq \tau \leq T$, $A_V(\tau,\tau) = 0$ everywhere on Ω.
(3) For all τ and t with $0 \leq \tau \leq t \leq T$, the process $t \mapsto A_V(\tau,t)$, $\tau \leq t \leq T$, is non-decreasing and continuous.
(4) For all $0 \leq \tau \leq \lambda \leq t \leq T$ and $x \in E$,

$$A_V(\tau,t) = A_V(\tau,\lambda) + A_V(\lambda,t)$$

$\mathbb{P}_{\tau,x}$-a.s.
(5) For all $0 \leq \tau \leq t \leq T$ and $x \in E$,

$$\mathbb{E}_{\tau,x} A_V(\tau,t) = N(V)(\tau,t,x).$$

For any $V \in \mathcal{P}_f^*$, we define the functional A_V associated with V by

$$A_V(\tau,t) = A_{V^+}(\tau,t) - A_{V^-}(\tau,t).$$

This definition makes sense since both terms on the right-hand side are finite.

Lemma 3.19 *Let $V \in \mathcal{P}_f^*$. Then the functional A_V defined by (3.146) satisfies the following condition: For all $0 \leq \tau \leq t \leq T$, the random variable $A_V(\tau,t)$ is measurable with respect to the σ-algebra*

$$\mathcal{F}_{t-}^\tau = \sigma\left(X_s : \tau \leq s < t\right).$$

Proof. It is easy to see that without loss of generality we can assume that $V \geq 0$. Let u_k be a sequence of numbers such that $\tau \leq u_k < t$ and $u_k \uparrow t$. Since the random variable $A_V(\tau, u_k)$ is $\mathcal{F}_{u_k}^\tau$-measurable, it is also $\mathcal{F}_{t-}^\tau$-measurable. Moreover, for all $\omega \in \Omega$ with $\int_\tau^t V(s, X_s(\omega))\, ds < \infty$, we have

$$\lim_{k \to \infty} \int_\tau^{u_k} V(s, X_s(\omega))\, ds = \int_\tau^t V(s, X_s(\omega))\, ds. \quad (3.147)$$

It is not hard to see that in order to prove Lemma 3.19, it suffices to show that the set

$$C_j = \left\{\omega : \int_\tau^t V(s, X_s(\omega))\, ds \leq j\right\} \tag{3.148}$$

belongs to the σ-algebra $\mathcal{F}_{t-}^\tau$ for all $j \geq 1$. This fact easily follows from the equality

$$C_j = \bigcap_{k=1}^\infty \left\{\omega : \int_\tau^{u_k} V(s, X_s(\omega))\, ds \leq j\right\}.$$

This completes the proof of Lemma 3.19. □

Our next goal is to construct the additive functional A_μ associated with a time-dependent measure μ from the non-autonomous Kato class $\mathcal{P}_m^*$. The first step in the construction is to approximate μ by a sequence of functions V_k in a special sense. We will need the potentials $N(V)$ and $N(\mu)$ and the functions $M(V)$ and $M(\mu)$ defined after Remark 3.2. Recall that for $V \in \widehat{\mathcal{P}}_f^*$ the function $M(V)$ is given by

$$M(V)(\tau, t) = \sup_{r : \tau \leq r \leq t} \sup_{x \in E} |N(V)(r, t, x)|, \quad 0 \leq \tau \leq t \leq T.$$

Similarly, for $\mu \in \widehat{\mathcal{P}}_m^*$ the function $M(\mu)$ is defined by

$$M(\mu)(\tau, t) = \sup_{r : \tau \leq r \leq t} \sup_{x \in E} |N(\mu)(r, t, x)|, \quad 0 \leq \tau \leq t \leq T.$$

Lemma 3.20 *Let P be a transition probability function, and let $V \in \mathcal{P}_f^*$. For $k \geq 1$, $0 \leq \tau \leq T$, and $x \in E$, put*

$$V_k(\tau, x) = kN(V)\left(\tau, \min\left(\tau + \frac{1}{k}, T\right), x\right). \tag{3.149}$$

Then the following conditions hold:

$$V_k \in \mathcal{P}_f^* \tag{3.150}$$

for all $k \geq 1$;

$$\lim_{k \to \infty} \sup_{0 \leq \tau \leq t \leq T} \sup_{x \in E} |N(V - V_k)(\tau, t, x)| = 0; \tag{3.151}$$

and

$$\lim_{t - \tau \downarrow 0} \sup_{k \geq 1} \sup_{x \in E} N(|V_k|)(\tau, t, x) = 0. \tag{3.152}$$

Lemma 3.21 *Let P be a transition probability function possessing a density p, and let $\mu \in \mathcal{P}_m^*$. For $k \geq 1$, $0 \leq \tau \leq T$, and $x \in E$, put*

$$V_k(\tau, x) = k N(\mu)\left(\tau, \min\left(\tau + \frac{1}{k}, T\right), x\right). \qquad (3.153)$$

Then conditions (3.150)-(3.152) in Lemma 3.20 hold with μ instead of V.

Proof. We will only prove Lemma 3.20. The proof of Lemma 3.21 is similar. It is not hard to see that

$$\begin{aligned}
N(V_k)(\tau, t, x) &= k \int_\tau^t Y(\tau, s) ds \int_s^{\min(s+\frac{1}{k}, T)} Y(s, u) V(u)(x) du \\
&= k \int_\tau^t ds \int_s^{\min(s+\frac{1}{k}, T)} Y(\tau, u) V(u)(x) du \\
&= k \int_\tau^{\min(t+\frac{1}{k}, T)} Y(\tau, u) V(u)(x) du \int_\tau^t \chi_{C_k(u)}(s) ds, \quad (3.154)
\end{aligned}$$

where $\chi_{C_k(u)}$ is the characteristic function of the set

$$C_k(u) = \left\{ s : s \leq u \leq \min\left(s + \frac{1}{k}, T\right) \right\}.$$

It follows from (3.154) that

$$\begin{aligned}
N(V_k)(\tau, t, x) = {}& k \int_\tau^{\min(\tau+\frac{1}{k}, T)} Y(\tau, u) V(u)(x) du \int_\tau^t \chi_{C_k(u)}(s) ds \\
& + \int_{\min(\tau+\frac{1}{k}, t)}^t Y(\tau, u) V(u)(x) du \\
& + k \int_t^{\min(t+\frac{1}{k}, T)} Y(\tau, u) V(u)(x) du \int_\tau^t \chi_{C_k(u)}(s) ds.
\end{aligned}$$
$$(3.155)$$

Next (3.155) gives

$$\begin{aligned}
|N(V - V_k)(\tau, t, x)| \leq {}& \int_\tau^{\min(\tau+\frac{1}{k}, t)} Y(\tau, u) |V(u)|(x) du \\
& + k \int_\tau^{\min(\tau+\frac{1}{k}, t)} Y(\tau, u) |V(u)|(x) du \int_\tau^t \chi_{C_k(u)}(s) ds \\
& + k \int_t^{\min(t+\frac{1}{k}, T)} Y(\tau, u) |V(u)|(x) du \int_\tau^t \chi_{C_k(u)}(s) ds.
\end{aligned}$$

Since the Lebesgue measure of the set $C_k(u)$ does not exceed $\frac{1}{k}$, we get

$$|N(V - V_k)(\tau, t, x)| \leq 2N(|V|)\left(\tau, \min\left(\tau + \frac{1}{k}, t\right), x\right)$$

$$+ Y(\tau, t) \int_t^{\min(t+\frac{1}{k}, T)} Y(t, u)|V(u)|(x) du$$

$$= 2N(|V|)\left(\tau, \min\left(\tau + \frac{1}{k}, t\right), x\right)$$

$$+ Y(\tau, t)N(|V|)\left(t, \min\left(t + \frac{1}{k}, T\right)\right)(x).$$

Therefore,

$$\sup_{x \in E} |N(V - V_k)(\tau, t, x)| \leq 2 \sup_{x \in E} N(|V|)\left(\tau, \min\left(\tau + \frac{1}{k}, t\right), x\right)$$

$$+ \sup_{x \in E} N(|V|)\left(t, \min\left(t + \frac{1}{k}, T\right), x\right).$$
(3.156)

Now it is clear that condition (3.151) in Lemma 3.20 follows from condition (3.156) and the definition of the class $\mathcal{P}_f^*$. Since

$$N(|V_k|)(\tau, t, x) \leq k \int_\tau^t Y(\tau, s) ds \int_s^{\min(s+\frac{1}{k}, T)} Y(s, u)|V(u)|(x) du$$

$$= k \int_\tau^t ds \int_s^{\min(s+\frac{1}{k}, T)} Y(\tau, u)|V(u)|(x) du$$

$$= k \int_\tau^{\min(t+\frac{1}{k}, T)} Y(\tau, u)|V(u)|(x) du \int_{\tau, t} \chi_{C_k(u)}(s) ds,$$

we get

$$N(|V_k|)(\tau, t, x) \leq k \min\left(t - \tau, \frac{1}{k}\right) N(|V|)\left(\tau, \min\left(t + \frac{1}{k}, T\right), x\right).$$
(3.157)

It is easy to see that (3.150) follows from (3.157).

It remains to show that condition (3.152) holds. Let $\epsilon > 0$. Then (3.157) and Lemma 3.2 imply that there exist $\delta_1 > 0$ and $k_0 > 1$ such that

$$\sup_{x \in E} N(|V_k|)(\tau, t, x) < \epsilon$$
(3.158)

for all $t - \tau < \delta_1$ and $k \geq k_0$. Moreover, since $V_k \in \mathcal{P}_f^*$, there exists $\delta_2 > 0$ such that $\delta_2 < \delta_1$ and (3.158) holds for all $t - \tau < \delta_2$ and $k \leq k_0$. Therefore, (3.158) holds for all $k \geq 1$ and $t - \tau < \delta_2$. This establishes equality (3.152) and completes the proof of Lemma 3.20. □

Remark 3.6 Suppose that the conditions in Lemma 3.21 hold. Then it is not hard to see that (3.151) implies

$$\lim_{k \to \infty} M(V_k)(\tau, t) = M(\mu)(\tau, t).$$

Moreover, (3.157) gives

$$\limsup_{k \to \infty} N(|V_k|)(\tau, t, x) \leq N(|\mu|)(\tau, t, x)$$

and

$$\limsup_{k \to \infty} M(|V_k|)(\tau, t) \leq M(|\mu|)(\tau, t).$$

In the next definition, we introduce a special type of approximation based on the formulas in Lemmas 3.20 and 3.21.

Definition 3.21 Let P be a transition probability function and let $V \in \mathcal{P}_f^*$. By definition, a sequence of functions $V_k \in \mathcal{P}_f^*$, $k \geq 1$, approaches a function V in the potential sense if

$$\lim_{k \to \infty} \sup_{0 \leq \tau \leq t \leq T} \sup_{x \in E} |N(V - V_k)(\tau, t, x)| \to 0,$$

and

$$\lim_{t - \tau \downarrow 0} \sup_{k \geq 1} \sup_{x \in E} N(|V_k|)(\tau, t, x) = 0.$$

Suppose that P possesses a density p and let $\mu \in \mathcal{P}_f^*$. By definition, a sequence of time-dependent measures $\mu_k \in \mathcal{P}_m^*$, $k \geq 1$, approaches a time-dependent measure μ in the potential sense if

$$\lim_{k \to \infty} \sup_{0 \leq \tau \leq t \leq T} \sup_{x \in E} |N(\mu - \mu_k)(\tau, t, x)| \to 0,$$

and

$$\lim_{t - \tau \downarrow 0} \sup_{k \geq 1} \sup_{x \in E} N(|\mu_k|)(\tau, t, x) = 0.$$

It follows from (3.149), (3.153), and from the definition of the class $\hat{\mathcal{P}}_f^*$ that the functions V_k in Lemmas 3.20 and 3.21 are bounded. Hence, for any function $V \in \mathcal{P}_f^*$ (any time-dependent measure $\mu \in \mathcal{P}_m^*$) there exists a sequence of bounded functions approximating the function V (the time-dependent measure μ) in the sense of Definition 3.21.

Lemma 3.22 *Let V_k be a sequence of Borel functions on $[0,T] \times E$ such that*

$$\lim_{t-\tau\downarrow 0} \sup_{k\geq 1} \sup_{x\in E} N\left(|V_k|\right)(\tau,t,x) = 0.$$

Then $\sup_k \|V_k\|_f < \infty$. The same result is true for time-dependent measures.

Proof. By the assumption in Lemma 3.22, there exists $\delta > 0$ such that for $t - \tau < \delta$ and $k \geq 1$,

$$\sup_{x\in E} \sup_{k\geq 1} N\left(|V_k|\right)(\tau,t,x) < 1. \tag{3.159}$$

Moreover, for all $0 \leq \tau < t \leq T$, there exists a partition $\tau = t_0 < \cdots < t_n = t$ of the interval $[\tau,t]$ such that $\max\{|t_{j+1} - t_j| : 0 \leq j \leq n-1\} < \delta$ and $n < \delta^{-1}T$. It follows that

$$N\left(|V_k|\right)(\tau,t,x) = \sum_{j=0}^{n-1} \int_{t_j}^{t_{j+1}} Y(\tau,s)\,|V_k|(x)ds$$

$$= \sum_{j=0}^{n-1} Y(\tau,t_j) \int_{t_j}^{t_{j+1}} Y(t_j,s)\,|V_k|(x)ds$$

$$= \sum_{j=0}^{n-1} Y(\tau,t_j)\, N\left(|V_k|\right)(t_j,t_{j+1})(x).$$

Therefore, (3.159) implies

$$\sup_{k\geq 1} \sup_{x\in E} N\left(|V_k|\right)(\tau,t,x) \leq \sum_{k\geq 1}^{n-1} N\left(|N_k|\right)(t_j,t_{j+1},x) \leq n < \frac{T}{\delta}.$$

This completes the proof of Lemma 3.22 in the case of functions. The case of time-dependent measures is similar. □

The next theorem concerns the existence of the additive functional A_μ associated with a time-dependent measure $\mu \in \mathcal{P}_m^*$.

Theorem 3.14 *Let $P \in \mathcal{PM}$ be a transition probability function possessing a density p, and let X_t be a corresponding progressively measurable Markov process. Then for every time-dependent nonnegative measure μ from the class $\mathcal{P}_m^*$, there exists a functional $A_\mu(\tau, t)$ of the process X_t such that*

(1) For all τ and t with $0 \leq \tau \leq t \leq T$, the random variable $A_\mu(\tau, t)$ is nonnegative and finite everywhere on Ω.

(2) For all τ and t with $0 \leq \tau \leq t \leq T$, the random variable $A_\mu(\tau, t)$ is $\mathcal{F}_{t-}^\tau$-measurable.

(3) For all $0 \leq \tau \leq T$ and $x \in E$, $A_\mu(\tau, \tau) = 0$ $\mathbb{P}_{\tau,x}$-a.s.

(4) For all τ and t with $0 \leq \tau \leq t \leq T$, the process $t \mapsto A_\mu(\tau, t)$, $\tau \leq t \leq T$, is non-decreasing and continuous everywhere on Ω.

(5) For all $0 \leq \tau \leq \lambda \leq t \leq T$ and $x \in E$,

$$A_\mu(\tau, t) = A_\mu(\tau, \lambda) + A_\mu(\lambda, t) \ \mathbb{P}_{\tau,x}\text{-a.s.}$$

(6) For all $0 \leq \tau \leq t \leq T$ and $x \in E$,

$$\mathbb{E}_{\tau,x} A_\mu(\tau, t) = N(\mu)(\tau, t, x).$$

Proof. Let V_k be the sequence of functions in formula (3.153) corresponding to the time-dependent measure μ. Our first goal is to show that

$$\lim_{k,j \to \infty} \sup_{\tau: 0 \leq \tau \leq T} \sup_{x \in E} \mathbb{E}_{\tau,x} \sup_{t: \tau \leq t \leq T} \left| A_{V_k}(\tau, t) - A_{V_j}(\tau, t) \right|^n = 0 \quad (3.160)$$

for every $n \geq 1$. This will allow us to define the functional A_μ as the limit of the sequence A_{V_k}.

Lemma 3.23 *Let $P \in \mathcal{PM}$ and $V \in \mathcal{P}_f^*$. Then for all $0 \leq \tau \leq t \leq T$, $x \in E$, and all integers $n \geq 2$,*

$$|\mathbb{E}_{\tau,x} A_V(\tau, t)^n| \leq n! N(|V|)(\tau, t, x) M(|V|)(\tau, t)^{n-2} M(V)(\tau, t). \quad (3.161)$$

Proof. It follows from the Markov property and the condition $V \in \mathcal{P}_f^*$ that

$$\mathbb{E}_{\tau,x} A_V(\tau, t)^2$$
$$= 2\mathbb{E}_{\tau,x} \int_\tau^t V(s, X_s) \, ds \int_s^t V(u, X_u) \, du$$
$$= 2\mathbb{E}_{\tau,x} \int_\tau^t V(s, X_s) \, ds \int_s^t \mathbb{E}_{\tau,x} \left(V(u, X_u) \mid \mathcal{F}_s^\tau \right) du$$

$$= 2\mathbb{E}_{\tau,x} \int_\tau^t V(s, X_s)\, ds \int_s^t \mathbb{E}_{s,z} V(u, X_u)\, du \Big|_{z=X_s}$$

$$\leq 2 \int_\tau^t ds\, Y(\tau, s)|V(s)|(x) ds \sup_{s:\tau\leq s\leq t} \sup_{y\in E} \left|\int_s^t Y(s, u)V(u)(y) du\right|. \quad (3.162)$$

Now it is clear that (3.162) implies (3.161) with $n = 2$.

Next, let $n > 2$ be any positive integer. By induction, we get

$$\mathbb{E}_{\tau,x} A_V(\tau, t)^n$$

$$= n!\mathbb{E}_{\tau,x} \int_\tau^t V(t_1, X_{t_1})\, dt_1 \int_{t_1}^t V(t_2, X_{t_2})\, dt_2 \cdots \int_{t_{n-1}}^t V(t_n, X_{t_n})\, dt_n$$

$$\leq n! \int_\tau^t Y(\tau, s)|V(s)|(x) ds \sup_{r:\tau\leq r\leq t} \sup_{y\in E} \left[\int_r^t Y(r, u)|V(u)|(y) du\right]^{n-2}$$

$$\sup_{r:\tau\leq r\leq t} \sup_{y\in E} \left|\int_r^t Y(r, u)V(u)(y) du\right|.$$

Now it is not hard to see that the previous estimate implies inequality (3.161).

This completes the proof of Lemma 3.23. $\square$

Corollary 3.1 *Suppose that the conditions in Lemma 3.23 are satisfied. Then for any odd integer $n \geq 3$,*

$$\mathbb{E}_{\tau,x} |A_V(\tau, t)|^n$$
$$\leq \sqrt{(n-1)!(n+1)!} N(|V|)(\tau, t, x) M(|V|)(\tau, t)^{n-2} M(V)(\tau, t). \quad (3.163)$$

Proof. If $n \geq 3$ is odd, then

$$\mathbb{E}_{\tau,x} |A_V(\tau, t)|^n \leq \left\{\mathbb{E}_{\tau,x} A_V(\tau, t)^{n-1}\right\}^{\frac{1}{2}} \left\{\mathbb{E}_{\tau,x} A_V(\tau, t)^{n+1}\right\}^{\frac{1}{2}}.$$

Now it is clear that (3.163) follows from Lemma 3.23. $\square$

Lemma 3.23 and Corollary 3.1 provide pointwise estimates for the expression $\mathbb{E}_{\tau,x} |A_V(\tau, t)|^n$. The next lemma shows that stronger uniform estimates hold.

Lemma 3.24 *Let $P \in \mathcal{PM}$ and $V \in \mathcal{P}_f^*$. Then for any τ with $0 \leq \tau \leq T$, any $\delta > 0$ such that $\tau + \delta \leq T$, and any even integer $n \geq 2$, the following estimate holds:*

$$\sup_{x\in E} \mathbb{E}_{\tau,x} \sup_{t:\tau\leq t\leq \tau+\delta} A_V(\tau, t)^n$$

$$\leq c_n M(|V|)(\tau, \tau + \delta)^{n-1} M(V)(\tau, \tau + \delta), \qquad (3.164)$$

where

$$c_n = 2^n \left[\left(\frac{n}{n-1} \right)^n n! + 1 \right]. \qquad (3.165)$$

Moreover, for any odd integer $n \geq 3$,

$$\sup_{x \in E} \mathbb{E}_{\tau,x} \sup_{t: \tau \leq t \leq \tau + \delta} |A_V(\tau, t)|^n$$
$$\leq c_n M(|V|)(\tau, \tau + \delta)^{n-1} M(V)(\tau, \tau + \delta) \qquad (3.166)$$

where $c_n = \sqrt{c_{n-1} c_{n+1}}$.

Remark 3.7 For $n = 1$, the following estimate holds:

$$\sup_{x \in E} \mathbb{E}_{\tau,x} \sup_{t: \tau \leq t \leq \tau + \delta} |A_V(\tau, t)|$$
$$\leq \{c_2 M(|V|)(\tau, \tau + \delta) M(V)(\tau, \tau + \delta)\}^{\frac{1}{2}}. \qquad (3.167)$$

Estimate (3.167) follows from (3.164) with $n = 2$.

Proof. We will prove estimate (3.164). Estimate (3.166) follows from (3.164) and Hölder's inequality.

Let $n \geq 2$ be an even integer, and let $V \in \mathcal{P}_f^*$. Given $x \in E$ and τ, δ, and t with $0 \leq \tau \leq t \leq \tau + \delta \leq T$, put

$$M_t = \mathbb{E}_{\tau,x} \left(A_V(\tau, \tau + \delta) \mid \mathcal{F}_t^\tau \right).$$

It follows from (3.161) that M_t is an $\mathcal{F}_t^\tau$-martingale from the space L^n. By the Markov property, for every t with $0 \leq \tau \leq t \leq \tau + \delta \leq T$,

$$M_t = A_V(\tau, t) + \int_t^{\tau + \delta} \mathbb{E}_{\tau,x} \left(V(s, X_s) \mid \mathcal{F}_t^\tau \right) ds$$
$$= A_V(\tau, t) + \int_t^{\tau + \delta} Y(t, s) V(s)(X_t) ds$$

$\mathbb{P}_{\tau,x}$-a.s. Hence, M_t is a modification of the functional

$$\widetilde{M}_t = A_V(\tau, t) + N(V)(t, \tau + \delta, X_t).$$

Let us fix a partition $\tau = t_0 < t_1 < \cdots < t_k = \tau + \delta$ of the interval $[\tau, \tau + \delta]$. By Doob's inequality (see Subsection 5.6), we get

$$\mathbb{E}_{\tau,x} \sup_{j: 0 \leq j \leq k} A_V(\tau, t_j)^n$$

$$\leq 2^n \mathbb{E}_{\tau,x} \sup_{j:0\leq j\leq k} M_{t_j}^n + 2^n \mathbb{E}_{\tau,x} \sup_{j:0\leq j\leq k} \left| N(V)\left(t_j, \tau+\delta, X_{t_j}\right)\right|^n$$

$$\leq 2^n \left(\frac{n}{n-1}\right)^n \mathbb{E}_{\tau,x} A_V(\tau, \tau+\delta)^n + 2^n \left[\sup_{z\in E} \sup_{s:\tau\leq s\leq \tau+\delta} |N(V)(s, \tau+\delta, z)|\right]^n$$

$$\leq 2^n \left(\frac{n}{n-1}\right)^n \mathbb{E}_{\tau,x} A_V(\tau, \tau+\delta)^n + 2^n M(V)(\tau, \tau+\delta)^n. \tag{3.168}$$

It follows from (3.161) and (3.168) that

$$\mathbb{E}_{\tau,x} \sup_{j:0\leq j\leq k} A_V(\tau, t_j)^n \leq c_n M(|V|)(\tau, \tau+\delta)^{n-1} M(V)(\tau, \tau+\delta) \tag{3.169}$$

for all $x \in E$ where c_n is defined by (3.165). Now choose a sequence of successive refinements of the partition $\tau = t_0 < t_1 < \cdots < t_k = \tau + \delta$ on the left-hand side of (3.169), for which the maximum length of the partition intervals tends to 0. Passing to the limit in (3.169) and using the monotone convergence theorem and the continuity of the functional $A_V(\tau, t)$ with respect to the variable t, we get estimate (3.164).

This completes the proof of Lemma 3.24. □

Let us continue the proof of Theorem 3.14. For a nonnegative time-dependent measure $\mu \in \mathcal{P}_m^*$, define the sequence V_k by (3.153). Then (3.164) and (3.166) imply

$$\sup_{x\in E} \mathbb{E}_{\tau,x} \sup_{t:\tau\leq t\leq \tau+\delta} \left| A_{V_k-V_j}(\tau, t)\right|^n$$

$$\leq c_n M\left(|V_k - V_j|\right)(\tau, \tau+\delta)^{n-1} M\left(V_k - V_j\right)(\tau, \tau+\delta). \tag{3.170}$$

It follows from Lemma 3.21 that formula (3.160) holds. Therefore,

$$\lim_{k,j\to\infty} \sup_{\tau:0\leq \tau\leq T} \sup_{x\in E} \mathbb{E}_{\tau,x} \sup_{t:\tau\leq t\leq T} \left|A_{V_k}(\tau, t) - A_{V_j}(\tau, t)\right| = 0,$$

and hence, there exists a functional $\widehat{A}_\mu(\tau, t)$ of the process X_t such that for all $0 \leq \tau \leq t \leq T$,

$$\lim_{k\to\infty} \sup_{x\in E} \mathbb{E}_{\tau,x} \sup_{u:\tau\leq u\leq t} \left|\widehat{A}_\mu(\tau, u) - A_{V_k}(\tau, u)\right| = 0. \tag{3.171}$$

The random variable $\widehat{A}_\mu(\tau, t)$ in (3.171) is measurable with respect to the σ-algebra $\bar{\mathcal{F}}_t^\tau$. It follows from (3.171) that there exists an increasing sequence n_k of positive integers such that

$$\lim_{k\to\infty} \sup_{u:\tau\leq u\leq t} \left|\widehat{A}_\mu(\tau, u) - A_{V_{n_k}}(\tau, u)\right| = 0 \tag{3.172}$$

$\mathbb{P}_{\tau,x}$-almost surely for all $x \in E$. The sequence n_k in (3.172) depends on τ and t. Moreover, it can be chosen independently of x. This can be shown using (3.171) and reasoning as in the standard proof of the fact that the convergence in measure implies the existence of an almost everywhere convergent subsequence (see, e.g., the proof of Proposition 18 on pages 95-96 in [Royden (1988)]).

We are finally ready to define the functional A_μ. For all $0 \leq \tau \leq t \leq T$, put

$$A_\mu(\tau,t) = \begin{cases} \lim_{k \to \infty} A_{V_{n_k}}(\tau,t), & \text{if } \lim_{i,j \to \infty} \sup_{u \in [\tau,t]} \left| A_{V_{n_i}}(\tau,t) - A_{V_{n_j}}(\tau,t) \right| = 0 \\ 0, & \text{otherwise.} \end{cases}$$
(3.173)

Then for all $k \geq 1$, $A_{V_{n_k}}$ is nondecreasing and continuous everywhere on Ω. Moreover, the random variable $A_{V_{n_k}}(\tau,t)$ is $\mathcal{F}_{t-}^\tau$-measurable. Indeed, for any $V \in \mathcal{P}_f^*$ we have

$$A_V(\tau,t) = \lim_{k \to \infty} \int_\tau^{t-\frac{1}{k}} V(s,X_s)\,ds,$$

and hence $A_V(\tau,t)$ is $\mathcal{F}_t^\tau$-measurable. It follows from the properties of the functionals $A_{V_{n_k}}$ mentioned above that the functional A_μ defined by (3.173) satisfies conditions (1)-(6) in Theorem 3.14.

This completes the proof of Theorem 3.14. $\square$

The next assertion concerns the uniqueness of the functional A_μ.

Lemma 3.25 *Let μ be a non-negative time-dependent measure from the class $\mathcal{P}_m^*$, and let A_1 and A_2 be two functionals satisfying conditions 1-5 in Theorem 3.14. Then for every $0 \leq \tau \leq T$ and $x \in E$, the processes $A_1(\tau,t)$ and $A_2(\tau,t)$ are indistinguishable.*

Proof. It is not hard to see, using the conditions in Theorem 3.14, that

$$\mathbb{E}_{\tau,x}\left[A_1(\tau,t) - A_2(\tau,t)\right]^2$$
$$= 2 \sum_{i,j=1}^2 (-1)^{i+j} \mathbb{E}_{\tau,x} \int_\tau^t \left[A_i(\tau,t) - A_i(\tau,s)\right] dA_j(\tau,s)$$
$$= 2 \sum_{i,j=1}^2 (-1)^{i+j} \mathbb{E}_{\tau,x} \int_\tau^t A_i(s,t) dA_j(\tau,s). \tag{3.174}$$

By condition (1) in Theorem 3.14, the random variable $A_i(t,s)$ is $\mathcal{F}_t^s$-measurable for $i = 1, 2$. Next, using the Markov property in (3.174), we obtain

$$\mathbb{E}_{\tau,x}\left[A_1(\tau,t) - A_2(\tau,t)\right]^2$$

$$= 2\sum_{i,j=1}^{2}(-1)^{i+j}\mathbb{E}_{\tau,x}\int_\tau^t \mathbb{E}_{z,s}A_i(s,t)\big|_{z=X_s}\,dA_j(\tau,s)$$

$$= 2\sum_{i,j=1}^{2}(-1)^{i+j}\mathbb{E}_{\tau,x}\int_\tau^t N_\mu(s,t,X_s)\,dA_j(\tau,s) = 0. \qquad (3.175)$$

It follows from (3.175) that for given τ and x, the process A_2 is a modification of A_1. Since both processes are continuous, they are indistinguishable. This completes the proof of Lemma 3.25. $\square$

Definition 3.22 Let $\mu \in \mathcal{P}_m^*$, and denote by μ^+ and μ^- the positive and the negative variation of the time-dependent measure μ, respectively. Then the functional A_μ is defined as follows:

$$A_\mu(\tau,t) = A_{\mu^+}(\tau,t) - A_{\mu^-}(\tau,t).$$

Lemma 3.26 *Let $P \in \mathcal{PM}$ be a transition probability function possessing a density p, and let $\mu \in \mathcal{P}_m^*$. Then estimates (3.164) and (3.166) hold with μ instead of V.*

Proof. We will prove estimate (3.164) for a time-dependent measure μ. It is clear that estimate (3.166) for μ follows from estimate (3.164).

Let $\mu \in \mathcal{P}_m^*$, and let V_k be the sequence constructed for μ in Lemma 3.21. Then $V_k \in \mathcal{P}_f^*$. Applying estimate (3.164) to the sequence V_k, we see that

$$\sup_{x\in E}\mathbb{E}_{\tau,x}\sup_{t:\tau\le t\le\tau+\delta}|A_{V_k}(\tau,t)|^n \le c_n M(|V_k|)(\tau,\tau+\delta)^{n-1}M(V_k)(\tau,\tau+\delta).$$
(3.176)

It follows from Remark 3.6 that

$$\limsup_{k\to\infty} M(|V_k|)(\tau,\tau+\delta) \le M(|\mu|)(\tau,\tau+\delta) \qquad (3.177)$$

and

$$M(V_k)(\tau,\tau+\delta) \to M(\mu)(\tau,\tau+\delta) \qquad (3.178)$$

as $k \to \infty$. Now using (3.170), (3.173), (3.176), (3.177), and (3.178), we see that (3.164) holds for μ.

This completes the proof of Lemma 3.26. □

Next, we turn our attention to additive functionals corresponding to time-dependent measures from non-autonomous Kato classes. The time-reversal and the fact that if $\widetilde{P}$ is a backward transition probability function, then

$$P(\tau, x; t, B) = \widetilde{P}(T-t, B; T-\tau, x) \tag{3.179}$$

is a transition probability function will be used in the definition of the additive functional. Recall that the following notation was used for the time reversed objects: $\widetilde{\mathbb{P}}^{t,y} = \mathbb{P}_{T-t,y}$, $\widetilde{X}^t = X_{T-t}$, and $\widetilde{\mathcal{F}}^t_\tau = \mathcal{F}^{T-t}_{T-\tau}$, where the forward objects correspond to the transition probability function P, while the backward ones are associated with the backward transition probability function $\widetilde{P}$ (see Section 1.4).

We will say that a transition probability function $\widetilde{P}$ belongs to the class $\widetilde{\mathcal{PM}}$ if there exists a backward Markov process $\widetilde{X}^t$ that is progressively measurable with respect to the filtration $\widetilde{\mathcal{F}}^t_\tau$ and has $\widetilde{P}$ as its backward transition function. It is not hard to see that the process X_t is $\mathcal{F}^\tau_t$-progressively measurable if and only if the time reversed process $\widetilde{X}^t$ is $\widetilde{\mathcal{F}}^t_\tau$-progressively measurable. Therefore,

$$\widetilde{P} \in \widetilde{\mathcal{PM}} \Longleftrightarrow P \in \mathcal{PM}.$$

For a function $V \in \mathcal{P}_f$, the functional $\widetilde{A}_V$ is defined by

$$\widetilde{A}_V(\tau, t) = \int_\tau^t V\left(s, \widetilde{X}^s\right) ds. \tag{3.180}$$

$\widetilde{\mathbb{P}}^{t,y}$-a.s. Now let $\mu \in \mathcal{P}_m$ and denote by $\widetilde{\mu}$ the time-reversal of μ given by $\widetilde{\mu}(t) = \mu(T-t)$. Then the functional $\widetilde{A}_\mu$ is defined as follows:

$$\widetilde{A}_\mu(\tau, t) = A_{\widetilde{\mu}}(T-t, T-\tau). \tag{3.181}$$

The functional $\widetilde{A}_\mu$ satisfies the following conditions:

(1) For all $\tau \leq t$, the random variable $\widetilde{A}_\mu(\tau, t)$ is $\widetilde{\mathcal{F}}^t_\tau$-measurable.
(2) For all t and $y \in E$, $\widetilde{A}_\mu(t, t) = 0$ $\widetilde{\mathbb{P}}^{t,y}$-a.s.
(3) For all $\tau \leq t$ and $y \in E$, the process $\tau \mapsto \widetilde{A}_\mu(\tau, t)$, $0 \leq \tau \leq t$, is non-increasing and continuous $\widetilde{\mathbb{P}}^{t,y}$-a.s.
(4) For all $\tau \leq \lambda \leq t$, $\widetilde{A}_\mu(\tau, t) = \widetilde{A}_\mu(\tau, \lambda) + \widetilde{A}_\mu(\lambda, t)$ $\widetilde{\mathbb{P}}^{t,y}$-a.s.
(5) For all $\tau \leq t$ and $y \in E$, $\widetilde{\mathbb{E}}^{t,y} \widetilde{A}_\mu(\tau, t) = \widetilde{N}(\mu)(\tau, t, y)$.

Note that the free propagator $U(t,\tau)$ associated with $\widetilde{P}$ is related to the backward free propagator $Y(\tau,t)$ corresponding to P by the formula

$$Y(T-t, T-\tau) = U(t,\tau)$$

(see formula (2.11) in Section 2.2).

3.10 Exponential Estimates for Additive Functionals

In this section we study the multiplicative functionals $\exp\{-\int_\tau^t V(s, X_s)\}ds$ and $\exp\{-A_\mu(t,\tau)\}$. These functionals are called the Kac functionals. The next result is a generalization of Khas'minski's Lemma to non-autonomous Kato classes. The exponential estimate in Lemma 3.27 is a useful result. For instance, Lemma 3.27 implies that the backward Feynman-Kac propagators Y_V and Y_μ with $V \in \mathcal{P}_f^*$ and $\mu \in \mathcal{P}_m^*$ are uniformly bounded on the space $L_{\mathcal{E}}^\infty$ (see Theorem 4.2 below). For the proof of Khas'minski's lemma in the autonomous case see [Simon (1982)].

Lemma 3.27 *The following assertions hold:*

(a) Let $P \in \mathcal{PM}$, $V \in \mathcal{P}_f^*$, and let the numbers τ and t with $0 \le \tau \le t \le T$ be such that $M(|V|)(\tau, t) < 1$. Then

$$\sup_{x \in E} \mathbb{E}_{\tau,x} \exp\left\{\int_\tau^t |V(s, X_s)|\, ds\right\} \le \frac{1}{1 - M(|V|)(\tau, t)}. \tag{3.182}$$

(b) Suppose that $P \in \mathcal{PM}$ has a density p. Let $\mu \in \mathcal{P}_m^*$, and let the numbers τ and t with $0 \le \tau \le t \le T$ be such that $M(|\mu|)(\tau, t) < 1$. Then

$$\sup_{x \in E} \mathbb{E}_{\tau,x} \exp\left\{A_{|\mu|}(\tau, t)\right\} \le \frac{1}{1 - M(|\mu|)(\tau, t)}. \tag{3.183}$$

Proof. It follows from estimate (3.161) that

$$\mathbb{E}_{\tau,x} \frac{A_{|V|}(\tau,t)^n}{n!} \le M(|V|)(\tau, t). \tag{3.184}$$

It is not hard to see that (3.184) implies (3.182). Moreover, reasoning as in the proof of Lemma 3.26, we see that (3.182) implies (3.183). □

The next lemmas contain more exponential estimates:

Lemma 3.28 *The following assertions hold:*

(a) Let $P \in \mathcal{PM}$, $V \in \mathcal{P}_f^*$, and let the numbers τ and t with $0 \leq \tau \leq t \leq T$ be such that $M(|V|)(t,\tau) < 1$. Then

$$\mathbb{E}_{\tau,x} \exp\{|A_V(\tau,t)|\} \leq 1 + \{2N(|V|)(\tau,t,x)M(V)(\tau,t)\}^{\frac{1}{2}}$$
$$+ \frac{2\sqrt{3}}{3} \frac{N(|V|)(\tau,t,x)M(V)(\tau,t)}{1 - M(|V|)(\tau,t)}. \quad (3.185)$$

(b) Suppose that $P \in \mathcal{PM}$ has a density p. Let $\mu \in \mathcal{P}_m^*$, and let the numbers τ and t with $0 \leq \tau \leq t \leq T$ be such that $M(|\mu|)(\tau,t) < 1$. Then estimate (3.185) holds with μ instead of V.

Proof. Estimate (3.185) follows from Lemma 3.23, Corollary 3.1, and the fact that

$$\sqrt{(m-1)!(m+1)!} \leq \frac{2\sqrt{3}}{3}(m!)$$

for all $m \geq 3$. The proof of part (b) is similar. Here we reason as in the proof of Lemma 3.26. □

Lemma 3.29 *The following assertions hold:*

(a) Let $P \in \mathcal{PM}$, $q \geq 1$, $1 < r < \infty$, $\frac{1}{r} + \frac{1}{r'} = 1$, $V \in \mathcal{P}_f^*$, and $W \in \mathcal{P}_f^*$. Let τ and t with $0 \leq \tau \leq t \leq T$ be such that $M(rq|W|)(\tau,t) < 1$ and $M(r'q|V-W|)(\tau,t) < 1$. Then

$$\mathbb{E}_{\tau,x} |\exp\{A_V(\tau,t)\} - \exp\{A_W(\tau,t)\}|^q$$
$$\leq \frac{1}{(1 - M(rq|W|)(\tau,t))^{\frac{1}{r}}}$$
$$\times \left\{ [2N(r'q|V-W|)(\tau,t,x) M(r'q(V-W))(\tau,t)]^{\frac{1}{2}} \right.$$
$$\left. + \frac{2\sqrt{3}}{3} \frac{N(r'q|V-W|)(\tau,t,x) M(r'q(V-W))(\tau,t)}{1 - M(r'q|V-W|)(\tau,t)} \right\}^{\frac{1}{r'}}. \quad (3.186)$$

(b) Suppose that $P \in \mathcal{PM}$ has a density p, and let $q \geq 1$, $1 < r < \infty$, $\frac{1}{r} + \frac{1}{r'} = 1$, $\mu \in \mathcal{P}_m^*$, and $\nu \in \mathcal{P}_m^*$. If the numbers τ and t with $0 \leq \tau \leq t \leq T$ are such that $M(rq|\nu|)(\tau,t) < 1$ and $M(r'q|\mu-\nu|)(\tau,t) < 1$, then

$$\mathbb{E}_{\tau,x} |\exp\{A_\mu(\tau,t)\} - \exp\{A_\nu(\tau,t)\}|^q$$

$$\leq \frac{1}{(1 - M(rq|\nu|)(\tau,t))^{\frac{1}{r}}} \left\{ [2N\left(r'q|\mu - \nu|\right)(\tau,t,x)M\left(r'q(\mu - \nu)\right)(\tau,t)]^{\frac{1}{2}} \right.$$
$$\left. + \frac{2\sqrt{3}}{3} \frac{N\left(r'q|\mu - \nu|\right)(\tau,t,x)M\left(r'q(\mu - \nu)\right)(\tau,t)}{1 - M\left(r'q|\mu - \nu|\right)(\tau,t)} \right\}^{\frac{1}{r'}}. \quad (3.187)$$

Proof. It is not hard to see that for all a and b with $-\infty < a < \infty$ and $b \geq 1$,

$$|e^a - 1|^b \leq e^{|a|b} - 1. \quad (3.188)$$

By (3.188) and Hölder's inequality,

$$\mathbb{E}_{\tau,x} |\exp\{A_V(\tau,t)\} - \exp\{A_W(\tau,t)\}|^q$$
$$\leq \{\mathbb{E}_{\tau,x} \exp\{A_{rqW}(\tau,t)\}\}^{\frac{1}{r}} \{\mathbb{E}_{\tau,x} |\exp\{A_{V-W}(\tau,t)\} - 1|^{r'q}\}^{\frac{1}{r'}}$$
$$\leq \{\mathbb{E}_{\tau,x} \exp\{A_{rq|W|}(\tau,t)\}\}^{\frac{1}{r}} \{\mathbb{E}_{\tau,x} \exp\{|A_{r'q(V-W)}(\tau,t)|\} - 1\}^{\frac{1}{r'}}.$$
$$(3.189)$$

Now it is clear that (3.186) follows from (3.182), (3.185) and (3.189). This completes the proof of part (a) of Lemma 3.29.

Next we will prove part (b) of Lemma 3.29. By Lemma 3.21, there exist the approximating sequences g_k (for μ) and h_k (for ν). It follows from the assumptions in part (b) of Lemma 3.29 that $M(rq|\nu|)(\tau,t) < 1$ and $M\left(r'q|\mu - \nu|\right)(\tau,t) < 1$. Therefore, there exists a sequence $\{k'\}$ of positive integers such that

$$M\left(rq|h_{k'}|\right)(\tau,t) < 1, \quad M\left(r'q|g_{k'} - h_{k'}|\right)(\tau,t) < 1,$$
$$\lim_{k' \to \infty} A_{g_{k'}}(\tau,t) = A_\mu(\tau,t), \quad \lim_{k' \to \infty} A_{h_{k'}}(\tau,t) = A_\nu(\tau,t),$$
$$\limsup_{k' \to \infty} M\left(|h_{k'}|\right)(\tau,t) \leq M(|\mu|)(\tau,t),$$
$$\limsup_{k' \to \infty} M\left(|g_{k'} - h_{k'}|\right)(\tau,t) \leq M(|\mu - \nu|)(\tau,t),$$
$$\limsup_{k' \to \infty} N\left(|g_{k'} - h_{k'}|\right)(\tau,t,x) \leq N(|\mu - \nu|)(\tau,t,x),$$

and

$$\lim_{k' \to \infty} M\left(g_{k'} - h_{k'}\right)(\tau,t) = M(\mu - \nu)(\tau,t).$$

It follows from (3.186) that

$$\mathbb{E}_{\tau,x} \left|\exp\left\{A_{g_{k'}}(\tau,t)\right\} - \exp\left\{A_{h_{k'}}(\tau,t)\right\}\right|^q$$
$$\leq \frac{1}{(1 - M(rq|h_{k'}|)(\tau,t))^{\frac{1}{r}}}$$
$$\left\{[2N(r'q|g_{k'} - h_{k'}|)(\tau,t,x) M(r'q(g_{k'} - h_{k'}))(\tau,t)]^{\frac{1}{2}}\right.$$
$$\left. + \frac{2\sqrt{3}}{3} \frac{N(r'q|g_{k'} - h_{k'}|)(\tau,t,x) M(r'q(g_{k'} - h_{k'}))(\tau,t)}{1 - M(r'q|g_{k'} - h_{k'}|)(\tau,t)}\right\}^{\frac{1}{r'}}. \quad (3.190)$$

Now using Fatou's Lemma in (3.190), we see that estimate (3.187) holds. This completes the proof of Lemma 3.29. □

It is not hard to see from formula (3.165) that $c_n \leq c2^n n!$. However, this inequality does not allow us to get an exponential estimate for the functional A_V from inequalities (3.164) and (3.166). Next, we will obtain such an estimate by modifying the proof of Lemma 3.24.

Lemma 3.30 *The following assertions hold:*

(a) *Let $P \in \mathcal{PM}$ and $V \in \mathcal{P}_f^*$. Then for every τ with $0 \leq \tau \leq T$ and every $\delta > 0$ such that $\tau + \delta \leq T$ and $M(|V|)(\tau, \tau + \delta) < 1$, the following estimate holds:*

$$\sup_{x \in E} \mathbb{E}_{\tau,x} \exp\left\{\sup_{t:\tau \leq t \leq \tau+\delta} |A_V(\tau,t)|\right\}$$
$$\leq \exp\{M(V)(\tau, \tau+\delta)\} \left(1 + c\{M(|V|)(\tau,\tau+\delta) M(V)(\tau,\tau+\delta)\}^{\frac{1}{2}}\right.$$
$$\left. + c\frac{M(|V|)(\tau,\tau+\delta) M(V)(\tau,\tau+\delta)}{1 - M(|V|)(\tau,\tau+\delta)}\right). \quad (3.191)$$

(b) *Suppose that $P \in \mathcal{PM}$ has a density p, and let $\mu \in \mathcal{P}_m^*$. Then for every τ with $0 \leq \tau \leq T$ and every $\delta > 0$ such that $\tau + \delta \leq T$ and $M(|\mu|)(\tau, \tau + \delta) < 1$, the following estimate holds:*

$$\sup_{x \in E} \mathbb{E}_{\tau,x} \exp\left\{\sup_{t:\tau \leq t \leq \tau+\delta} |A_\mu(\tau,t)|\right\}$$
$$\leq \exp\{M(\mu)(\tau, \tau+\delta)\} \left(1 + c\{M(|\mu|)(\tau,\tau+\delta) M(\mu)(\tau,\tau+\delta)\}^{\frac{1}{2}}\right.$$
$$\left. + c\frac{M(|\mu|)(\tau,\tau+\delta) M(\mu)(\tau,\tau+\delta)}{1 - M(|\mu|)(\tau,\tau+\delta)}\right).$$

Proof. In the proof of Lemma 3.30, we will use the same notation as in the proof of Lemma 3.24. By Doob's inequality (see Subsection 5.6), for every $n \geq 2$,

$$\mathbb{E}_{\tau,x} \sup_{j:0\leq j\leq k} \left|M_{t_j}\right|^n \leq \left(\frac{n}{n-1}\right)^n \mathbb{E}_{\tau,x} \left|A_V(\tau,\tau+\delta)\right|^n$$

$$\leq \left(\frac{n}{n-1}\right)^n n! M(|V|)(\tau,\tau+\delta)^{n-1} M(V)(\tau,\tau+\delta).$$

Next, dividing the previous inequality by $n!$, adding up the resulting inequalities, and using (3.164) and the equality

$$M_t = A_V(\tau,t) + N(V)(t,t+\delta,X_t)$$

we get

$$\mathbb{E}_{\tau,x} \exp\left\{\sup_{j:0\leq j\leq k} \left|M_{t_j}\right|\right\}$$

$$\leq 1 + \mathbb{E}_{\tau,x} \sup_{j:0\leq j\leq k} \left|M_{t_j}\right| + c\frac{M(|V|)(\tau,\tau+\delta)M(V)(\tau,\tau+\delta)}{1 - M(|V|)(\tau,\tau+\delta)}$$

$$\leq 1 + \mathbb{E}_{\tau,x} \sup_{j:0\leq j\leq k} \left|A_V(\tau,t_j)\right| + \mathbb{E}_{\tau,x} \sup_{j:0\leq j\leq k} \left|N(V)(t_j,\tau+\delta,X_{t_j})\right|$$

$$+ c\frac{M(|V|)(\tau,\tau+\delta)M(V)(\tau,\tau+\delta)}{1 - M(|V|)(\tau,\tau+\delta)}$$

$$\leq 1 + \left\{\mathbb{E}_{\tau,x} \sup_{j:0\leq j\leq k} \left|A_V(\tau,t_j)\right|^2\right\}^{\frac{1}{2}}$$

$$+ M(V)(\tau,\tau+\delta) + c\frac{M(|V|)(\tau,\tau+\delta)M(V)(\tau,\tau+\delta)}{1 - M(|V|)(\tau,\tau+\delta)}$$

$$\leq 1 + \{c_2 M(|V|)(\tau,\tau+\delta)M(V)(\tau,\tau+\delta)\}^{\frac{1}{2}}$$

$$+ M(V)(\tau,\tau+\delta) + c\frac{M(|V|)(\tau,\tau+\delta)M(V)(\tau,\tau+\delta)}{1 - M(|V|)(\tau,\tau+\delta)}. \tag{3.192}$$

We also have

$$\mathbb{E}_{\tau,x} \exp\left\{\sup_{j:0\leq j\leq k} \left|M_{t_j}\right|\right\}$$

$$\geq \mathbb{E}_{\tau,x} \exp\left\{\sup_{j:0\leq j\leq k} \left|A_V(\tau,t_j)\right| - \sup_{j:0\leq j\leq k} \left|N(V)(t_j,\tau+\delta,X_{t_j})\right|\right\}$$

$$\geq \exp\{-M(V)(\tau, \tau+\delta)\} \mathbb{E}_{\tau,x} \exp\left\{\sup_{j:0\leq j\leq k} |A_V(\tau, t_j)|\right\}. \tag{3.193}$$

It follows from (3.192) and (3.193) that

$$\mathbb{E}_{\tau,x} \exp\left\{\sup_{j:0\leq j\leq k} |A_V(\tau, t_j)|\right\}$$
$$\leq \exp\{M(V)(\tau, \tau+\delta)\} \left[1 + \{c_2 M(|V|)(\tau, \tau+\delta) M(V)(\tau, \tau+\delta)\}^{\frac{1}{2}}\right.$$
$$\left. + M(V)(\tau, \tau+\delta) + c\frac{M(|V|)(\tau, \tau+\delta) M(V)(\tau, \tau+\delta)}{1 - M(|V|)(\tau, \tau+\delta)}\right].$$

Therefore,

$$\mathbb{E}_{\tau,x} \exp\left\{\sup_{j:0\leq j\leq k} |A_V(\tau, t_j)|\right\}$$
$$\leq \exp\{M(V)(\tau, \tau+\delta)\} \left(1 + c\{M(|V|)(\tau, \tau+\delta) M(V)(\tau, \tau+\delta)\}^{\frac{1}{2}}\right.$$
$$\left. + c\frac{M(|V|)(\tau, \tau+\delta) M(V)(\tau, \tau+\delta)}{1 - M(|V|)(\tau, \tau+\delta)}\right). \tag{3.194}$$

Finally, consider a sequence of successive refinements of the partition $\tau = t_0 < t_1 < \cdots < t_k = \tau + \delta$ of the interval $[\tau, \tau+\delta]$ such that the maximum length of the partition intervals tends to 0. It follows from the monotone convergence theorem and the continuity of the functional $A_V(\tau, t)$ that one can pass to the limit in (3.194). This establishes estimate (3.191). The proof of part (b) of Lemma 3.30 is similar. Here we reason as in the proof of part (b) of Lemma 3.29. □

Let $\widetilde{P}$ be a backward transition probability function. Recall that for a Borel function V on $[0, T] \times E$ we denoted by $\widetilde{A}_V$ the additive functional associated with V (see formula (3.180)). Similarly, for a time-dependent measure μ we denoted by $\widetilde{A}_\mu$ the additive functional corresponding to μ (see formula (3.180)). It is not difficult to see that all the results for the functionals A_V and A_μ obtained in this section can be reformulated for the functionals $\widetilde{A}_V$ and $\widetilde{A}_\mu$.

3.11 Probabilistic Description of Non-Autonomous Kato Classes

A probabilistic characterization of non-autonomous Kato classes can be obtained from property (5) of the functionals A_V, A_μ, $\widetilde{A}_V$, and $\widetilde{A}_\mu$ in Definitions 3.4 and 3.5. For all $\tau \le t$ and $x \in E$, we have

$$\mathbb{E}_{\tau,x} A_V(\tau,t) = N(V)(\tau,t,x), \quad \widetilde{\mathbb{E}}^{t,x} \widetilde{A}_V(\tau,t) = \widetilde{N}(V)(\tau,t,x),$$

$$\mathbb{E}_{\tau,x} A_\mu(\tau,t) = N(\mu)(\tau,t,x), \text{ and } \widetilde{\mathbb{E}}^{t,x} \widetilde{A}_\mu(\tau,t) = \widetilde{N}(\mu)(\tau,t,x).$$

It follows that the following lemmas hold.

Lemma 3.31 *Let P be a transition probability function. Then*

$$V \in \widehat{\mathcal{P}}_f^* \iff \sup_{0 \le \tau \le t \le T} \sup_{x \in E} \mathbb{E}_{\tau,x} A_{|V|}(\tau,t) < \infty.$$

Let $V \in \widehat{\mathcal{P}}_f^*$. Then

$$V \in \mathcal{P}_f^* \iff \lim_{t-\tau \to 0+} \sup_{x \in E} \mathbb{E}_{\tau,x} A_{|V|}(\tau,t) = 0.$$

Suppose that the transition probability function P has a density p. Then

$$\mu \in \widehat{\mathcal{P}}_m^* \iff \sup_{0 \le \tau \le t \le T} \sup_{x \in E} \mathbb{E}_{\tau,x} A_{|\mu|}(\tau,t) < \infty.$$

Let $\mu \in \widehat{\mathcal{P}}_m^*$. Then

$$\mu \in \mathcal{P}_m^* \iff \lim_{t-\tau \to 0+} \sup_{x \in E} \mathbb{E}_{\tau,x} A_{|\mu|}(\tau,t) = 0.$$

Lemma 3.32 *Let $\widetilde{P}$ be a backward transition probability function. Then*

$$V \in \widehat{\mathcal{P}}_f \iff \sup_{0 \le \tau \le t \le T} \sup_{x \in E} \widetilde{\mathbb{E}}^{t,x} \widetilde{A}_{|V|}(\tau,t) < \infty.$$

Let $V \in \widehat{\mathcal{P}}_f$. Then

$$V \in \mathcal{P}_f \iff \lim_{t-\tau \to 0+} \sup_{x \in E} \widetilde{\mathbb{E}}^{t,x} \widetilde{A}_{|V|}(\tau,t) = 0.$$

Suppose that the backward transition probability function $\widetilde{P}$ has a density $\widetilde{p}$. Then

$$\mu \in \widehat{\mathcal{P}}_m \iff \sup_{0 \le \tau \le t \le T} \sup_{x \in E} \widetilde{\mathbb{E}}^{t,x} \widetilde{A}_{|\mu|}(\tau,t) < \infty.$$

Let $\mu \in \widehat{\mathcal{P}}_m$. Then

$$\mu \in \mathcal{P}_m \iff \lim_{t-\tau \to 0+} \sup_{x \in E} \widetilde{\mathbb{E}}^{t,x} \widetilde{A}_{|\mu|}(\tau, t) = 0.$$

3.12 Notes and Comments

(a) The reader, interested in additive and multiplicative functionals of time-homogeneous Markov processes, may consult [Blumenthal and Getoor (1968); Revuz and Yor (1991)]. The additive functional A_μ considered in Sections 3.1 and 3.9 was studied in [Gulisashvili (2004b); Gulisashvili (2004c)]. It is known that under certain restrictions on measures and processes there is a correspondence between additive functionals and measures (the Revuz correspondence, see [Revuz and Yor (1991); Beznea and Boboc (2000); Beznea and Boboc (2004)]). For more information on additive and multiplicative functionals associated with measures and the corresponding Schrödinger semigroups see [Blanchard and Ma (1990a); Blanchard and Ma (1990b); Albeverio, Blanchard, and Ma; Albeverio and Ma (1991); Glover, Rao, and Song (1993); Glover, Rao, Šikić, and Song (1994)].

(b) The exponential estimates for the additive functionals A_V and A_μ (see Section 3.10) were obtained in [Gulisashvili (2004b); Gulisashvili (2004c)]. The approximation in the potential sense discussed in Section 3.9 was defined in [Gulisashvili (2004b); Gulisashvili (2004c)].

(c) Non-autonomous Kato classes were studied in [Sturm (1994); Qi Zhang (1996); Qi Zhang (1997); Schnaubelt and Voigt (1999); Nagasawa (2000); Räbiger, Rhandi, Schnaubelt, and Voigt (2000); Gulisashvili (2002a); Gulisashvili (2002b); Gulisashvili (2004a); Gulisashvili (2004b); Gulisashvili (2004c); Gulisashvili (2005)]. Our presentation of the non-autonomous Kato classes follows [Gulisashvili (2004b); Gulisashvili (2004c)].

(d) For more information on fundamental solutions of parabolic initial and final value problems in the case of differential operators in non-divergence form see [Dressel (1940); Dressel (1946); Friedman (1964); Il'in, Kalashnikov, and Oleinik (1962); Ladyženskaja, Solonnikov, and Ural'ceva (1968); Porper and Eidel'man (1984); Eidel'man (1969); Eidel'man and Zhitarashu (1998)].

(e) For the results concerning weak fundamental solutions and the Gaussian estimates in the case of differential operators in divergence form see [Nash (1958); Aronson (1967); Aronson (1968); Fabes (1993); Fabes and Stroock (1986); Liskevich and Semenov (2000); Porper and Eidel'man (1984); Semenov (1999); Qi Zhang (1996); Qi Zhang (1997)].

(f) The relations between fundamental solutions and transition densities are discussed in [Porper and Eidel'man (1984)].

(g) There exists a rich literature on stochastic integrals and stochastic differential equations, for example, [Friedman (1975); Friedman (1976); Chung and Williams (1990); Ikeda and Watanabe (1989); Métivier and Pellaumail (1980); Métivier, M. and Viot, M. (1987); Lamberton and Lapeyre (1996); Øksendal (1998); Protter (2005)].

(h) More information on the Itô formula can be found in [Bhattacharya and Waymire (1990); Durrett (1996); Chung and Williams (1990); Ikeda and Watanabe (1989); Karatzas and Shreve (1991); Lamberton and Lapeyre (1996); Revuz and Yor (1991); Stroock (2003)].

(i) The martingale problem was introduced and studied by Stroock and Varadhan (see [Stroock and Varadhan (1979)]). Liggett used the martingale problem to study interacting particle systems (see [Liggett (2005)]).

(j) Diffusion processes are discussed in [Itô and McKean (1965); Dynkin (2002); Eberle (1999); Revuz and Yor (1991); Ikeda and Watanabe (1989); Nagasawa (1993); Rogers and Williams (2000a); Rogers and Williams (2000b); Stroock (1987); Stroock and Varadhan (1979); Carlen (1984)].

Chapter 4
Feynman-Kac Propagators

4.1 Schrödinger Semigroups with Kato Class Potentials

Feynman-Kac propagators are two-parameter analogues of Schrödinger semigroups. In this section we gather several known results concerning Schrödinger semigroups with Kato class potentials. Some of these results are formulated without proof. We refer the reader to [Simon (1982)] for more information on Schrödinger semigroups.

Let V be a Lebesgue measurable function on $\mathbb{R}^d$, and consider the following initial value problem for the perturbed heat equation:

$$\begin{cases} \dfrac{\partial u(t,x)}{\partial t} - \dfrac{1}{2}\Delta u(t,x) + V(x)u(t,x) = 0 \\ u(\tau, x) = g(x), \end{cases} \quad (4.1)$$

where $0 \leq \tau < t \leq T$ and $x \in \mathbb{R}^d$. In (4.1), g is a Lebesgue measurable function on $\mathbb{R}^d$, and the symbol Δ stands for the Laplace operator defined by $\Delta = \sum_{i=1}^{d} \dfrac{\partial^2}{\partial x_i^2}$. It is known that under certain restrictions on the potential function V, there exists the Schrödinger semigroup

$$S(t) = e^{-tH}, \quad t \geq 0,$$

where H is the Schrödinger operator given by $H = -\frac{1}{2}\Delta + V$. The Schrödinger semigroup is bounded on the space L^p with $1 \leq p \leq \infty$. Moreover, the function u defined by $u(t,x) = S(t-\tau)g(x)$ is a solution to initial value problem (4.1). It is also known that the Feynman-Kac formula holds for the Schrödinger semigroup, that is,

$$u(t,x) = e^{-(t-\tau)H}g(x) = \mathbb{E}_x g(B_{t-\tau}) \exp\left\{-\int_0^{t-\tau} V(B_s)\,ds\right\}, \quad (4.2)$$

where B_t is Brownian motion. One can also consider the following final value problem for the perturbed backward heat equation:

$$\begin{cases} \dfrac{\partial u(\tau,x)}{\partial \tau} + \dfrac{1}{2}\Delta u(\tau,x) - V(x)u(\tau,x) = 0 \\ u(t,x) = f(x) \end{cases} \quad (4.3)$$

where $0 \leq \tau < t \leq T$, $x \in \mathbb{R}^d$, and f is a Lebesgue measurable function on $\mathbb{R}^d$. The Feynman-Kac formula in this case is as follows:

$$u(\tau,x) = e^{-(t-\tau)H}f(x) = \mathbb{E}_x g(B_{t-\tau}) \exp\left\{-\int_0^{t-\tau} V(B_s)\,ds\right\}. \quad (4.4)$$

If $V = 0$ in (4.1) and (4.3), then we get the classical heat equation and the backward heat equation. In this case we have

$$u(t,x) = \mathbb{E}_x g(B_{t-\tau}) \quad (4.5)$$

and

$$u(\tau,x) = \mathbb{E}_x f(B_{t-\tau}). \quad (4.6)$$

Formulae (4.5) and (4.6) can be derived from the following two facts:

(1) The equality $\dfrac{\partial g_d}{\partial s} = \dfrac{1}{2}\Delta g_d$ holds for the Gaussian density g_d.
(2) The formula

$$\mathbb{P}_x[B_t \in A] = \int_A g_d(t,y)\,dy, \quad t > 0, \ A \in \mathcal{B}_{\mathbb{R}^d},$$

holds for one-dimensional distributions of Brownian motion.

Recall that we discussed Brownian motion in Subsection 1.12.1.

Definition 4.1 For $1 \leq p < \infty$, the local uniform L^p-space $L^p_{\text{loc},u}$ consists of all Lebesgue measurable functions V on $\mathbb{R}^d$ for which

$$\sup_{x \in \mathbb{R}} \int_{y:|x-y|<1} |V(y)|^p\,dy < \infty.$$

Definition 4.2 A locally integrable function V on the space $\mathbb{R}^d$ with $d \geq 3$ belongs to the Kato class K_d, provided that

$$\lim_{\alpha \downarrow 0} \sup_{x \in \mathbb{R}^d} \int_{y:|x-y|\leq \alpha} \frac{|V(y)|}{|x-y|^{d-2}} dy = 0.$$

If $d = 2$, then the Kato condition is

$$\lim_{\alpha \downarrow 0} \sup_{x \in \mathbb{R}^d} \int_{y:|x-y|\leq \alpha} |V(y)| \ln \frac{1}{|x-y|} dy = 0.$$

For $d = 1$, the Kato class coincides with the space $L^1_{\text{loc},u}$.

It is clear from Definition 4.2 that the Kato condition utilizes the convolution of the function $|V|$ with the truncated Riesz potential kernel. For $d \geq 3$, the Kato class K_d, equipped with the norm

$$\|V\|_{K_d} = \sup_{x \in \mathbb{R}^d} \int_{y:|x-y|\leq 1} \frac{|V(y)|}{|x-y|^{d-2}} dy,$$

is a Banach space (see [Voigt (1986)]). Similar result holds for the classes K_1 and K_2.

The next assertion provides several equivalent characterizations of the Kato class. They are expressed in terms of Brownian motion and the heat semigroup

$$e^{\frac{1}{2}t\Delta} f(x) = \int_{\mathbb{R}^d} f(y) g_d(t, x-y) dy = \mathbb{E}_x \left[f(B_t) \right]. \tag{4.7}$$

Theorem 4.1 *Let V be a locally integrable function on $\mathbb{R}^d$. Then the following are equivalent:*

(1) The function V belongs to the Kato class K_d.
(2) The equality

$$\lim_{t \downarrow 0} \sup_{x \in \mathbb{R}^d} \mathbb{E}_x \int_0^t |V(B_s)| \, ds = 0 \tag{4.8}$$

holds.
(3) The equality

$$\lim_{t \downarrow 0} \sup_{x \in \mathbb{R}^d} \int_0^t e^{\frac{1}{2}s\Delta} |V|(x) ds = 0 \tag{4.9}$$

holds.

Proof. The equivalence of conditions (2) and (3) in Theorem 4.1 follows from (4.7). Next, we will prove the equivalence of conditions (1) and (3) for $d \geq 3$. The cases where $d = 1$ and $d = 2$ are similar.

Let us compare the truncated Riesz potential kernel $\dfrac{\chi_{B(0,\sqrt{t})}(x)}{|x|^{d-2}}$ and the kernel

$$\int_0^t \frac{1}{(2\pi s)^{\frac{d}{2}}} \exp\left\{-\frac{|x|^2}{2s}\right\} ds = \frac{1}{2\pi^{\frac{d}{2}}} \frac{1}{|x|^{d-2}} \int_{|x|^2/(2t)}^{\infty} u^{\frac{d}{2}-2} e^{-u} du,$$

appearing in condition (3). We will show that the following estimates hold:

$$\frac{1}{2\pi^{\frac{d}{2}}} \frac{1}{|x|^{d-2}} \int_{|x|^2/(2t)}^{\infty} u^{\frac{d}{2}-2} e^{-u} du \leq \frac{c_d}{|x|^{d-2}}, \tag{4.10}$$

$$\frac{1}{2\pi^{\frac{d}{2}}} \frac{1}{|x|^{d-2}} \int_{|x|^2/(2t)}^{\infty} u^{\frac{d}{2}-2} e^{-u} du \geq \frac{c'_d \chi_{B(0,\sqrt{t})}(x)}{|x|^{d-2}}, \tag{4.11}$$

and

$$\frac{1}{2\pi^{\frac{d}{2}}} \frac{1}{|x|^{d-2}} \int_{|x|^2/(2t)}^{\infty} u^{\frac{d}{2}-2} e^{-u} du \leq c''_d e^{-\frac{1}{4t}|x|^2} \quad \text{for } |x| \geq \sqrt{t}. \tag{4.12}$$

Inequality (4.10) is obvious since $d > 2$. Inequality (4.11) is a consequence of the inequality

$$\int_{\frac{1}{2}}^{\infty} u^{\frac{d}{2}-2} e^{-u} du > 0.$$

Finally, inequality (4.12) can be obtained from the estimate

$$\int_a^{\infty} u^{\frac{d}{2}-2} e^{-u} du \leq e^{-\frac{1}{2}a} \int_a^{\infty} u^{\frac{d}{2}-2} e^{-\frac{1}{2}u} du, \quad a > 0.$$

It is not hard to see that the equivalence of conditions (1) and (3) in Theorem 4.1 can be derived from estimates (4.10)-(4.12).

This completes the proof of Theorem 4.1. □

Next, we will formulate several assertions concerning the Kato class and Schrödinger semigroups. The following inclusions hold for the Kato class K_d:

$$L^p_{\text{loc,u}} \subset K_d \subset L^1_{\text{loc,u}}, \quad p > \frac{d}{2}.$$

The Schrödinger semigroup e^{-tH} with $V \in K_d$ is bounded on the space L^p with $1 \leq p \leq \infty$ and strongly continuous on the space L^p with $1 \leq p < \infty$. It is (L^p-L^q)-smoothing, that is,

$$e^{-tH} : L^p \to L^q \text{ for all } 1 \leq p \leq q \leq \infty \text{ and } t > 0.$$

Moreover, e^{-tH} is a strongly continuous semigroup on the spaces C_0 and BUC.

The Kato class can be defined for more general homogeneous Markov processes (see Sections 3.1 and 3.2 in [Chung and Zhao (1995)]). Definition 4.3 below is based on the probabilistic characterization (4.8) of the Kato class K_d.

Definition 4.3 Let $P \in \mathcal{PM}$ be a time-homogeneous transition probability function, and let $(X_t, \mathcal{F}_t, \mathbb{P}_x)$ be a corresponding progressively measurable time-homogeneous Markov process on $(\Omega, \mathcal{F})$ with state space $(E, \mathcal{E})$. Then it is said that a Borel function V on E belongs to the Kato class J if

$$\lim_{t \downarrow 0} \sup_{x \in E} \mathbb{E}_x \int_0^t |V(X_s)| ds = 0.$$

Let X_t be a time-homogeneous Markov process. Then the semigroup associated with X_t is defined by

$$S(t)f(x) = \mathbb{E}_x f(X_t).$$

Chung and Zhao studied the semigroup $S(t)$ and its perturbations by functions from the Kato class J (see [Chung and Zhao (1995)]).

4.2 Feynman-Kac Propagators

Let $P \in \mathcal{PM}$ be a transition probability function, and let X_t be a corresponding progressively measurable Markov process. Recall that for a function $V \in \mathcal{P}_f^*$, the Kac functional M_V is defined as follows:

$$M_V(\tau, t) = \exp\left\{-\int_\tau^t V(s, X_s) ds\right\}$$

where $0 \leq \tau \leq t \leq T$. The progressive measurability of the process X_t implies the $\mathcal{F}_t^\tau$-measurability of the random variable $M_V(\tau, t)$. If P has a density p, and $\mu \in \mathcal{P}_m^*$ is a time-dependent measure, then the Kac functional

M_μ is given by

$$M_\mu(\tau,t) = \exp\{-A_\mu(\tau,t)\}, \qquad (4.13)$$

where A_μ is the additive functional associated with μ (see Theorem (3.14)).

Now we are finally ready to define backward Feynman-Kac propagators. Recall that the class $\mathcal{PM}$ of transition functions was introduced in Definition 3.20.

Definition 4.4

(a) Let $P \in \mathcal{PM}$ be a transition probability function, and let $V \in \mathcal{P}_f^*$. Then the family of operators given by

$$Y_V(\tau,t)g(x) = \mathbb{E}_{\tau,x} g(X_t)\exp\{-\int_\tau^t V(s,X_s)\,ds\}, \ 0 \le \tau \le t \le T,$$

is called the backward Feynman-Kac propagator associated with the transition probability function P and the function V.

(b) Let $P \in \mathcal{PM}$ be a transition probability function possessing a density p, and let $\mu \in \mathcal{P}_m^*$. Then the family of operators defined by

$$Y_\mu(\tau,t)g(x) = \mathbb{E}_{\tau,x} g(X_t)\exp\{-A_\mu(\tau,t)\}, \ 0 \le \tau \le t \le T,$$

is called the backward Feynman-Kac propagator associated with the transition probability function P and the time-dependent measure μ.

Our next goal is to introduce forward Feynman-Kac propagators. Here we need the definitions from Sections 1.4 and 3.3. For a backward transition probability function $\widetilde{P} \in \widetilde{\mathcal{PM}}$, a progressively measurable backward Markov process $\widetilde{X}^t$ associated with $\widetilde{P}$, and a function $V \in \mathcal{P}_f$, the functional $\widetilde{A}_V$ is defined by

$$\widetilde{A}_V(\tau,t) = \int_\tau^t V\left(s, \widetilde{X}^s\right) ds$$

$\widetilde{\mathbb{P}}^{t,y}$-a.s. In the case of a time-dependent measure $\mu \in \mathcal{P}_m$, the functional $\widetilde{A}_\mu$ is given by

$$\widetilde{A}_\mu(\tau,t) = A_{\widetilde{\mu}}(T-t, T-\tau).$$

Here $\widetilde{\mu}$ is the time-reversal of the time-dependent measure μ, defined by $\widetilde{\mu}(t) = \mu(T-t)$ (see Section 3.9).

Definition 4.5

(a) Let $\widetilde{P} \in \widetilde{\mathcal{PM}}$ be a backward transition probability function, and let $V \in \mathcal{P}_f$. Then the family of operators given by

$$U_V(t,\tau)g(x) = \widetilde{\mathbb{E}}^{t,x} g\left(\widetilde{X}_t\right) \exp\{-\int_\tau^t V\left(s, \widetilde{X}_s\right) ds\}, \quad 0 \leq \tau \leq t \leq T,$$

is called the Feynman-Kac propagator associated with the backward transition probability function $\widetilde{P}$ and the function V.

(b) Let $\widetilde{P} \in \widetilde{\mathcal{PM}}$ be a backward transition probability function possessing a density $\widetilde{p}$, and let $\mu \in \mathcal{P}_m$. Then the family of operators defined by

$$U_\mu(\tau,t)g(x) = \widetilde{\mathbb{E}}^{t,x} g\left(\widetilde{X}_\tau\right) \exp\left\{-\widetilde{A}_\mu(t,\tau)\right\}, \quad 0 \leq \tau \leq t \leq T,$$

is called the Feynman-Kac propagator associated with the backward transition probability function $\widetilde{P}$ and the time-dependent measure μ.

4.3 The Behavior of Feynman-Kac Propagators in L^p-Spaces

The backward Feynman-Kac propagators Y_V and Y_μ with $V \in \mathcal{P}_f^*$ and $\mu \in \mathcal{P}_m^*$ inherit various properties of the free backward propagator Y. In this section and in Section 4.4, we study the relations between Y, Y_V, and Y_μ. We follow [Gulisashvili (2004b); Gulisashvili (2004c)] in our discussion of the inheritance problem.

The first result in the present section concerns the L^∞-boundedness of backward propagators.

Theorem 4.2 *Let $P \in \mathcal{PM}$. Then the following assertions hold:*

(a) *For any $V \in \mathcal{P}_f^*$, Y_V is a backward propagator on $L_\mathcal{E}^\infty$.*

(b) *If P has a density p, then for every $V \in \mathcal{P}_f^*$, Y_V is a backward propagator on L^∞.*

(c) *If P has a density p, then for every $\mu \in \mathcal{P}_m^*$, Y_μ is a backward propagator on L^∞.*

Proof. Let $g \in L_\mathcal{E}^\infty$. Then it is not hard to prove that the function $Y_V(\tau,t)g$ is Borel measurable. By Lemma 3.27,

$$\sup_{x \in E} |Y_V(\tau,t)g(x)| \leq \|g\|_\infty \sup_{x \in E} \mathbb{E}_{\tau,x} \exp\left\{A_{|V|}(\tau,t)\right\} \leq \|g\|_\infty \frac{1}{1 - M(|V|)(\tau,t)}$$

for all τ, t with $M(|V|)(\tau - t) < 1$. Therefore, $Y_V(\tau, t) \in L(L_{\mathcal{E}}^{\infty}, L_{\mathcal{E}}^{\infty})$ for all τ and t such that the difference $t - \tau$ is small. By the Markov property, we see that Y_V is a backward propagator on $L_{\mathcal{E}}^{\infty}$. This establishes part (a) of Theorem 4.2. The proof of parts (b) and (c) is similar. □

Remark 4.1 Under the conditions in part (b) of Theorem 4.2, the following estimate holds:

$$\|Y_V(\tau, t)\|_{\infty \to \infty} \leq \exp\left\{\alpha\left(\frac{t-\tau}{\delta} + 1\right)\right\},$$

where $\delta > 0$ is any number such that

$$\rho(\delta) = \sup\{M(|V|)(\eta, \lambda) : \lambda - \eta < \delta\} < 1$$

and $\alpha = \ln \dfrac{1}{1 - \rho(\delta)}$. Similar estimates hold under the conditions in parts (a) and (c) of Theorem 4.2.

The next result concerns approximations in the potential sense and the uniform convergence of Feynman-Kac propagators.

Theorem 4.3 Let $P \in \mathcal{PM}$ and $V \in \mathcal{P}_f^*$. Suppose that a sequence $V_k \in \mathcal{P}_f^*$, $k \geq 1$, approaches V in the potential sense. Then

$$\lim_{k \to \infty} \sup_{(\tau, t): 0 \leq \tau \leq t \leq T} \|Y_V(\tau, t) - Y_{V_k}(\tau, t)\|_{L_{\mathcal{E}}^{\infty} \to L_{\mathcal{E}}^{\infty}} = 0.$$

Suppose that $P \in \mathcal{PM}$ has a density p and let $\mu \in \mathcal{P}_m^*$. Suppose also that a sequence $\mu_k \in \mathcal{P}_m^*$, $k \geq 1$, approaches μ in the potential sense. Then

$$\lim_{k \to \infty} \sup_{(\tau, t): 0 \leq \tau \leq t \leq T} \|Y_\mu(\tau, t) - Y_{\mu_k}(\tau, t)\|_{\infty \to \infty} = 0.$$

Proof. We will prove only the second part of Theorem 4.3. The proof of the first part is similar.

Let μ and μ_k be such as in the formulation of Theorem 4.3, and let $f \in L^{\infty}$. Then, by part (b) of Lemma 3.29 with $q = 1$ and $r = 2$, there exists $\delta_0 > 0$ such that

$$|Y_\mu(\tau, t)f(x) - Y_{\mu_k}(\tau, t)f(x)|$$
$$\leq \alpha \|f\|_\infty \left(\{M(\mu - \mu_k)(\tau, t)\}^{\frac{1}{2}} + M(\mu - \mu_k)(\tau, t)\right) \quad (4.14)$$

for all τ and t with $t - \tau < \delta_0$ and all $x \in E$. In (4.14), the constant α does not depend on x, t, τ, and k. It follows from (4.14) that

$$\lim_{k \to \infty} \sup_{(\tau,t):0 \leq t-\tau<\delta_0} \|Y_\mu(\tau,t) - Y_{\mu_k}(\tau,t)\|_{L^\infty \to L^\infty_\mathcal{E}} = 0. \tag{4.15}$$

Next we will get rid of the restriction $t - \tau < \delta_0$ in (4.15). Consider a partition $0 = t_0 < t_1 < t_2 < \cdots < t_n = T$ of the interval $[0,T]$ such that $t_{j+1} - t_j < \delta_0$ for all j with $0 \leq j \leq n-1$. Then the estimate in (4.14) holds, provided that t and τ belong to the same interval $[t_j, t_{j+1}]$, $0 \leq j \leq n-1$. Next, we will show how to complete the proof of the second part of Theorem 4.3, using the previous assertion and the properties of backward propagators. We will consider only a special case, where $t_j \leq \tau \leq t_{j+1} \leq t \leq t_{j+2}$ with $0 \leq j \leq n-2$, and leave the rest as an exercise for the reader. By the uniform boundedness of the propagators Y_{μ_k}, $k \geq 1$, on the space L^∞ (this follows from Remark 4.1), we have

$$\|Y_\mu(\tau,t) - Y_{\mu_k}(\tau,t)\|_{\infty \to \infty}$$
$$\leq \|Y_\mu(\tau,t_{j+1}) Y_\mu(t_{j+1},t) - Y_{\mu_k}(\tau,t_{j+1}) Y_{\mu_k}(t_{j+1},t)\|_{\infty \to \infty}$$
$$\leq \|Y_\mu(\tau,t_{j+1}) Y_\mu(t_{j+1},t) - Y_\mu(\tau,t_{j+1}) Y_{\mu_k}(t_{j+1},t)\|_{\infty \to \infty}$$
$$+ \|Y_\mu(\tau,t_{j+1}) Y_{\mu_k}(t_{j+1},t) - Y_{\mu_k}(\tau,t_{j+1}) Y_{\mu_k}(t_{j+1},t)\|_{\infty \to \infty}$$
$$\leq \alpha \|Y_\mu(t_{j+1},t) - Y_{\mu_k}(t_{j+1},t)\|_{\infty \to \infty}$$
$$+ \alpha \|Y_\mu(\tau,t_{j+1}) - Y_{\mu_k}(\tau,t_{j+1})\|_{\infty \to \infty}.$$

Since $t - t_{j+1} < \delta_0$ and $t_{j+1} - \tau < \delta_0$, we can apply the special case of (4.15) that has already been established. It follows that equality (4.15) holds for all $t_j \leq \tau \leq t_{j+1} \leq t \leq t_{j+2}$ with $0 \leq j \leq n-2$.

This completes the proof of Theorem 4.3. □

Corollary 4.1 *Let $P \in \mathcal{PM}$, $V \in \mathcal{P}_f^*$, and suppose that V_k is defined by (3.149). Then*

$$\lim_{k \to \infty} \sup_{(\tau,t):0 \leq \tau \leq t \leq T} \|Y_V(\tau,t) - Y_{V_k}(\tau,t)\|_{L^\infty_\mathcal{E} \to L^\infty_\mathcal{E}} = 0.$$

Let $P \in \mathcal{PM}$, $\mu \in \mathcal{P}_m^*$, and suppose that V_k is defined by (3.153). Suppose also that P has a density p. Then

$$\lim_{k \to \infty} \sup_{(\tau,t):0 \leq \tau \leq t \leq T} \|Y_\mu(\tau,t) - Y_{V_k}(\tau,t)\|_{\infty \to \infty} = 0.$$

Since the sequence V_k defined by (3.149) approaches V in the potential sence and, similarly, the sequence V_k defined by (3.153) approaches μ in the potential sense (see Lemmas 3.20 and 3.21), Corollary 4.1 follows from Theorem 4.3.

The next lemma will be important in the sequel.

Lemma 4.1 *Let $P \in \mathcal{PM}$. Then the following assertions hold:*

(a) For any $V \in \mathcal{P}_f^*$,

$$\lim_{t-\tau \downarrow 0} \|Y_V(\tau,t) - Y(\tau,t)\|_{L_{\mathcal{E}}^\infty \to L_{\mathcal{E}}^\infty} = 0. \quad (4.16)$$

(b) If P has a density p, then for every $V \in \mathcal{P}_f^*$,

$$\lim_{t-\tau \downarrow 0} \|Y_V(\tau,t) - Y(\tau,t)\|_{\infty \to \infty} = 0. \quad (4.17)$$

(c) If P has a density p, then for every $\mu \in \mathcal{P}_m^*$,

$$\lim_{t-\tau \downarrow 0} \|Y_\mu(\tau,t) - Y(\tau,t)\|_{\infty \to \infty} = 0. \quad (4.18)$$

Proof. It follows from part (a) of Lemma 3.27 and from the definition of the class $\mathcal{P}_f^*$ that

$$\limsup_{t-\tau \downarrow 0} \|Y_V(\tau,t) - Y(\tau,t)\|_{\infty \to \infty} \le \limsup_{t-\tau \downarrow 0} \sup_{x \in E} \left(\mathbb{E}_{\tau,x} \exp\left\{A_{|V|}(\tau,t)\right\} - 1 \right)$$

$$\le \limsup_{t-\tau \downarrow 0} \sup_{x \in E} \frac{M(|V|)(\tau,t)}{1 - M(|V|)(\tau,t)} = 0.$$

This implies equality (4.17). The proof of (4.16) and (4.18) is similar. In the proof of (4.18), we use part (b) of Lemma 3.27. □

The next result provides sufficient conditions for the boundedness of backward Feynman-Kac propagators on the space L^s.

Theorem 4.4 *Suppose that $1 < s < \infty$, $1 \le r < s$, and $P \in \mathcal{PM}$. Then the following assertions hold:*

(a) Let $V \in \mathcal{P}_f^*$, and suppose that the free backward propagator Y satisfies the condition $Y(\tau,t) \in L(L_{\mathcal{E}}^r, L_{\mathcal{E}}^r)$ for all $0 \le \tau \le t \le T$. Then Y_V is a backward propagator on $L_{\mathcal{E}}^s$. If, in addition, Y is uniformly bounded on $L_{\mathcal{E}}^r$ and strongly continuous on $L_{\mathcal{E}}^s$, then Y_V is a strongly continuous backward propagator on $L_{\mathcal{E}}^s$.

(b) If P has a density p, and if $Y(\tau,t) \in L(L^r, L^r)$ for all $0 \le \tau \le t \le T$, then Y_V is a backward propagator on L^s. If, in addition, Y is uniformly bounded on L^r and strongly continuous on L^s, then Y_V is a strongly continuous backward propagator on L^s.

(c) Suppose that P has a density p, and let $\mu \in \mathcal{P}_m^*$. If $Y(\tau,t) \in L(L^r, L^r)$ for all $0 \leq \tau < t \leq T$, then Y_μ is a backward propagator on L^s. If, in addition, Y is uniformly bounded on L^r and strongly continuous on L^s, then Y_μ is a strongly continuous backward propagator on L^s.

Remark 4.2 We do not know whether Theorem 4.4 holds for $r = s$. In the case of the heat semigroup, Theorem 4.4 may fail for $s = 1$. This was established in [Gulisashvili (2005)].

Proof. We will prove only part (b) of Theorem 4.4. Parts (a) and (c) are similar. Assume that the conditions in part (b) hold, and let $g \in L^s$. By Hölder's inequality,

$$|Y_V(\tau,t)g(x)| \leq \left\{\mathbb{E}_{\tau,x} |g(X_t)|^{\frac{s}{r}}\right\}^{\frac{r}{s}} \left\{\mathbb{E}_{\tau,x} \exp\left\{\frac{s}{s-r} A_{|V|}(\tau,t)\right\}\right\}^{\frac{s-r}{s}}$$

$$= \left\{Y(\tau,t)|g|^{\frac{s}{r}}(x)\right\}^{\frac{r}{s}} \left\{Y_{\frac{s}{s-r}|V|}(\tau,t)1(x)\right\}^{\frac{s-r}{s}}$$

$$\leq \left\|Y_{\frac{s}{s-r}|V|}(\tau,t)\right\|_{\infty \to \infty}^{\frac{s-r}{s}} \left\{Y(\tau,t)|g|^{\frac{s}{r}}(x)\right\}^{\frac{r}{s}} \quad (4.19)$$

for all t and τ with $0 \leq \tau \leq t \leq T$. It follows from Remark 4.1 that

$$\left\|Y_{\frac{s}{s-r}|V|}\tau,t)\right\|_{\infty \to \infty} \leq c \quad (4.20)$$

where $c \geq 1$ depends on s, r, and V. Next (4.19) and (4.20) give

$$|Y_V(\tau,t)g(x)| \leq c\left\{Y(\tau,t)|g|^{\frac{s}{r}}(x)\right\}^{\frac{r}{s}}. \quad (4.21)$$

Now we see that (4.21) implies

$$\|Y_V(\tau,t)\|_{s \to s} \leq c \|Y(\tau,t)\|_{r \to r}^{\frac{r}{s}}. \quad (4.22)$$

Therefore, Y_V is a propagator on L^s.

Remark 4.3 Inequality (4.22) provides an estimate for the norm of the backward Feynman-Kac propagator on the space L^s.

Let us return to the proof of part (b) of Theorem 4.4. The next result will be needed in the proof.

Theorem 4.5 Let $1 < s < \infty$, $1 \leq r < s$, and $P \in \mathcal{PM}$. Then the following conditions are satisfied:

(a) Let $V \in \mathcal{P}_f^*$, and suppose that the free backward propagator Y is uniformly bounded on $L_{\mathcal{E}}^r$. Then

$$\lim_{t-\tau\downarrow 0}\|Y_V(\tau,t)-Y(\tau,t)\|_{L_{\mathcal{E}}^s\to L_{\mathcal{E}}^s}=0. \qquad (4.23)$$

(b) Suppose that P has a density p, and Y is uniformly bounded on L^r. Then for any $V \in \mathcal{P}_f^*$,

$$\lim_{t-\tau\downarrow 0}\|Y_V(\tau,t)-Y(\tau,t)\|_{s\to s}=0. \qquad (4.24)$$

(c) Suppose that P has a density p, and Y is uniformly bounded on L^r. Then for any $\mu \in \mathcal{P}_m^*$,

$$\lim_{t-\tau\downarrow 0}\|Y_\mu(\tau,t)-Y(\tau,t)\|_{s\to s}=0. \qquad (4.25)$$

Proof. We begin with the proof of part (b) of Theorem 4.5. It follows from part (b) of Theorem 4.4 that Y_V is a backward propagator on the space L^s. Let $g \in L^s$. By Hölder's inequality and inequality (3.188), we obtain

$$|Y_V(\tau,t)g(x) - Y(\tau,t)g(x)|$$

$$\leq \left\{\mathbb{E}_{\tau,x}|g(X_t)|^{\frac{s}{r}}\right\}^{\frac{r}{s}}\left\{\mathbb{E}_{\tau,x}|\exp\{A_V(\tau,t)\}-1|^{\frac{s}{s-r}}\right\}^{\frac{s-r}{s}}$$

$$\leq \left\{Y(\tau,t)|g|^{\frac{s}{r}}(x)\right\}^{\frac{r}{s}}\left\{\mathbb{E}_{\tau,x}\exp\left\{\frac{s}{s-r}A_{|V|}(\tau,t)\right\}-1\right\}^{\frac{s-r}{s}}. \qquad (4.26)$$

It follows from (4.26), part (a) of Lemma 3.27, the definition of the class $\mathcal{P}_f^*$, and the uniform boundedness of Y on L^r that

$$\limsup_{t-\tau\downarrow 0}\|Y_V(\tau,t)-Y(\tau,t)\|_{s\to s}$$

$$\leq a(s,r,V)\limsup_{t-\tau\downarrow 0}\sup_{x\in E}\left\{\mathbb{E}_{\tau,x}\exp\left\{\frac{s}{s-r}A_{|V|}(\tau,t)\right\}-1\right\}^{\frac{s-r}{s}}$$

$$\leq c(s,r,V)\limsup_{t-\tau\downarrow 0}\left[\frac{\frac{s}{s-r}M(|V|)(\tau,t)}{1-\frac{s}{s-r}M(|V|)(\tau,t)}\right]^{\frac{s-r}{s}}=0.$$

This implies equality (4.24). Equalities (4.23) and (4.25) can be obtained similarly. Part (b) of Lemma 3.27 is used in the proof of equality (4.25) instead of part (a) of Lemma 3.27. □

Let us continue the proof of part (b) of Theorem 4.4. Suppose that the free backward propagator Y is locally uniformly bounded on L^r and strongly continuous on L^s. We have already shown above that Y_V is a backward propagator on L^s. Moreover, Y_s is uniformly bounded (see estimate (4.22)). Therefore, in order to prove the strong continuity of Y_V, it suffices to establish the separate strong continuity (see Theorem 2.1).

Let $0 \leq \tau \leq t \leq T$, and suppose that $t' \geq t$ and $g \in L^s$. Then

$$\|Y_V(\tau, t') g - Y_V(\tau, t)g\|_s$$
$$= \|Y_V(\tau, t)(Y_V(t, t') g - g)\|_s \leq M \|Y_V(t, t') g - g\|_s$$
$$\leq M \|g\|_s \|Y_V(t, t') - Y(t, t')\|_s + M \|Y(t, t') g - g\|_s.$$

It follows from Theorem 4.5 and from the strong continuity of Y that

$$\lim_{t' \downarrow t} \|Y_V(\tau, t') g - Y_V(\tau, t)g\|_s = 0. \tag{4.27}$$

Similarly, we get

$$\lim_{t' \uparrow t} \|Y_V(\tau, t') g - Y_V(\tau, t)g\|_s = 0. \tag{4.28}$$

Suppose that $\tau' \leq \tau$. Then

$$\|Y_V(\tau', t) g - Y_V(\tau, t)g\|_s = \|(Y_V(\tau, \tau') - I) Y_V(\tau, t)g\|_s$$
$$\leq \|Y_V(\tau', \tau) - Y(\tau', \tau)\|_{s \to s} \|Y_V(\tau, t)g\|_s$$
$$+ \|Y(\tau', \tau) Y_V(\tau, t)g - Y_V(\tau, t)g\|_s.$$

It follows from (4.22), Theorem 4.5, and from the strong continuity of Y that

$$\lim_{\tau' \uparrow \tau} \|Y_V(\tau', t) g - Y_V(\tau, t)g\|_s = 0. \tag{4.29}$$

Finally, let $\tau < \tau' < t$, and let λ be such that $\tau' < \lambda < t$. Then

$$\|Y_V(\tau', t) g - Y_V(\tau, t)g\|_s = \|(Y_V(\tau', \lambda) - Y_V(\tau, \lambda)) Y_V(\lambda, t)g\|_s$$
$$\leq \|(Y(\tau', \lambda) - Y(\tau, \lambda)) Y_V(\lambda, t)g\|_s$$
$$+ \|Y_V(\tau', \lambda) - Y(\tau', \lambda)\|_{s \to s} \|Y_V(\lambda, t)g\|_s$$
$$+ \|Y_V(\tau, \lambda) - Y(\tau, \lambda)\|_{s \to s} \|Y_V(\lambda, t)g\|_s$$
$$\leq \|(Y(\tau', \lambda) - Y(\tau, \lambda)) Y_V(\lambda, t)g\|_s$$
$$+ M \|Y_V(\tau', \lambda) - Y(\tau', \lambda)\|_{s \to s} \|g\|_s$$
$$+ M \|Y_V(\tau, \lambda) - Y(\tau, \lambda)\|_{s \to s} \|g\|_s$$

$$= I_1 + I_2 + I_3. \tag{4.30}$$

For every $\epsilon > 0$, fix λ such that $\tau < \lambda < t$ and $I_2 + I_3 \leq \frac{1}{2}\epsilon$ for all τ' with $\tau < \tau' < \lambda$. This can be done using Theorem 4.5. Then the strong continuity of Y implies the existence of $\delta > 0$ such that $I_1 \leq \frac{1}{2}\epsilon$ for all τ' with $\tau < \tau' \leq \tau + \delta < \lambda$. Hence, (4.30) gives

$$\lim_{\tau' \downarrow \tau} \|Y_V(\tau', t)g - Y_V(\tau, t)g\|_s = 0, \tag{4.31}$$

and it follows from (4.27), (4.28), (4.29), and (4.31) that Y_V is separately strongly continuous.

This completes the proof of Theorem 4.4. □

The next result is an (L^s-L^q)-smoothing theorem for backward Feynman-Kac propagators.

Theorem 4.6 *Let $1 < s < q \leq \infty$, $1 \leq r < s$, and $P \in \mathcal{PM}$. Then the following conditions hold:*

(a) *Let $V \in \mathcal{P}_f^*$, and suppose that $Y(\tau, t) \in L\left(L_{\mathcal{E}}^r, L_{\mathcal{E}}^{\frac{rq}{s}}\right)$ for all $0 \leq \tau < t \leq T$. Then $Y_V(\tau, t) \in L(L_{\mathcal{E}}^s, L_{\mathcal{E}}^q)$ for all $0 \leq \tau < t \leq T$.*
(b) *If P has a density p, $V \in \mathcal{P}_f^*$, and $Y(\tau, t) \in L\left(L^r, L^{\frac{rq}{s}}\right)$ for all $0 \leq \tau < t \leq T$, then $Y_V(\tau, t) \in L(L^s, L^q)$.*
(c) *If P has a density p, $\mu \in \mathcal{P}_m^*$, and $Y(\tau, t) \in L\left(L^r, L^{\frac{rq}{s}}\right)$ for all $0 \leq \tau < t \leq T$, then $Y_\mu(\tau, t) \in L(L^s, L^q)$.*

Proof. We will prove part (b) of Theorem 4.6 in the case $q \neq \infty$. The proof in the case $q = \infty$ is similar. Let $g \in L^s$. By estimate (4.21),

$$\|Y_V(\tau, t)g\|_q \leq c \left\{ \int_E \left\{ Y(\tau, t)|g|^{\frac{s}{r}}(x) \right\}^{\frac{rq}{s}} dx \right\}^{\frac{1}{q}},$$

and hence the assumptions in Theorem 4.6 imply that

$$\|Y_V(\tau, t)g\|_q \leq c |Y(\tau, t)|_{r \to \frac{rq}{s}}^{\frac{r}{s}} \|g\|_s. \tag{4.32}$$

Now it is clear that part (b) of Theorem 4.6 follows from (4.32). The proof of parts (a) and (c) is similar. □

4.4 Feller, Feller-Dynkin, and BUC-Property of Feynman-Kac Propagators

In this section we study the behavior of backward Feynman-Kac propagators on the spaces of continuous functions. Recall that a backward BC-propagator is called a backward Feller propagator. A backward C_0-propagator is called a backward Feller-Dynkin propagator. If a backward $L_{\mathcal{E}}^{\infty}$-propagator Q is such that

$$Q(\tau, t) \in L(L_{\mathcal{E}}^{\infty}, BC), \ 0 \leq \tau < t \leq T,$$

then it is said that Q satisfies the strong Feller condition. If a backward $L_{\mathcal{E}}^{\infty}$-propagator Q is such that

$$Q(\tau, t) \in L(L_{\mathcal{E}}^{\infty}, BUC), \ 0 \leq \tau < t \leq T,$$

then it is said that Q satisfies the strong BUC-condition.

Theorem 4.7 *Let $P \in \mathcal{PM}$ and $V \in \mathcal{P}_f^*$. Then the following assertions hold:*

(a) If Y satisfies the strong Feller condition, then Y_V also satisfies the same condition.

(b) If Y satisfies the strong BUC-condition, then Y_V also satisfies the same condition.

Proof. We will prove part (a) of Theorem 4.7. The proof of Part (b) is similar. The next lemma will be used in the proof of part (a).

Lemma 4.2 *The following assertion holds for $P \in \mathcal{PM}$:*

(a) Let $V \in \mathcal{P}_f^*$. Then for all $x, x' \in E$, $0 \leq \tau < t \leq T$, $g \in L_{\mathcal{E}}^{\infty}$, and $\lambda > 0$ with $\tau + \lambda < t$,

$$|Y_V(\tau,t)g(x') - Y_V(\tau,t)g(x)|$$
$$\leq 2 \left\| Y_V(\tau, \tau+\lambda) - Y(\tau, \tau+\lambda) \right\|_{\infty \to \infty} \left\| Y_V(\tau+\lambda, t)g \right\|_{\infty}$$
$$+ |Y(\tau, \tau+\lambda) Y_V(\tau+\lambda, t) g(x') - Y(\tau, \tau+\lambda) Y_V(\tau+\lambda, t) g(x)|.$$
(4.33)

(b) Suppose that P has a density p. Then inequality (4.33) holds for all $g \in L^{\infty}$.

(c) Suppose that P has a density p. If $\mu \in \mathcal{P}_m^*$, then for all $x \in E$, $x' \in E$, $0 \le \tau < t \le T$, $g \in L^\infty$, and $\lambda > 0$ with $\tau + \lambda < t$,

$$|Y_\mu(\tau,t)g(x') - Y_\mu(\tau,t)g(x)|$$
$$\le 2\, \|Y_\mu(\tau,\tau+\lambda) - Y(\tau,\tau+\lambda)\|_{\infty\to\infty} \|Y_\mu(\tau+\lambda,t)g\|_\infty$$
$$+ |Y(\tau,\tau+\lambda)Y_\mu(\tau+\lambda,t)g(x') - Y(\tau,\tau+\lambda)Y_\mu(\tau+\lambda,t)g(x)|\,.$$

Proof. We will prove only part (a) of Lemma 4.2. The proof of parts (b) and (c) is similar. We have

$$|Y_V(\tau,t)g(x') - Y_V(\tau,t)g(x)|$$
$$= |Y_V(\tau,\tau+\lambda)Y_V(\tau+\lambda,t)g(x') - Y_V(\tau,\tau+\lambda)Y_V(\tau+\lambda,t)g(x)|$$
$$\le |Y_V(\tau,\tau+\lambda)Y_V(\tau+\lambda,t)g(x') - Y(\tau,\tau+\lambda)Y_V(\tau+\lambda,t)g(x')|$$
$$+ |Y(\tau,\tau+\lambda)Y_V(\tau+\lambda,t)g(x') - Y(\tau,\tau+\lambda)Y_V(\tau+\lambda,t)g(x)|$$
$$+ |Y_V(\tau,\tau+\lambda)Y_V(\tau+\lambda,t)g(x) - Y(\tau,\tau+\lambda)Y_V(\tau+\lambda,t)g(x)|$$
$$\le 2\,\|Y_V(\tau,\tau+\lambda) - Y(\tau,\tau+\lambda)\|_{\infty\to\infty} \|Y_V(\tau+\lambda,t)g\|_\infty$$
$$+ |Y(\tau,\tau+\lambda)Y_V(\tau+\lambda,t)g(x') - Y(\tau,\tau+\lambda)Y_V(\tau+\lambda,t)g(x)|\,.$$

This completes the proof of Lemma 4.2. □

Let us go back to the proof of part (a) of Theorem 4.7. Suppose that Y satisfies the strong Feller condition, and let $g \in L_{\mathcal{E}}^\infty$. Since Y_V is a uniformly bounded backward $L_{\mathcal{E}}^\infty$-propagator (see Remark 4.1), we have

$$\|Y_V(\tau,t)\|_{\infty\to\infty} < M \tag{4.34}$$

for all $0 \le \tau \le t \le T$. It follows from (4.34) and Theorem 4.5 that for every $\epsilon > 0$ there exists $\lambda > 0$ such that $\tau + \lambda < t$ and

$$2\,\|Y_V(\tau,\tau+\lambda) - Y(\tau,\tau+\lambda)\|_{\infty\to\infty}\|Y_V(\tau+\lambda,t)g\|_\infty < \frac{\epsilon}{2}. \tag{4.35}$$

Moreover, for a number λ such as above and any fixed $x \in E$, there exists $\delta > 0$ such that

$$|Y(\tau,\tau+\lambda)Y_V(\tau+\lambda,t)g(x') - Y(\tau,\tau+\lambda)Y_V(\tau+\lambda,t)g(x)| < \frac{\epsilon}{2} \tag{4.36}$$

for all x' with $\rho(x',x) < \delta$. This follows from (4.34) and from the assumption that Y is a backward strong Feller propagator. Now it is easy to see that part (a) of Theorem 4.7 follows from (4.35), (4.36), and Lemma 4.2.

This completes the proof of Theorem 4.7. □

The next three theorems are the inheritance results in the case of the Feller property, the Feller-Dynkin property, and the BUC-property. Since it is not known whether these properties are inherited by Y_V and Y_μ from Y, we impose an additional restriction on the free backward propagator Y.

Theorem 4.8 *Let $P \in \mathcal{PM}$, $V \in \mathcal{P}_f^*$, and suppose that Y satisfies the strong Feller condition. Then the following assertions hold:*

(a) If Y is a backward Feller-Dynkin propagator, then the same is true for Y_V.

(b) If Y is a strongly continuous backward Feller-Dynkin propagator, then the same is true for Y_V.

Proof. (a) Let $g \in C_0$. Then, by part (a) of Theorem 4.7,

$$Y_V(\tau, t) g \in BC \text{ for all } 0 \leq \tau \leq t \leq T.$$

For every $\epsilon > 0$, there exists a compact set K_ϵ such that $|g(x)| < \epsilon$ for all $x \in E \setminus K_\epsilon$. Moreover, Urysohn's Lemma implies that there exists a continuous function g_ϵ on E with compact support such that $g_\epsilon(x) = 1$ for all $x \in K_\epsilon$. It follows that

$$|Y_V(\tau, t) g(x)| \leq |Y_V(\tau, t) g g_\epsilon(x)| + |Y_V(\tau, t) g (1 - g_\epsilon)(x)|$$

$$\leq c_{\tau, t} \left[\left\{ Y(\tau, t) |g g_\epsilon|^2 (x) \right\}^{\frac{1}{2}} + \epsilon \right].$$

Now it is clear that part (a) of Theorem 4.8 holds.

(b) Let Y be a strongly continuous Feller-Dynkin propagator. It follows from part (a) of Theorem 4.8 that Y_V is a Feller-Dynkin propagator. By adapting the proof of the strong continuity of Y_V on the space L^s in Theorem 4.4 to the case of the space C_0, we see that Y_V is strongly continuous on C_0. □

If the free backward propagator Y satisfies the strong BUC-condition, then part (b) of Theorem 4.7 implies that Y_V is a backward BUC-propagator. Moreover, the following assertion holds.

Theorem 4.9 *Let $P \in \mathcal{PM}$, $V \in \mathcal{P}_f^*$, and suppose that Y satisfies the strong BUC-condition. If Y is a strongly continuous backward BUC-propagator, then the same is true for Y_V.*

Proof. It follows from Theorem 4.7 that Y_V is a backward BUC-propagator. Now the strong continuity of the Feynman-Kac propagator Y_V on the space BUC can be obtained by reasoning as in the proof of the strong continuity of Y_V on L^s in Theorem 4.4. Here we use the space BC instead of the space L^s. □

The next theorem provides sufficient conditions for the continuity of the function $(\tau, x) \to Y_V(\tau,t)g(x)$ on the set $[0,t) \times E$. Let ξ denote the topology on the space BC generated by the uniform convergence of functions on compact subsets of the space E.

Theorem 4.10 *Let $P \in \mathcal{PM}$, and suppose that Y satisfies the following conditions:*

(i) Y is a backward strong Feller propagator.
(ii) For every function $h \in BC$ such that $h = Y_V(r,s)g$ with $0 \le r < s \le T$ and $g \in BC$, the mapping $(u,v) \mapsto Y(u,v)h$ of the space $\{(u,v) : 0 \le u \le v \le T\}$ into the space (BC, ξ) is continuous.

Then for any $V \in \mathcal{P}_f^*$, $t \in (0,T]$, and $g \in L_\mathcal{E}^\infty$, the function $(\tau, x) \mapsto Y_V(\tau,t)g(x)$ is continuous on the space $[0,t) \times E$.

Proof. Suppose that the conditions in Theorem 4.10 hold. Using part (a) of Theorem 4.7, we see that Y_V is a backward strong Feller propagator. Given $t \in (0,T]$ and $g \in L_\mathcal{E}^\infty$, fix $x \in E$ and τ with $0 \le \tau < t$. Suppose that τ' is close to τ and $x' \in U(x)$, where $U(x)$ is a relatively compact neighborhood of x. Then we have

$$|Y_V(\tau',t)g(x') - Y_V(\tau,t)g(x)|$$
$$\le |Y_V(\tau',t)g(x') - Y_V(\tau,t)g(x')| + |Y_V(\tau,t)g(x') - Y_V(\tau,t)g(x)|$$
$$= J_1 + J_2. \qquad (4.37)$$

Since Y_V is a backward strong Feller propagator,

$$\lim_{x' \to x} J_2 = 0. \qquad (4.38)$$

Next, we will estimate the quantity $\sup_{x' \in U(x)} J_1$. Let us first suppose that $\tau' < \tau$. Then we have

$$\sup_{x' \in U(x)} J_1 \le \sup_{x' \in U(x)} |(Y_V(\tau',\tau) - I) Y_V(\tau,t)g(x')|$$
$$\le \sup_{x' \in U(x)} |(Y_V(\tau',\tau) - Y(\tau',\tau)) Y_V(\tau,t)g(x')|$$
$$+ \sup_{x' \in U(x)} |(Y(\tau',\tau) - I) Y_V(\tau,t)g(x')|$$
$$\le M |Y_V(\tau',\tau) - Y(\tau',\tau)|_{\infty \to \infty} |g|_\infty$$

$$+ \sup_{x' \in U(x)} |(Y(\tau', \tau) - I) Y_V(\tau, t) g(x')|. \qquad (4.39)$$

Put $h = Y_V(\tau, t)g$. Then for any small $\varepsilon > 0$ we have

$$h = Y_V(\tau, t - \varepsilon) Y_V(t - \varepsilon, t) g = Y_V(\tau, t - \varepsilon) h_1.$$

Since condition (ii) in Theorem 4.10 holds, the function h_1 belongs to the space BC. It follows from (4.39), condition (ii) in Theorem 4.10, and Lemma 4.1 that

$$\lim_{\tau' \uparrow \tau} \sup_{x' \in U(x)} J_1 = 0. \qquad (4.40)$$

Next, suppose that $\tau < \tau'$. Then for every λ with $\tau' < \lambda < t$,

$$\sup_{x' \in U(x)} J_1 \leq \sup_{x' \in U(x)} |(Y_V(\tau', \lambda) - Y_V(\tau, \lambda)) Y_V(\lambda, t) g(x')|$$

$$\leq \sup_{x' \in U(x)} |(Y_V(\tau', \lambda) - Y(\tau', \lambda)) Y_V(\lambda, t) g(x')|$$

$$+ \sup_{x' \in U(x)} |(Y_V(\tau, \lambda) - Y(\tau, \lambda)) Y_V(\lambda, t) g(x')|$$

$$+ \sup_{x' \in U(x)} |(Y(\tau', \lambda) - Y(\tau, \lambda)) Y_V(\lambda, t) g(x')|$$

$$\leq M |g|_\infty |Y_V(\tau', \lambda) - Y(\tau', \lambda)|_{\infty \to \infty}$$

$$+ M |g|_\infty |Y_V(\tau, \lambda) - Y(\tau, \lambda)|_{\infty \to \infty}$$

$$+ \sup_{x' \in U(x)} |(Y(\tau', \lambda) - Y(\tau, \lambda)) Y_V(\lambda, t) g(x')|$$

$$= C_1 + C_2 + C_3. \qquad (4.41)$$

Using Lemma 4.1, we see that the following statement holds: for every $\varepsilon > 0$ there exists $\lambda \in (\tau, t)$ such that if $\tau < \tau' < \lambda$, then $C_1 + C_2 < \frac{1}{2}\varepsilon$. Moreover,

$$Y_V(\lambda, t)g = Y(\lambda, t - \delta) Y_V(t - \delta, t) g = Y(\lambda, t - \delta) h \qquad (4.42)$$

where $h \in BC$. Now condition (ii) in Theorem 4.10 and (4.42) imply that there exists $\eta > 0$ such that $\tau < \tau' < \tau + \eta < \lambda$ and $C_3 \leq \frac{1}{2}\varepsilon$. Hence, (4.41) gives

$$\lim_{\tau' \downarrow \tau} \sup_{x' \in U(x)} J_1 = 0. \qquad (4.43)$$

Now it is clear that Theorem 4.10 follows from (4.37), (4.38), (4.40), and (4.43). □

Corollary 4.2 Let $P \in \mathcal{PM}$, and suppose that Y is a backward strong Feller propagator. Suppose also that for every $g \in BC$, the mapping $(u,v) \mapsto Y(u,v)g$ of the space $\{(u,v) : 0 \le u \le v \le T\}$ into the space (BC, ξ) is continuous. Then for any $V \in \mathcal{P}_f^*$, $t \in (0,T]$, and $g \in L_\mathcal{E}^\infty$, the function $(\tau, x) \to Y_V(\tau, t)g(x)$ is continuous on the set $[0,t) \times E$.

Corollary 4.3 Let $P \in \mathcal{PM}$, and suppose that Y is a strongly continuous backward BUC-propagator. Suppose also that Y possesses the strong BUC-property. Then for any $V \in \mathcal{P}_f^*$, $t \in (0,T]$, and $g \in L_\mathcal{E}^\infty$, the function $(\tau, x) \to Y_V(\tau, t)g(x)$ is continuous on the set $[0,t) \times E$.

It is not hard to see that Corollaries 4.2 and 4.3 follow from Theorem 4.10. The proof is left as an exercise for the reader.

Remark 4.4 If a transition probability function $P \in \mathcal{PM}$ has a density p, then Theorems 4.7–4.10 and Corollaries 4.2 and 4.3 hold for any time-dependent measure μ from the class $\mathcal{P}_m^*$. The proofs of these results for $\mu \in \mathcal{P}_m^*$ are similar to the proofs for $V \in \mathcal{P}_f^*$.

4.5 Integral Kernels of Feynman-Kac Propagators

Backward Feynman-Kac propagators Y_V with $V \in \mathcal{P}_f^*$ possess integral kernels. This will be shown in this section. The kernels of the backward Feynman-Kac propagators Y_V and Y_μ are measures. These measures are absolutely continuous with respect to the reference measure m provided that the transition probability function P has a density p.

Let P be a transition probability function from the class $\mathcal{PM}$, and let X_t be a corresponding progressively measurable process. Recall that the free backward propagator Y associated with P is the family of integral operators on the space $L_\mathcal{E}^\infty$ defined by

$$Y(\tau, t)f(x) = \int_E f(y) P(\tau, x; t, dy)$$

for all $0 \le \tau < t \le T$, $x \in E$, and $f \in L_\mathcal{E}^\infty$. In general, the kernel of such a propagator is a Borel measure. If the transition probability function P possesses a density p, then

$$Y(\tau, t)f(x) = \int_E f(y) p(\tau, x; t, y) dy$$

for all $0 \leq \tau < t \leq T$, $x \in E$, and $f \in L_{\mathcal{E}}^{\infty}$. Therefore, in this case, the kernel of Y is a Borel function.

Let V be a function from the Kato class $\mathcal{P}_f^*$. Then the corresponding backward Feynman-Kac propagator Y_V is a uniformly bounded family of linear operators on the space $L_{\mathcal{E}}^{\infty}$. This is also true for the backward Feynman-Kac propagator Y_μ associated with a time-dependent measure μ from the class $\mathcal{P}_m^*$, provided that the transition probability function P has a density p (see Theorem 4.2 and Remark 4.1). We have already mentioned above that the free backward propagator Y is a family of integral operators. It will be shown below that if the transition density p is strictly positive, then the backward Feynman-Kac propagators Y_V and Y_μ associated with $V \in \mathcal{P}_f^*$ and $\mu \in \mathcal{P}_m^*$ are also families of integral operators. In the formulation of the next theorem, the symbol $\mu_{t,y}^{\tau,x}$ stands for the pinned measure on the σ-algebra $\mathcal{F}_{t-}^\tau = \sigma\left(X_s : \tau \leq s < t\right)$. This measure pins the process X_t at the point x at time τ and at the point y at time t (see the end of Section 1.11 for the definition of the pinned measure).

Theorem 4.11 *Let P be a transition probability function from the class $\mathcal{PM}$, and let X_t be a corresponding progressively measurable Markov process. Then the following assertions hold:*

(1) If $V \in \mathcal{P}_f^*$, then the function

$$P_V(\tau, x; t, A) = Y_V(\tau, t)\chi_A(x) \qquad (4.44)$$

where $0 \leq \tau < t \leq T$, $x \in E$, and $A \in \mathcal{E}$, is a transition function. Moreover,

$$P_V(\tau, x; t, A)$$
$$= \lim_{u \uparrow t} \mathbb{E}_{\tau, x} \exp\left\{-\int_\tau^u V(s, X_s)\,ds\right\} P(u, X_u; t, A) \qquad (4.45)$$

for all $0 \leq \tau < t \leq T$, $x \in E$, and $A \in \mathcal{E}$. The backward Feynman-Kac propagator Y_V can be represented as follows:

$$Y_V(\tau, t)f(x) = \int_E f(y) P_V(\tau, x; t, dy) \qquad (4.46)$$

for all $0 \leq \tau < t \leq T$, $x \in E$, and $f \in L_{\mathcal{E}}^{\infty}$.

(2) If the transition probability function P has a density p and if $\mu \in \mathcal{P}_m^*$, then the function

$$P_\mu(\tau, x; t, A) = Y_\mu(\tau, t)\chi_A(x) \qquad (4.47)$$

where $0 \leq \tau < t \leq T$, $x \in E$, and $A \in \mathcal{E}$, is a transition function. For all $0 \leq \tau < t \leq T$ and $x \in E$, the measure $A \mapsto P_\mu(\tau, x; t, A)$ is absolutely continuous with respect to the reference measure m. The Radon–Nikodym derivative p_μ of this measure is a transition density, and for all $0 \leq \tau < t \leq T$, $x \in E$, and $f \in L_\mathcal{E}^\infty$, the following formula holds:

$$Y_\mu(\tau, t)f(x) = \int_E f(y) p_\mu(\tau, x; t, y)\, dy. \tag{4.48}$$

Moreover, if the transition probability function P has a strictly positive density p, then for all τ and t with $0 \leq \tau \leq t \leq T$ and $x \in E$,

$$\begin{aligned} p_\mu(\tau, x; t, y) &= \lim_{u \uparrow t} \mathbb{E}_{\tau, x} \exp\left\{-A_\mu(\tau, u)\right\} p(u, X_u; t, y) \\ &= \int_\Omega \exp\left\{-A(\tau, t)\right\} d\mu_{t,y}^{\tau, x} \end{aligned} \tag{4.49}$$

for m-almost all $y \in E$.

Proof. Let $V \in \mathcal{P}_f^*$, and let P_V be the function defined by (4.44). Then for fixed τ, t, and A, the function $x \mapsto P_V(\tau, x; t, A)$ is a Borel function on E. This follows from the fact that Y_V is a backward propagator on the space $L_\mathcal{E}^\infty$ (see Theorem 4.2). Next fix τ, t, and x. Since

$$Y_V(\tau, t)\chi_A(x) = \mathbb{E}_{\tau, x} \chi_A(X_t) \exp\left\{-\int_\tau^t V(s, X_s)\, ds\right\}, \tag{4.50}$$

the function $A \mapsto P_V(\tau, x; t, A)$ is a Borel measure on $\mathcal{E}$. Moreover, the fact that the function P_V satisfies the Chapman-Kolmogorov equation follows from the properties of Y_V.

Our next goal is to prove the equality in (4.45). By (4.50), the Markov property, and the properties of conditional expectations,

$$\begin{aligned} P_V(\tau, x; t, A) &= Y_V(\tau, t)\chi_A(x) \\ &= \mathbb{E}_{\tau, x} \chi_A(X_t) \exp\left\{-\int_\tau^t V(s, X_s)\, ds\right\} \\ &= \lim_{u \uparrow t} \mathbb{E}_{\tau, x} \chi_A(X_t) \exp\left\{-\int_\tau^u V(s, X_s)\, ds\right\} \\ &= \lim_{u \uparrow t} \mathbb{E}_{\tau, x} \exp\left\{-\int_\tau^u V(s, X_s)\, ds\right\} \mathbb{E}_{\tau, x}\left[\chi_A(X_t) \mid \mathcal{F}_u^\tau\right] \\ &= \lim_{u \uparrow t} \mathbb{E}_{\tau, x} \exp\left\{-\int_\tau^u V(s, X_s)\, ds\right\} \mathbb{E}_{u, X_u}\left[\chi_A(X_t)\right] \end{aligned}$$

$$= \lim_{u \uparrow t} \mathbb{E}_{\tau,x} \exp\left\{-\int_\tau^u V(s, X_s)\, ds\right\} P(u, X_u; t, A).$$

Therefore, formula (4.45) holds. Moreover, if $f \in L_{\mathcal{E}}^\infty$ and $f \geq 0$, then

$$Y_V(\tau, t) f(x) = \int_0^\infty Y_V(\tau, t) \chi_{\{f > \lambda\}}(x)\, d\lambda. \qquad (4.51)$$

It follows from (4.51) that the function P_V is the integral kernel of the backward Feynman-Kac propagator Y_V.

This completes the proof of part (1) of Theorem 4.11.

Next, we will prove part (2) of Theorem 4.11. Suppose that P has a density p, and let $\mu \in \mathcal{P}_m^*$. We have

$$Y_\mu(\tau, t) \chi_A(x) = \mathbb{E}_{\tau,x} \chi_A(X_t) \exp\{-A_\mu(\tau, t)\}. \qquad (4.52)$$

It follows from formula (4.52) that for all $0 \leq \tau < t \leq T$ and $x \in E$, the function

$$A \mapsto P_\mu(\tau, x; t, A) \qquad (4.53)$$

is a Borel measure on $\mathcal{E}$. Since Y_μ is a backward propagator on the space $L_{\mathcal{E}}^\infty$ (see Theorem 4.2), the set function $x \mapsto P_\mu(\tau, x; t, A)$ is a Borel function on E for all $0 \leq \tau < t \leq T$ and $A \in \mathcal{E}$. Moreover, P_μ satisfies the Chapman-Kolmogorov equation. By the continuity of the functional $A_\mu(\tau, t)$ with respect the variable t, the Markov property, and the properties of conditional expectations, we get

$$\begin{aligned}
P_\mu(\tau, x; t, A) &= Y_\mu(\tau, t) \chi_A(x) \\
&= \lim_{u \uparrow t} \mathbb{E}_{\tau,x} \exp\{-A_\mu(\tau, u)\} \mathbb{E}_{\tau,x} \left[\chi_A(X_t) \mid \mathcal{F}_u^\tau\right] \\
&= \lim_{u \uparrow t} \mathbb{E}_{\tau,x} \exp\{-A_\mu(\tau, u)\} \mathbb{E}_{\tau,x} \left[\chi_A(X_t) \mid \mathcal{F}_u^\tau\right] \\
&= \lim_{u \uparrow t} \mathbb{E}_{\tau,x} \exp\{-A_\mu(\tau, u)\} \mathbb{E}_{u, X_u}[\chi_A(X_t)] \\
&= \lim_{u \uparrow t} \mathbb{E}_{\tau,x} \exp\{-A_\mu(\tau, u)\} Y(u, t) \chi_A(X_u) \\
&= \lim_{u \uparrow t} \mathbb{E}_{\tau,x} \exp\{-A_\mu(\tau, u)\} P(u, X_u; t, A) \\
&= \lim_{u \uparrow t} \int_A dy\, \mathbb{E}_{\tau,x} \exp\{-A_\mu(\tau, u)\} p(u, X_u; t, y) \qquad (4.54)
\end{aligned}$$

for all $0 \leq \tau < t \leq T$, $x \in E$, and $A \in \mathcal{E}$. It follows from (4.54) that the measure in (4.53) is the setwise limit of the measures

$$A \mapsto \int_A dy \mathbb{E}_{\tau,x} \exp\{-A_\mu(\tau,u)\} p(u, X_u; t, y). \tag{4.55}$$

Since the measures in (4.55) are absolutely continuous with respect to m, the measure in (4.53) is also absolutely continuous with respect to m. By the Radon–Nikodym theorem, the transition function P_μ has a density p_μ. It is not difficult to prove that p_μ satisfies the Chapman-Kolmogorov equation for transition densities. By (4.47) and the formula

$$Y_\mu(\tau,t) f(x) = \int_0^\infty Y_\mu(\tau,t) \chi_{\{f > \lambda\}}(x) d\lambda,$$

where f is a nonnegative function from $L_\mathcal{E}^\infty$, we see that formula (4.48) holds.

Our next goal is to establish the equalities in (4.49). We have

$$\exp\{-A_\mu(\tau,t)\} \leq \exp\{A_{\mu^-}(\tau,t)\} \tag{4.56}$$

for all $0 \leq \tau \leq t \leq T$. Since $\mu^- \in \mathcal{P}_m^*$, the family of functions defined by

$$\{y \mapsto \mathbb{E}_{\tau,x} \exp\{-A_\mu(\tau,u)\} p(u, X_u; t, y) : \tau \leq u < t\} \tag{4.57}$$

is a bounded subset of the space $L^1(m)$. This can be established using Theorem 4.2 and Remark 4.1. Moreover, since for all $u \in [\tau, t)$ the random variable $A_\mu(\tau, u)$ is $\mathcal{F}_u^\tau$-measurable (see Theorem 3.14), Lemma 1.25 implies that

$$\mathbb{E}_{\tau,x} \exp\{-A_\mu(\tau,u)\} p(u, X_u; t, y)$$
$$= \int_\Omega \exp\{-A_\mu(\tau,u)\} d\mu_{t,y}^{\tau,x}. \tag{4.58}$$

Next, let u_k be a sequence of numbers such that $\tau < u_k < t$ and $u_k \uparrow t$ as $k \to \infty$. It follows from $\mu^- \in \mathcal{P}_f^*$, formula (4.58) with $u = u_k$, the monotone convergence theorem, and Remark 4.1 that

$$\int_E dy \int_\Omega \exp\{A_{\mu^-}(\tau,t)\} d\mu_{t,y}^{\tau,x} = \lim_{k \to \infty} \int_E dy \int_\Omega \exp\{A_{\mu^-}(\tau,u_k)\} d\mu_{t,y}^{\tau,x}$$
$$= \lim_{k \to \infty} \int_E dy \mathbb{E}_{\tau,x} \exp\{A_{\mu^-}(\tau,u_k)\} p(u_k, X_{u_k}; t, y)$$
$$= \sup_{k \geq 1} \mathbb{E}_{\tau,x} \exp\{A_{\mu^-}(\tau,u_k)\} \leq \mathbb{E}_{\tau,x} \exp\{A_{\mu^-}(\tau,t)\} < \infty. \tag{4.59}$$

Therefore,
$$\int_\Omega \exp\{A_{\mu^-}(\tau,t)\}\, d\mu_{t,y}^{\tau,x} < \infty \tag{4.60}$$

for m-almost all $y \in E$. Next (4.56), (4.58), (4.60), and the dominated convergence theorem give

$$\lim_{k\to\infty} \mathbb{E}_{\tau,x} \exp\{-A_\mu(\tau,u_k)\} p(u_k, X_{u_k}; t, y)$$
$$= \int_\Omega \exp\{-A_\mu(\tau,t)\}\, d\mu_{t,y}^{\tau,x} \tag{4.61}$$

for m-almost all $y \in E$. Note that the exceptional set in (4.61) does not depend on the sequence u_k. Taking this into account, we see that (4.61) implies

$$\lim_{u\uparrow t} \mathbb{E}_{\tau,x} \exp\{-A_\mu(\tau,u)\} p(u, X_u; t, y)$$
$$= \int_\Omega \exp\{-A_\mu(\tau,t)\}\, d\mu_{t,y}^{\tau,x} \tag{4.62}$$

for m-almost all $y \in E$. Moreover, it follows from (4.54), (4.56), (4.58), (4.59), (4.61), (4.62), and the dominated convergence theorem that

$$\int_A p_\mu(\tau,x;t,y)\, dy = \lim_{k\to\infty} \int_A \mathbb{E}_{\tau,x} \exp\{-A_\mu(\tau,u_k)\} p(u_k, X_{u_k}; t, y)\, dy$$
$$= \int_A \lim_{k\to\infty} \mathbb{E}_{\tau,x} \exp\{-A_\mu(\tau,u_k)\} p(u_k, X_{u_k}; t, y)\, dy$$
$$= \int_A \lim_{u\uparrow t} \mathbb{E}_{\tau,x} \exp\{-A_\mu(\tau,u)\} p(u, X_u; t, y)\, dy$$
$$= \int_A \int_\Omega \exp\{-A_\mu(\tau,t)\}\, d\mu_{t,y}^{\tau,x}$$

for any Borel subset A of the space E. Therefore,

$$p_\mu(\tau,x;t,y) = \lim_{k\to\infty} \mathbb{E}_{\tau,x} \exp\{-A_\mu(\tau,u)\} p(u, X_u; t, y)$$
$$= \int_\Omega \exp\{-A_\mu(\tau,t)\}\, d\mu_{t,y}^{\tau,x}$$

for m-almost all $y \in E$. This establishes formula (4.49).

The proof of Theorem 4.11 is thus completed. □

4.6 Feynman-Kac Propagators and Howland Semigroups

In Section 2.4, we discussed Howland semigroups associated with propagators and backward propagators and considered special examples of Howland semigroups on the space $L_{\mathcal{B}}^{\infty}([0,T], L_{\mathcal{E}}^{\infty})$ corresponding to free propagators and free backward propagators on the space $L_{\mathcal{E}}^{\infty}$. Let us recall that for a free backward propagator, $Y(\tau, t)f(x) = E_{\tau,x}f(X_t)$, associated with a transition probability function P, the Howland semigroup $S_Y(t)$ is given by the following formula:

$$S_Y(t)F(\tau, x) = E_{\tau,x} F\left((\tau + t) \wedge T, X_{(\tau+t)\wedge T}\right)$$

for all $\tau \in [0,t]$, $t \in [0,T]$, $x \in E$, and $F \in L_{\mathcal{B}}^{\infty}([0,T], L_{\mathcal{E}}^{\infty})$. Similarly, let $\widetilde{P}$ be a backward transition probability function, and let U be the corresponding free propagator on the space $L_{\mathcal{E}}^{\infty}$. Then the Howland semigroup $S_U(\tau)$ is defined by

$$S_U(\tau)F(t, x) = \widetilde{\mathbb{E}}^{t,x} F\left((t - \tau) \vee 0, X_{(t-\tau)\vee 0}\right)$$

for all $\tau \in [0,T]$, $t \in [0,T]$, $x \in E$, and $F \in L_{\mathcal{B}}^{\infty}([0,T], L_{\mathcal{E}}^{\infty})$.

Next, assume that the transition probability function P belongs to the class $\mathcal{PM}$, and let V be a Borel function on $[0,T] \times E$ from the non-autonomous Kato class $\mathcal{P}_f^*$. Recall that the backward Feynman-Kac propagator Y_V is defined on the space $L_{\mathcal{E}}^{\infty}$ by the formula

$$Y_V(\tau, t)f(x) = \mathbb{E}_{\tau,x} f(X_t) \exp\left\{-\int_\tau^t V(s, X_s)\, ds\right\},$$

where $0 \leq \tau \leq t \leq T$, $x \in E$, and $f \in L_{\mathcal{E}}^{\infty}$. Similarly, if P has a density p, then for every time-dependent measure μ from the class $\mathcal{P}_m^*$, the backward Feynman-Kac propagator Y_μ is defined on the space $L_{\mathcal{E}}^{\infty}$ by the formula

$$Y_\mu(\tau, t)f(x) = \mathbb{E}_{\tau,x} f(X_t) \exp\{-A_\mu(\tau, t)\}$$

where $0 \leq \tau \leq t \leq T$, $x \in E$, and $f \in L_{\mathcal{E}}^{\infty}$ (see Definition 4.4 and Theorem 4.2). For a backward transition function $\widetilde{P} \in \widetilde{\mathcal{M}}$ and $V \in \mathcal{P}_f$, the Feynman-Kac propagator U_V is given by

$$U_V(t, \tau)f(x) = \widetilde{\mathbb{E}}^{t,x} f\left(\widetilde{X}_\tau\right) \exp\left\{-\int_\tau^t V\left(s, \widetilde{X}_s\right) ds\right\}$$

where $0 \leq \tau \leq t \leq T$, $x \in E$, and $f \in L_{\mathcal{E}}^{\infty}$. If $\widetilde{P}$ has a density $\widetilde{p}$, then for every $\mu \in \mathcal{P}_f$, the Feynman-Kac propagator U_μ is given by the following formula:

$$U_\mu(t,\tau)f(x) = \widetilde{\mathbb{E}}^{t,x} f\left(\widetilde{X}_\tau\right) \exp\left\{-\widetilde{A}_\mu(t,\tau)\right\}$$

where $0 \leq \tau \leq t \leq T$, $x \in E$, and $f \in L_{\mathcal{E}}^\infty$ (see Definition 4.5).

Since Y_V and Y_μ with $V \in \mathcal{P}_f^*$ and $\mu \in \mathcal{P}_m^*$ are backward propagators on the space $L_{\mathcal{E}}^\infty$, the corresponding Howland semigroups are defined by the following formulas:

$$S_{Y_V}(t) F(\tau, x)$$
$$= \mathbb{E}_{\tau,x} F\left((\tau+t) \wedge T, X_{(\tau+t) \wedge T}\right) \exp\left\{-\int_\tau^{(\tau+t) \wedge T} V(s, X_s) \, ds\right\}$$

and

$$S_{Y_\mu}(t) F(\tau, x) = \mathbb{E}_{\tau,x} F\left((\tau+t) \wedge T, X_{(\tau+t) \wedge T}\right) \exp\left\{-A_\mu(\tau, (\tau+t) \wedge T)\right\}$$

where $0 \leq \tau \leq t \leq T$, $x \in E$, and $F \in L_\mathcal{B}^\infty([0,T], L_\mathcal{E}^\infty)$. Moreover, since U_V and U_μ with $V \in \mathcal{P}_f$ and $\mu \in \mathcal{P}_m$ are propagators on the space $L_\mathcal{E}^\infty$, the corresponding Howland semigroups are defined by the following formulas:

$$S_{U_V}(\tau) F(t, x)$$
$$= \widetilde{\mathbb{E}}^{t,x} F\left((t-\tau) \vee 0, X_{(t-\tau) \vee 0}\right) \exp\left\{-\int_{(t-\tau) \vee 0}^t V\left(s, \widetilde{X}_s\right) ds\right\}$$

and

$$S_{U_\mu}(\tau) F(t,x) = \widetilde{\mathbb{E}}^{t,x} F\left((t-\tau) \vee 0, X_{(t-\tau) \vee 0}\right) \exp\left\{-A_\mu(t, (t-\tau) \vee 0)\right\}$$

where $0 \leq \tau \leq t \leq T$, $x \in E$, and $F \in L_\mathcal{B}^\infty([0,T], L_\mathcal{E}^\infty)$.

The next assertion shows that the non-autonomous Kato class $\mathcal{P}_f^*$ coincides with the Kato class J associated with the Howland semigroup $S_Y(t)$ (see Definitions 3.4 and 4.3).

Lemma 4.3 *Let P be a transition probability function from the class $\mathcal{PM}$, and let X_t be a corresponding progressively measurable Markov process. Let $\widehat{X}_t$ be a space-time process associated with X_t. Then $\mathcal{P}_f^* = J$, where $\mathcal{P}_f^*$ is the non-autonomous Kato class corresponding to the process X_t, and J is the Kato class associated with the process $\widehat{X}_t$.*

Proof. Let V be a Borel function on $[0,T] \times E$. By Definition 3.4,

$$V \in \mathcal{P}_f^* \iff \lim_{t-\tau \downarrow 0} \sup_{x \in E} \mathbb{E}_{\tau,x} \int_\tau^t |V(s, X_s)| \, ds = 0.$$

It follows that

$$V \in \mathcal{P}_f^* \iff \lim_{u \downarrow 0} \sup_{(\tau,x) \in [0,T] \times E} \mathbb{E}_{\tau,x} \int_\tau^{(u+\tau) \wedge T} |V(s, X_s)|\, ds = 0. \quad (4.63)$$

On the other hand, Definition 4.3 shows that

$$V \in J \iff \lim_{t \downarrow 0} \sup_{(\tau,x) \in [0,T] \times E} \mathbb{E}_{(\tau,x)} \int_0^t \left|V\left(\widehat{X}_s\right)\right| ds = 0.$$

It is not difficult to see from the previous equivalence that

$$V \in J \iff \lim_{u \downarrow 0} \sup_{(\tau,x) \in [0,T] \times E} \mathbb{E}_{\tau,x} \int_\tau^{(u+\tau) \wedge T} |V(s, X_s)|\, ds = 0. \quad (4.64)$$

Now it is clear that Lemma 4.3 follows from (4.63) and (4.64). □

In the book by Chung and Zhao [Chung and Zhao (1995)] various results concerning the inheritance of properties of free semigroups by Feynman-Kac semigroups are discussed. One may be tempted to derive the inheritance results obtained in Sections 4.3 and 4.4 of the present book from the similar results for Feynman-Kac semigroups established in Section 3.2 of [Chung and Zhao (1995)]. An encouraging motivation for this approach is the coincidence of the non-autonomous backward Kato class $\mathcal{P}_f^*$ and the Kato class J associated with the space-time process $\widehat{X}_t$ (see Lemma 4.3). However, this approach often fails. Next, we will give two examples illustrating the restricted applicability of Howland semigroups to the study of the inheritance problem for propagators. Let us first compare Theorem 4.7 in Section 4.4 of the present book and Proposition 3.12 in [Chung and Zhao (1995)]. Both results concern the inheritance of the strong Feller property. It is assumed in Proposition 3.12 in [Chung and Zhao (1995)] that the transition probability function P has a density p that is a symmetric bounded and time-homogeneous. It is also assumed that the function V belongs to the Kato class J (see condition (15) on page 70 in [Chung and Zhao (1995)]). Chung and Zhao established that under such restrictions, the Feynman-Kac semigroup inherits the strong Feller property from the free semigroup associated with the transition probability density p. On the other hand, Theorem 4.7 in the present book states that if P is a non-homogeneous transition probability function, then the backward Feynman-Kac propagator Y_V associated with a function V from the non-autonomous Kato class $\mathcal{P}_f^*$ inherits the strong Feller property from the free backward propagator Y. Let us try to obtain a special case of Theorem 4.7 using Proposition

3.12 in [Chung and Zhao (1995)]. Suppose that p is a transition probability density such that the corresponding free backward propagator,

$$Y(\tau, t)f(x) = \int_E f(y)p(\tau, x; t, y)dy, \qquad (4.65)$$

possesses the strong Feller property. This means that for every bounded Borel measurable function f and all $0 \leq \tau < t \leq T$, the function $x \mapsto Y(\tau, t)f(x)$ belongs to the space BC of bounded continuous functions on E. The Howland semigroup $S_Y(t)$ associated with the free backward propagator Y has the following form:

$$S_Y(t)F(\tau, x) = Y(\tau, (\tau + t) \wedge T)F((\tau + t) \wedge T)(x) \qquad (4.66)$$

for all $t > 0$, $(\tau, x) \in [0, T] \times E$, and all bounded Borel functions F on $[0, T] \times E$. Even if we forget that the transition probability function for the Howland semigroup has a singular component, and it is not necessarily symmetric and bounded, we still need to establish that the semigroup $S_Y(t)$ possesses the strong Feller property, that is, the function on the right-hand side of (4.66) is continuous on the space $[0, T] \times E$ for every $t > 0$. However, the validity of the previous assertion is not clear, since we only know the continuity of the function $x \mapsto Y(\tau, t)f(x)$ in the variable x, and no continuity assumption in the variable τ is imposed.

Our next example concerns the strong continuity of semigroups and propagators on the space L^p. Even if the free propagator Y is strongly continuous on the space $L^p(E)$, it is not clear how to prove that the corresponding Howland semigroup $S_Y(t)$ is strongly continuous on the space $L^p([0, T] \times E)$. More precisely, it is difficult to expect that the strong continuity of $S_Y(t)$, that is, the condition

$$\lim_{t \downarrow 0} \int \int_{[0,T] \times E} |F(\tau, x) - Y(\tau, (\tau + t) \wedge T)F((\tau + t) \wedge T)(x)|^p \, d\tau dx = 0$$

for all $F \in L^p([0, T] \times E)$, can be obtained from the strong continuity of the free backward propagator Y on the space $L^p(E)$.

4.7 Duhamel's Formula for Feynman-Kac Propagators

Duhamel's formula is an important link between the free backward propagator Y and the backward Feynman-Kac propagators Y_V and Y_μ. This formula shows that backward Feynman-Kac propagators can be obtained by solving a Volterra type integral equation. It will be established below that for a function $V \in \mathcal{P}_f^*$ and a time dependent measure $\mu \in \mathcal{P}_m^*$, Duhamel's formula holds pointwise.

Theorem 4.12

(a) Let $P \in \mathcal{PM}$, $f \in L_{\mathcal{E}}^\infty$, and $V \in \mathcal{P}_f^*$. Fix $t \in (0,T]$, and define a function on $[0,t] \times E$ by $u(\tau,x) = Y_V(\tau,t)f(x)$. Then for every $\tau \in [0,t]$, the function $x \mapsto u(\tau,x)$ belongs to the space $L_{\mathcal{E}}^\infty$. Moreover, the following Volterra type integral equation holds:

$$u(\tau,x) = Y(\tau,t)f(x) - \int_\tau^t Y(\tau,s)\left[V(s)u(s)\right](x)ds \qquad (4.67)$$

for all $x \in E$ and $0 \leq \tau \leq t \leq T$.

(b) Suppose that $P \in \mathcal{PM}$ has a density p. Let $f \in L_{\mathcal{E}}^\infty$, $\mu \in \mathcal{P}_m^*$, and $t \in (0,T]$, and define a function on $[0,t] \times E$ by $u(\tau,x) = Y_\mu(\tau,t)f(x)$. Then for every $\tau \in [0,t]$, the function $x \mapsto u(\tau,x)$ belongs to the space $L_{\mathcal{E}}^\infty$. Moreover, the following Volterra type integral equation holds:

$$u(\tau,x) = Y(\tau,t)f(x) - \int_\tau^t Y(\tau,s)\left[\mu(s)u(s)\right](x)ds \qquad (4.68)$$

for all $x \in E$ and $0 \leq \tau \leq t \leq T$.

Remark 4.5 The equations in (4.67) and (4.68) can be rewritten as follows:

$$Y_V(\tau,t)f(x) = Y(\tau,t)f(x) - \int_\tau^t Y(\tau,s)\left[V(s)Y_V(s,t)f\right](x)ds \qquad (4.69)$$

and

$$Y_\mu(\tau,t)f(x) = Y(\tau,t)f(x) - \int_\tau^t Y(\tau,s)\left[V(s)Y_\mu(s,t)f\right](x)ds. \qquad (4.70)$$

Proof. (a) Let $V \in \mathcal{P}_f^*$. By Theorem 4.2, for every $\tau \in [0,t]$ the function $x \mapsto u(\tau,x)$ belongs to the space $L_{\mathcal{E}}^\infty$. Moreover, under the assumptions in part (a) of Theorem 4.12, we have

$$\int_\tau^t Y(\tau,s)\left[V(s)Y_V(s,t)f\right](x)ds$$
$$= \int_\tau^t \mathbb{E}_{\tau,x} V(s,X_s)\mathbb{E}_{s,X_s} f(X_t) \exp\left\{-\int_s^t V(\lambda,X_\lambda)d\lambda\right\}ds,$$

where X_t is a progressively measurable process with P as its transition function. By the Markov property,

$$\int_\tau^t Y(\tau,s)\left[V(s)Y_V(s,t)f\right](x)ds$$

$$= \int_\tau^t \mathbb{E}_{\tau,x} V(s,X_s) \mathbb{E}_{\tau,x}(f(X_t)\exp\left\{-\int_s^t V(\lambda,X_\lambda)d\lambda\right\}|\mathcal{F}_s)ds$$

$$= \int_\tau^t \mathbb{E}_{\tau,x} f(X_t) V(s,X_s) \exp\left\{-\int_s^t V(\lambda,X_\lambda)d\lambda\right\} ds$$

$$= \int_\tau^t \mathbb{E}_{\tau,x} f(X_t) \frac{\partial}{\partial s} \exp\left\{-\int_s^t V(\lambda,X_\lambda)d\lambda\right\} ds$$

$$= \mathbb{E}_{\tau,x} f(X_t) - \mathbb{E}_{\tau,x} f(X_t) \exp\left\{-\int_\tau^t V(\lambda,X_\lambda)d\lambda\right\}$$

$$= Y(\tau,t)f(x) - Y_V(\tau,t)f(x).$$

This gives part (a) of Theorem 4.12.

(b) Suppose that the assumptions in part (b) of Theorem 4.12 hold, and let $\mu \in \mathcal{P}^*$. In the proof of part (b) we employ the approximation result from Section 3.9 (Lemma 3.21). Let V_k be the sequence of functions defined by (3.153). It follows from Lemma 3.21 and Theorem 4.3 that

$$\lim_{k\to\infty} \sup_{0\le\tau\le t\le T} \sup_{x\in E} |Y_\mu(\tau,t)f(x) - Y_{V_k}(\tau,t)f(x)| = 0 \qquad (4.71)$$

for all $f \in L_{\mathcal{E}}^\infty$. The functions V_k belong to the class $\mathcal{P}_f^*$ (see Lemma 3.21). Therefore, (4.69) gives

$$Y_{V_k}(\tau,t)f(x) = Y(\tau,t)f(x) - \int_\tau^t Y(\tau,s)\left[V_k(s)Y_{V_k}(s,t)f\right](x)ds. \qquad (4.72)$$

It follows from the properties of backward propagators that

$$\int_\tau^t Y(\tau,s)\left[V_k(s)Y_{V_k}(s,t)f\right](x)ds$$

$$= \int_\tau^t Y(\tau,s)k \int_s^{(s+\frac{1}{k})\wedge T} Y(s,\lambda)\left[\mu(\lambda)Y_{V_k}(s,t)f\right](x)d\lambda ds$$

$$\int_\tau^t k \int_s^{(s+\frac{1}{k})\wedge T} Y(\tau,\lambda)\left[\mu(\lambda)Y_{V_k}(s,t)f\right](x)d\lambda ds$$

$$= \int_\tau^{(t+\frac{1}{k})\wedge T} d\lambda k \int_{(\lambda-\frac{1}{k})\vee\tau}^\lambda Y(\tau,\lambda)\left[\mu(\lambda)Y_{V_k}(s,t)f\right](x)ds$$

$$= \int_{\tau}^{(t+\frac{1}{k})\wedge T} d\lambda Y(\tau,\lambda)\left[\mu(\lambda)k\int_{(\lambda-\frac{1}{k})\vee\tau}^{\lambda} Y_{V_k}(s,t)f ds\right](x). \quad (4.73)$$

Next, passing to the limit as $k\to\infty$ in (4.73) and using (4.71) and the definition of the class $\mathcal{P}_m^*$, we get

$$\lim_{k\to\infty}\int_{\tau}^{t} Y(\tau,s)\left[V_k(s)Y_{V_k}(s,t)f\right](x)ds = \int_{\tau}^{t} Y(\tau,\lambda)\left[\mu(\lambda)Y_\mu(\lambda,t)f\right](x)d\lambda. \quad (4.74)$$

Now it is clear that part (b) of Theorem 4.12 follows from (4.71), (4.72), and (4.74).

This completes the proof of Theorem 4.12. $\square$

Duhamel's formula shows that backward Feynman-Kac propagators generate solutions to Volterra type integral equations. Next, we will see that under certain restrictions, backward Feynman-Kac propagators generate solutions to final value problems. We will first reason informally. It will be assumed in the remaining part of this section that all functions are differentiable as many times as needed.

Let t and τ be such that $0 < \tau < t \leq T$. Fix γ with $0 \leq \gamma < \tau$ and apply the operator $Y(\gamma,\tau)$ to Duhamel's formula. Then, using the properties of backward propagators, we get

$$Y(\gamma,\tau)Y_\mu(\tau,t)f(x) = Y(\gamma,t)f(x) - \int_{\tau}^{t} Y(\gamma,s)\left[\mu(s)Y_\mu(s,t)f\right](x)ds \quad (4.75)$$

where $f \in L_{\mathcal{E}}^\infty$. Differentiating the equation in (4.75) from the right with respect to the variable τ on the interval (γ,t), we obtain

$$\frac{\partial^+}{\partial\tau}\left[Y(\gamma,\tau)Y_\mu(\tau,t)f(x)\right] = Y(\gamma,\tau)\left[\mu(\tau)Y_\mu(\tau,t)f\right](x).$$

It follows that

$$\left(\frac{\partial^+}{\partial\tau}Y(\gamma,\tau)\right)Y_\mu(\tau,t)f(x) + Y(\gamma,\tau)\left(\frac{\partial^+}{\partial\tau}Y_\mu(\tau,t)f(x)\right)$$
$$= Y(\gamma,\tau)\left[\mu(\tau)Y_\mu(\tau,t)f\right](x). \quad (4.76)$$

Our next goal is to apply Theorem 2.11 to the first term on the left-hand side of equality (4.76). By Theorem 2.11, if

$$Y_\mu(\tau_2,t)f \in D_+^{M,*}(\tau_1) \quad (4.77)$$

for all $0 < \tau_1 < \tau_2 < t$, then

$$Y(\gamma,\tau)A_+^M(\tau)Y_\mu(\tau,t)f(x) + Y(\gamma,\tau)\left(\frac{\partial^+}{\partial \tau}Y_\mu(\tau,t)f(x)\right)$$
$$= Y(\gamma,\tau)\left[\mu(\tau)Y_\mu(\tau,t)f\right](x). \tag{4.78}$$

The symbol $D_+^{M,*}(\tau_1)$ in (4.77) stands for the set defined by

$$D_+^{M,*}(\tau_1) = \bigcap_{t:\tau_1 < t < T} D^w\left(A_+^M(t)\right),$$

where $D^w\left(A_+^M(t)\right)$ denotes the subspace of the space $L_\mathcal{E}^\infty$ consisting of all functions for which the limit in formula (2.31) exists. Passing to the limit as $\gamma \uparrow \tau$ in (4.78), we see that if condition (4.77) holds, then the function $u(\tau,x) = Y_\mu(\tau,t)f(x)$ is a solution to the following final value problem:

$$\begin{cases} \frac{\partial^+}{\partial \tau}u(\tau,x) + A_+^M(\tau)u(\tau)(x) - \mu(\tau)u(\tau,x) = 0, \\ u(t,x) = f(x). \end{cases} \tag{4.79}$$

In the next section we will explain what can be done if the differentiability conditions in the reasoning above are not satisfied. It will be shown that under certain restrictions there exist viscosity solutions to final value problem (4.79).

4.8 Feynman-Kac Propagators and Viscosity Solutions

Viscosity solutions to partial differential equations were introduced in [Crandall and Lions (1983)]. See also [Crandall, Ishii, and Lions (1992)] for more information on viscosity solutions.

The main results in this section are Theorems 4.15 and 4.16. These theorems provide sufficient conditions for the solvability of the final value problem in (4.79) in the viscosity sense. We will first establish several preliminary results.

Theorem 4.13 *Let $\mu \in \mathcal{P}_m^*$, $f \in L^\infty$, and fix t such that $0 < t \leq T$. Then the following assertions hold:*

(a) Suppose that ψ is a bounded continuous function on $[0,T] \times E$, and let $(\tau_0,x_0) \in [0,t) \times E$ and $\delta > 0$ with $\tau_0 + \delta < t$ be such that

$$Y_\mu(\tau_0,t)f(x_0) - \psi(\tau_0,x_0) = \min_{(\tau,x)\in[\tau_0,\tau_0+\delta]\times E}(Y_\mu(\tau,t)f(x) - \psi(\tau,x)).$$

Then for every $0 < \epsilon < \delta$,

$$\frac{Y(\tau_0, \tau_0 + \epsilon)\psi(\tau_0 + \epsilon)(x_0) - \psi(\tau_0, x_0)}{\epsilon}$$
$$-\frac{1}{\epsilon}\int_{\tau_0}^{\tau_0+\epsilon} Y(\tau_0, s)\left[\mu(s)Y_\mu(s,t)f\right](x_0)ds \le 0. \quad (4.80)$$

(b) Suppose that ψ is a bounded continuous function on $[0,T] \times E$, and let $(\tau_0, x_0) \in [0,t) \times E$ and $\delta > 0$ with $\tau_0 + \delta < t$ be such that

$$Y_\mu(\tau_0, t)f(x_0) - \psi(\tau_0, x_0) = \max_{(\tau,x)\in[\tau_0,\tau_0+\delta]\times E}(Y_\mu(\tau,t)f(x) - \psi(\tau,x))$$

for some $(\tau_0, x_0) \in [0,t) \times E$. Then for every $0 < \epsilon < \delta$,

$$\frac{Y(\tau_0, \tau_0 + \epsilon)\psi(\tau_0 + \epsilon)(x_0) - \psi(\tau_0, x_0)}{\epsilon}$$
$$-\frac{1}{\epsilon}\int_{\tau_0}^{\tau_0+\epsilon} Y(\tau_0, s)\left[\mu(s)Y_\mu(s,t)f\right](x_0)ds \ge 0. \quad (4.81)$$

Proof. We will prove only part (a) of Theorem 4.13. The proof of part (b) is similar. Let ψ be any bounded Borel function on $[0,T] \times E$, and let M be any real number. Put

$$G(\tau, x) = Y_\mu(\tau, t)f(x) - \psi(\tau, x) - M. \quad (4.82)$$

Lemma 4.4 *Let $\mu \in \mathcal{P}_m^*$, and let $\epsilon > 0$ be such that $\tau + \epsilon < t$. Then the following equality holds for the function G defined by (4.82):*

$$G(\tau, x) - Y(\tau, \tau + \epsilon)G(\tau + \epsilon)(x)$$
$$= Y(\tau, \tau + \epsilon)\psi(\tau + \epsilon)(x) - \psi(\tau, x) - \int_\tau^{\tau+\epsilon} Y(\tau, s)\left[\mu(s)Y_\mu(s,t)f\right](x)ds. \quad (4.83)$$

Proof. We have

$$Y_\mu(\tau, t)f(x) - Y(\tau, \tau + \epsilon)Y_\mu(\tau + \epsilon, t)f(x)$$
$$= (Y_\mu(\tau, \tau + \epsilon) - Y(\tau, \tau + \epsilon))Y_\mu(\tau + \epsilon, t)f(x). \quad (4.84)$$

Using formula (4.68) in (4.84), we get

$$Y_\mu(\tau, t)f(x) - Y(\tau, \tau + \epsilon)Y_\mu(\tau + \epsilon, t)f(x)$$
$$= -\int_\tau^{\tau+\epsilon} Y(\tau, s)\left[\mu(s)Y_\mu(s, \tau + \epsilon)Y_\mu(\tau + \epsilon, t)f\right](x)ds$$

$$= -\int_{\tau}^{\tau+\epsilon} Y(\tau,s)\,[\mu(s)Y_\mu(s,t)f]\,(x)ds. \tag{4.85}$$

Now it is clear that (4.83) follows from (4.85).
This completes the proof of Lemma 4.4. □

Let us return to the proof of Theorem 4.13. Suppose that ψ is a function such as in the formulation of Theorem 4.13, and put

$$M = \min_{(\tau,x)\in[\tau_0,\tau_0+\delta]\times E}(Y_\mu(\tau,t)f(x) - \psi(\tau,x)).$$

Define the function G by (4.82). Then, Lemma 4.4 with $\tau = \tau_0$, $x = x_0$, and $\epsilon > 0$ such that $\tau_0 + \epsilon < t$ implies that

$$G(\tau_0, x_0) - Y(\tau_0, \tau_0+\epsilon)\,G(\tau_0+\epsilon)(x_0)$$
$$= Y(\tau_0, \tau_0+\epsilon)\,\psi(\tau_0+\epsilon)(x_0) - \psi(\tau_0, x_0)$$
$$- \int_{\tau_0}^{\tau_0+\epsilon} Y(\tau_0,s)\,[\mu(s)Y_\mu(s,t)f]\,(x_0)ds. \tag{4.86}$$

Next, dividing (4.86) by ϵ and using the facts that $G(\tau_0,x_0) = 0$ and $G(\tau_0+\epsilon,y) \geq 0$ for all $\epsilon < \delta$ and $y \in E$, we get estimate (4.80).
This completes the proof of Theorem 4.13. □

The next theorem is a local version of Theorem 4.13.

Theorem 4.14 *Let $\mu \in \mathcal{P}_m^*$, $f \in L^\infty$, and fix t such that $0 < t \leq T$. Then the following assertions hold:*

(a) Suppose that ψ is a bounded continuous function on $[0,T] \times E$ and (τ_0, x_0) is a point in $[0,t) \times E$. Suppose also that there exists $\delta > 0$ with $\tau_0 + \delta < t$ and a relatively compact neighborhood Q of x_0 in E such that

$$Y_\mu(\tau_0,t)f(x_0) - \psi(\tau_0,x_0) = \min_{(\tau,x)\in[\tau_0,\tau_0+\delta]\times\bar{Q}}(Y_\mu(\tau,t)f(x) - \psi(\tau,x)),$$

where $\bar{Q}$ denotes the closure of Q in E. Then for every $0 < \epsilon < \delta$,

$$\frac{Y(\tau_0,\tau_0+\epsilon)\,\psi(\tau_0+\epsilon)(x) - \psi(\tau_0,x_0)}{\epsilon}$$
$$- \frac{1}{\epsilon}\int_{\tau_0}^{\tau_0+\epsilon} Y(\tau_0,s)\,[\mu(s)Y_\mu(s,t)f]\,(x_0)ds$$
$$\leq \frac{\alpha}{\epsilon}\int_{E\setminus\bar{Q}} p(\tau_0,x_0;\tau_0+\epsilon,y)\,dy, \tag{4.87}$$

where $\alpha > 0$ and $\delta > 0$ do not depend on ϵ.

(b) Suppose that ψ is a bounded continuous function on $[0, T] \times E$, and let (τ_0, x_0) be a point in $[0, t) \times E$. Suppose also that there exists a relatively compact neighborhood Q of (τ_0, x_0) in $[0, t) \times E$ such that
$$Y_\mu(\tau_0, t) f(x_0) - \psi(\tau_0, x_0) = \max_{(\tau, x) \in \bar{Q}} (Y_\mu(\tau, t) f(x) - \psi(\tau, x)).$$

Then for every small $\epsilon > 0$,
$$\frac{Y(\tau_0, \tau_0 + \epsilon) \psi(\tau_0 + \epsilon)(x) - \psi(\tau_0, x_0)}{\epsilon}$$
$$- \frac{1}{\epsilon} \int_{\tau_0}^{\tau_0 + \epsilon} Y(\tau_0, s) [\mu(s) Y_\mu(s, t) f](x_0) ds$$
$$\geq -\frac{\alpha}{\epsilon} \int_{E \setminus \bar{Q}} p(\tau_0, x_0; \tau_0 + \epsilon, y) \, dy, \qquad (4.88)$$

where $\alpha > 0$ and $\delta > 0$ do not depend on ϵ.

Proof. We will only prove part (a) of Theorem 4.14. The proof of part (b) is similar. Suppose that the conditions in part (a) of Theorem 4.14 are satisfied. Define G by formula (4.82) with ψ as in the formulation of Theorem 4.14 and with the number M given by
$$M = \min_{(\tau, x) \in [\tau_0, \tau_0 + \delta] \times \bar{Q}} (Y_\mu(\tau, t) f(x) - \psi(\tau, x)).$$

Then, using equality (4.83) with $\tau = \tau_0$, $x = x_0$, and $\epsilon > 0$ such that $\epsilon < \delta$ and taking into account that $G(\tau_0, x_0) = 0$, $G(\tau, x) \geq 0$ for $(\tau, x) \in [\tau_0, \tau_0 + \delta] \times \bar{Q}$, and $|G(\tau, x)| \leq \alpha$, we get
$$\frac{Y(\tau_0, \tau_0 + \epsilon) \psi(\tau_0 + \epsilon)(x_0) - \psi(\tau_0, x_0)}{\epsilon}$$
$$- \frac{1}{\epsilon} \int_{\tau_0}^{\tau_0 + \epsilon} Y(\tau_0, s) [\mu(s) Y_\mu(s, t) f](x_0) ds$$
$$= -Y(\tau_0, \tau_0 + \epsilon) \chi_{\bar{Q}} G(\tau_0 + \epsilon)(x_0) - Y(\tau_0, \tau_0 + \epsilon) \chi_{E \setminus \bar{Q}} G(\tau_0 + \epsilon)(x_0)$$
$$\leq \frac{\alpha}{\epsilon} \int_{E \setminus \bar{Q}} p(\tau_0, x_0; \tau_0 + \epsilon, y) \, dy.$$

Therefore, estimate (4.87) holds.

This completes the proof of Theorem 4.14. □

Our next goal is to make several simplifications in inequalities (4.80), (4.81), (4.87), and (4.88). The following lemma concerns the first term on the left-hand side of estimates (4.80), (4.81), (4.87), and (4.88). For $h \in BC$ and $(\tau, x) \in [0, T] \times E$, we put

$$A_+(\tau)h(x) = \lim_{\epsilon \to 0+} \frac{Y(\tau, \tau + \epsilon)h(x) - h(x)}{\epsilon}, \quad (4.89)$$

provided that the limit in (4.89) exists and is finite. We will say that a bounded continuous function ψ on $[0, T] \times E$ is differentiable from the right at $\tau_0 \in [0, T)$ uniformly with respect to $y \in E$, if there exists a function $D_1^+ \psi(\tau_0, \cdot) \in BC$ such that

$$\lim_{\epsilon \to 0+} \sup_{y \in E} \left| \frac{\psi(\tau_0 + \epsilon, y) - \psi(\tau_0, y)}{\epsilon} - D_1^+ \psi(\tau_0, y) \right| = 0. \quad (4.90)$$

Lemma 4.5 *Suppose that the free backward propagator Y satisfies the conditions in Theorem 4.10, and let ψ be a bounded continuous function on $[0, T] \times R^n$. Let $\tau_0 \in [0, t)$ and $x_0 \in E$ be such that ψ is differentiable from the right at $\tau_0 \in [0, T)$ uniformly with respect to $y \in E$, and $A_+(\tau_0) \psi(\tau_0)(x_0)$ exists and is finite. Then*

$$\lim_{\epsilon \to 0+} \frac{Y(\tau_0, \tau_0 + \epsilon) \psi(\tau_0 + \epsilon)(x_0) - \psi(\tau_0, x_0)}{\epsilon}$$
$$= D_1^+ \psi(\tau_0, x_0) + [A_+(\tau_0) \psi(\tau_0)](x_0). \quad (4.91)$$

Proof. We have

$$\frac{Y(\tau_0, \tau_0 + \epsilon) \psi(\tau_0 + \epsilon)(x_0) - \psi(\tau_0, x_0)}{\epsilon}$$
$$= Y(\tau_0, \tau_0 + \epsilon) \left\{ \frac{\psi(\tau_0 + \epsilon) - \psi(\tau_0)}{\epsilon} - D_1^+ \psi(\tau_0) \right\}(x_0)$$
$$+ [Y(\tau_0, \tau_0 + \epsilon) D_1^+ \psi(\tau_0)](x_0) + \left[\frac{Y(\tau_0, \tau_0 + \epsilon) - I}{\epsilon} \psi(\tau_0) \right](x_0)$$
$$= I_1 + I_2 + I_3. \quad (4.92)$$

Since Y is a family of contraction operators on L^∞,

$$|I_1| \leq \sup_{y \in E} \left| \frac{\psi(\tau_0 + \epsilon, y) - \psi(\tau_0, y)}{\epsilon} - D_1^+ \psi(\tau_0, y) \right|.$$

It follows from (4.90) that

$$\lim_{\epsilon \to 0} I_1 = 0. \quad (4.93)$$

Since $D_1^+\psi(\tau_0,\cdot) \in BC$, and the conditions in Theorem 4.10 are satisfied, we get

$$\lim_{\epsilon \to 0} I_2 = D_1^+\psi(\tau_0, x_0). \tag{4.94}$$

Finally, (4.89) implies that

$$\lim_{\epsilon \to 0} I_3 = [A_+(\tau_0)\psi(\tau_0)](x_0). \tag{4.95}$$

Now it is clear that Lemma 4.5 follows from (4.92)-(4.95). □

Next, we turn our attention to the second term on the right-hand side of estimates (4.80), (4.81), (4.87), and (4.88). For $\mu \in \mathcal{P}_m^*$, consider its Radon-Nikodym-Lebesgue decomposition $d\mu(s) = V(s)dm + d\lambda(s)$, where $\lambda(s)$ is the singular part of $\mu(s)$ with respect to m. It is clear that $V \in \mathcal{P}_f^*$ and $\lambda \in \mathcal{P}_m^*$. Let $x_0 \in E$ and $\tau_0 \in [0, t)$ be given, and suppose that C_k with $-\infty < k < \infty$ is a strictly increasing sequence of Borel sets of positive measure m such that $x_0 \in C_k$ for all k, $\mathrm{diam}(C_k) + m(C_k) \to 0$ as $k \to -\infty$, and $\bigcup_{k=0}^{\infty} C_k = E$. For every integer j, put

$$\gamma_j(s) = \sup_{y \in E \setminus C_j} p(\tau_0, x_0; s, y),$$

and define the majorant p^* of p with respect to the family $\{C_k\}$ as follows:

$$p^*(\tau_0, x_0; s, z) = \gamma_j(s)$$

where j is the unique integer such that $z \in C_{j+1} \setminus C_j$. Let us also recall that the function $Y_\mu(s,t)f(x)$ is bounded on $[0,t] \times E$. Moreover, it is continuous on $[0,t) \times E$, by Theorem 4.10 and Remark 4.4. The following conditions will be used in the sequel:

$$\sup_{s:\tau_0 \leq s \leq \tau_0+\delta} \frac{1}{m(C_k)} \int_{C_k} |V(s,y) - V(\tau_0,x_0)|dy \to 0 \tag{4.96}$$

as $k \to -\infty$, where $\delta > 0$ is a number such that $\tau_0 + \delta < t$;

$$\sup_{k:k \geq j} \sup_{s:\tau_0 \leq s \leq \tau_0+\delta} \frac{1}{m(C_k)} \int_{C_k} |V(s,y)|dy \leq M_{1,j} \tag{4.97}$$

for all $j \in \mathbb{Z}$;

$$\sup_{s:\tau_0 \leq s \leq \tau_0+\delta} \frac{|\lambda(s)|(C_k)}{m(C_k)} \to 0 \tag{4.98}$$

as $k \to -\infty$;

$$\sup_{k:k\geq j} \sup_{s:\tau_0\leq s\leq \tau_0+\delta} \frac{|\lambda(s)|(C_k)}{m(C_k)} \leq M_{2,j} \qquad (4.99)$$

for all $j \in \mathbb{Z}$;

$$\sup_{s:\tau_0\leq s\leq \tau_0+\delta} \int_{C_k} p^*(\tau_0, x_0; s, z)\, dz \leq M_{3,k} \qquad (4.100)$$

for all $k \in \mathbb{Z}$, and

$$\lim_{s\to \tau_0+} \int_{E\setminus C_k} p^*(\tau_0, x_0; s, z)\, dz = 0 \qquad (4.101)$$

for all $k \in \mathbb{Z}$.

Remark 4.6 Condition (4.96) means that x_0 is a Lebesgue point of the function $V(\tau, \cdot)$ uniformly with respect to τ near τ_0. Condition (4.97) resembles a uniform local integrability condition for V. Similarly, condition (4.98) is a differentiability condition for the singular part λ of μ, while (4.99) is a uniform local integrability condition for λ. Conditions (4.100) and (4.101) are expressed in terms of the majorant p^* of the transition density p. They are based on the integrability condition for the majorant of an approximation of the identity (see [Stein and Weiss (1971)], Theorem 1.25). Condition (4.100) concerns the local integrability of p^*, while (4.101) is a stochastic continuity condition for p^*.

Lemma 4.6 *Let $\mu \in \mathcal{P}_m^*$, and assume that the free backward propagator Y satisfies the conditions in Theorem 4.10. Let $\tau_0 \in [0, t)$, $x_0 \in E$, and let $\{C_k\}$ be such that conditions (4.96)-(4.101) hold. Then*

$$\lim_{\epsilon\to 0+} \frac{1}{\epsilon}\int_{\tau_0}^{\tau_0+\epsilon} Y(\tau_0, s)\, [\mu(s)Y_\mu(s, t)f)](x_0)ds = V(\tau_0, x_0)\, Y_\mu(\tau_0, t)\, f(x_0). \qquad (4.102)$$

Proof. Put

$$D(s, y) = V(s, y)Y_\mu(s, t)f(y) \text{ and } d\nu(s) = Y_\mu(s, t)f(y)d\lambda(s).$$

Since the function $Y_\mu(s, t)f(y)$ is continuous on $[0, t) \times E$ and bounded on $[0, t] \times E$, it follows from (4.96) that

$$\sup_{s:\tau_0\leq s\leq \tau_0+\delta} \frac{1}{m(C_k)} \int_{C_k} |D(s, y) - D(\tau_0, x_0)|\, dy \to 0 \qquad (4.103)$$

as $k \to -\infty$. In (4.103), $\delta > 0$ is a number such that $\tau_0 + \delta < t$. Moreover, (4.97)–(4.99) imply

$$\sup_{k: k \geq j} \sup_{s: \tau_0 \leq s \leq \tau_0 + \delta} \frac{1}{m(C_k)} \int_{C_k} |D(s, y)| dy \leq M_{4,j} \qquad (4.104)$$

for all $j \in \mathbb{Z}$;

$$\sup_{s: \tau_0 \leq s \leq \tau_0 + \delta} \frac{|\nu(s)|(C_k)}{m(C_k)} \to 0 \qquad (4.105)$$

as $k \to -\infty$; and

$$\sup_{k: k \geq j} \sup_{s: \tau_0 \leq s \leq \tau_0 + \delta} \frac{|\nu(s)|(C_k)}{m(C_k)} \leq M_{5,j} \qquad (4.106)$$

for all $j \in \mathbb{Z}$.

Our next goal is to show that

$$\lim_{\epsilon \to 0+} \frac{1}{\epsilon} \int_{\tau_0}^{\tau_0 + \epsilon} Y(\tau_0, s) \left[|D(s) - D(\tau_0, x_0)| \right](x_0) ds = 0 \qquad (4.107)$$

and

$$\lim_{\epsilon \to 0+} \frac{1}{\epsilon} \int_{\tau_0}^{\tau_0 + \epsilon} Y(\tau_0, s) |\nu(s)|(x_0) ds = 0. \qquad (4.108)$$

Put

$$T(s, y) = |D(s, y) - D(\tau_0, x_0)|.$$

Then we have

$$\begin{aligned} Y(\tau_0, s) T(s)(x_0) &= \int_E T(s, y) p(\tau_0, x_0; s, y) \, dy \\ &\leq \int_E T(s, y) p^*(\tau_0, x_0; s, y) \, dy \\ &= \int_0^\infty d\lambda \int_{\{y: p^*(\tau_0, x_0; s, y) \geq \lambda\}} T(s, y) dy. \end{aligned} \qquad (4.109)$$

Since $\gamma_k(s)$ is a non-increasing sequence, it follows from (4.109) that there exists $\delta > 0$ such that

$$Y(\tau_0, s) T(s)(x_0) \leq \sum_{k \in \mathbb{Z}} (\gamma_k(s) - \gamma_{k+1}(s)) \int_{C_k} T(s, y) dy. \qquad (4.110)$$

It is not hard to see that for any $j \in \mathbb{Z}$, (4.110) gives

$$Y(\tau_0, s) T(s)(x_0)$$
$$\leq \sup_{s:\tau_0 \leq s \leq \tau_0 + \delta} \sum_{k=-\infty}^{j} (\gamma_k(s) - \gamma_{k+1}(s)) m(C_k) \frac{1}{m(C_k)} \int_{C_k} T(s, y) dy$$
$$+ \sum_{k=j+1}^{\infty} (\gamma_k(s) - \gamma_{k+1}(s)) m(C_k) \frac{1}{m(C_k)} \int_{C_k} T(s, y) dy = J_1(j) + J_2(j, s).$$
(4.111)

We have

$$J_1(j) \leq \left\{ \sup_{s:\tau_0 \leq s \leq \tau_0+\delta} \sup_{k:-\infty < k \leq j} \frac{1}{m(C_k)} \int_{C_k} |D(s, y) - D(\tau_0, x_0)| dy \right\}$$
$$\times \sup_{s:\tau_0 \leq s \leq \tau_0+\delta} \int_{C_j} p^*(\tau_0, x_0; s, y) \, dy. \tag{4.112}$$

Moreover, for every $j \in \mathbb{Z}$,

$$J_2(j, s) \leq (M_{4,j+1} + |D(\tau_0, x_0)|) \sum_{k=j+1}^{\infty} (\gamma_k(s) - \gamma_{k+1}(s)) m(C_k)$$
$$\leq (M_{4,j+1} + |D(\tau_0, x_0)|) \left[\gamma_{j+1}(s) m(C_{j+1}) + \sum_{k=j+2}^{\infty} \gamma_k(s) m(C_k \setminus C_{k-1}) \right]$$
$$\leq (M_{4,j+1} + |D(\tau_0, x_0)|) \left[\frac{m(C_{j+1})}{m(C_{j+1} \setminus C_j)} + 1 \right] \int_{E \setminus C_j} p^*(\tau_0, x_0; s, y) \, dy$$
$$\leq (M_{4,j+1} + |D(\tau_0, x_0)|) \left[\frac{m(C_{j+1})}{m(C_{j+1} \setminus C_j)} + 1 \right] \int_{E \setminus C_j} p^*(\tau_0, x_0; s, y) \, dy.$$
(4.113)

It follows from (4.111), (4.112), and (4.113) that

$$Y(\tau_0, s) [|D(s) - D(\tau_0, x_0)|] (x_0)$$
$$\leq \left\{ \sup_{s:\tau_0 \leq s \leq \tau_0+\delta} \sup_{k:-\infty < k \leq j} \frac{1}{m(C_k)} \int_{C_k} |D(s, y) - D(\tau_0, x_0)| dy \right\}$$
$$\times \sup_{s:\tau_0 \leq s \leq \tau_0+\delta} \int_{C_j} p^*(\tau_0, x_0; s, y) \, dy$$
$$+ (M_{4,j+1} + |D(\tau_0, x_0)|) \left[\frac{m(C_{j+1})}{m(C_{j+1} \setminus C_j)} + 1 \right] \int_{E \setminus C_j} p^*(\tau_0, x_0; s, y) \, dy$$

for all $j \in \mathbb{Z}$. Now it is not difficult to show that conditions (4.103) and (4.104) imply

$$\lim_{s \to \tau_0+} Y(\tau_0, s) \left[|D(s) - D(\tau_0, x_0)| \right](x_0) = 0.$$

This gives equality (4.107). The proof of equality (4.108) is similar. Here we use (4.105) and (4.106) instead of (4.103) and (4.104). It is clear that (4.107) and (4.108) imply (4.102).

This completes the proof of Lemma 4.6. □

Now we are ready to formulate the main results of the present section. The first of them concerns viscosity solutions in the case of global maxima or minima.

Theorem 4.15 *Let $\mu \in \mathcal{P}_m^*$, $f \in L^\infty$, $0 < t \leq T$, and suppose that the transition density p is such that the corresponding free backward propagator Y satisfies the conditions in Theorem 4.10. Then the following two assertions hold:*

Let $(\tau_0, x_0) \in [0, t) \times E$, $\delta > 0$ with $\tau_0 + \delta < t$, and ψ be such that:

(1) ψ is a bounded continuous function on $[0, T] \times E$;
(2) ψ is differentiable from the right at τ_0 uniformly with respect to $y \in E$;
(3) $A_+(\tau_0) \psi(\tau_0)(x_0)$ exists and is finite;
(4) There exists a sequence of sets C_k such that conditions (4.96)–(4.99) hold;
(5) The equality

$$Y_\mu(\tau_0, t) f(x_0) - \psi(\tau_0, x_0) = \min_{(\tau, x) \in [\tau_0, \tau_0 + \delta] \times E} (Y_\mu(\tau, t) f(x) - \psi(\tau, x))$$

holds.

Then

$$D_1^+ \psi(\tau_0, x_0) + [A_+(\tau_0) \psi(\tau_0)](x_0) - V(\tau_0, x_0) Y_\mu(\tau_0, t) f(x_0) \leq 0.$$

Suppose that conditions 1–4 in part (a) of Theorem 4.15 are satisfied. Suppose also that

$$Y_\mu(\tau_0, t) f(x_0) - \psi(\tau_0, x_0) = \max_{(\tau, x) \in [\tau_0, \tau_0 + \delta] \times E} (Y_\mu(\tau, t) f(x) - \psi(\tau, x)).$$

Then

$$D_1^+ \psi(\tau_0, x_0) + [A_+(\tau_0) \psi(\tau_0)](x_0) - V(\tau_0, x_0) Y_\mu(\tau_0, t) f(x_0) \geq 0.$$

It is clear that Theorem 4.15 follows from Theorem 4.13, Lemma 4.5, and Lemma 4.6.

Our next result concerns viscosity solutions in the case of local maxima and minima.

Theorem 4.16 *Let $\mu \in \mathcal{P}_m^*$, $f \in L^\infty$, $0 < t \le T$, and suppose that the transition density p is such that the corresponding free backward propagator Y satisfies the conditions in Theorem 4.10. Then the following two assertions hold:*

Let $(\tau_0, x_0) \in [0, t) \times E$, $\delta > 0$ with $\tau_0 + \delta < t$, and ψ be such that:

(1) ψ is a bounded continuous function on $[0, T] \times E$;
(2) ψ is differentiable from the right at τ_0 uniformly with respect to $y \in E$;
(3) $A_+(\tau_0) \psi(\tau_0)(x_0)$ exists and is finite;
(4) There exists a sequence of sets C_k such that conditions (4.96)–(4.99) hold;
(5) The equality

$$\lim_{\epsilon \to 0+} \frac{1}{\epsilon} \int_{E \setminus \bar{Q}} p(\tau_0, x_0; \tau_0 + \epsilon, y)\, dy = 0 \quad (4.114)$$

holds for every relatively compact neighborhood Q of x_0;
(6) There exists a relatively compact neighborhood Q of (τ_0, x_0) in $[0, t) \times E$ such that

$$Y_\mu(\tau_0, t) f(x_0) - \psi(\tau_0, x_0) = \min_{(\tau, x) \in [\tau_0, \tau_0 + \delta] \bar{Q}} (Y_\mu(\tau, t) f(x) - \psi(\tau, x)).$$

Then

$$D_1^+ \psi(\tau_0, x_0) + [A_+(\tau_0) \psi(\tau_0)](x_0) - V(\tau_0, x_0) Y_\mu(\tau_0, t) f(x_0) \le 0.$$

Suppose that conditions 1–5 in part (a) are satisfied. Suppose also that there exists a relatively compact neighborhood Q of (τ_0, x_0) in $[0, t) \times E$ such that

$$Y_\mu(\tau_0, t) f(x_0) - \psi(\tau_0, x_0) = \max_{(\tau, x) \in [\tau_0, \tau_0 + \delta] \bar{Q}} (Y_\mu(\tau, t) f(x) - \psi(\tau, x)).$$

Then

$$D_1^+ \psi(\tau_0, x_0) + [A_+(\tau_0) \psi(\tau_0)](x_0) - V(\tau_0, x_0) Y_\mu(\tau_0, t) f(x_0) \ge 0.$$

Theorem 4.16 follows from Theorem 4.14, Lemma 4.5, and Lemma 4.6.

Example 4.1 Let E be d-dimensional Euclidean space $\mathbb{R}^d$. We will assume that the reference measure m coincides with the Lebesgue measure m_d on $\mathbb{R}^d$. Suppose that p is a fundamental solution of a second order parabolic partial differential equation with time-dependent coefficients such as in Sections 3.6 and 3.7. Then p is a transition probability density (see Theorems 3.2 and 3.5). Recall that the density p in these theorems satisfies the upper Gaussian estimate; that is,

$$p(\tau, x; t, y) \leq \alpha_1 g_d \left(\alpha_2(t-\tau), x-y\right), \qquad (4.115)$$

where α_1 and α_2 are positive constants. In estimate (4.115), g_d stands for the d-dimensional Gaussian density given by

$$g_d(s, z) = \frac{1}{(2\pi s)^{\frac{d}{2}}} \exp\left\{-\frac{|z|^2}{2s}\right\}.$$

Define the radial majorant of the transition density p by

$$p^*(\tau, x; t, y) = \sup_{z: |z-x| \geq |y-x|} p(\tau, x; t, z).$$

It is clear that if estimate (4.115) holds for p, then

$$p^*(\tau, x; t, y) \leq \alpha_1 g_d \left(\alpha_2(t-\tau), x-y\right),$$

and hence, conditions (4.100) and (4.101) hold for p^*. It is not difficult to prove that condition (4.114) also holds. We will assume that the sets C_k in the formulation of Theorem 4.15 and Theorem 4.16 are given by $C_k = B(x_0, r_k)$ where $r_k \downarrow 0$ as $k \to -\infty$ and $r_k \uparrow \infty$ as $k \to \infty$. The next assertion follows from Theorem 4.15.

Corollary 4.4 *Let p be a transition probability density on $\mathbb{R}^d$ such that estimate (4.115) holds for p. Let $\mu \in \mathcal{P}^*$, $f \in L^\infty$, $0 < t \leq T$, and suppose that Y satisfies the conditions in Theorem 4.10. Then the following two assertions hold:*

Let $(\tau_0, x_0) \in [0, t) \times E$, $\delta > 0$ with $\tau_0 + \delta < t$, and ψ be such that:

(1) ψ is a bounded continuous function on $[0, T] \times E$;
(2) ψ is differentiable from the right at τ_0 uniformly with respect to $y \in E$;
(3) $A_+(\tau_0) \psi(\tau_0)(x_0)$ exists and is finite;
(4) Conditions (4.96)-(4.99) hold with $C_k = B(x_0, r_k)$ where r_k are such that $r_k \downarrow 0$ as $k \to -\infty$ and $r_k \uparrow \infty$ as $k \to \infty$;

(5) There exists a relatively compact neighborhood Q of x_0 in E such that

$$Y_\mu(\tau_0, t) f(x_0) - \psi(\tau_0, x_0) = \min_{(\tau,x) \in [\tau_0, \tau_0+\delta] \times \bar{Q}} (Y_\mu(\tau, t) f(x) - \psi(\tau, x)).$$

Then

$$D_1^+ \psi(\tau, x) + [A_+(\tau)\psi(\tau)](x) - V(\tau_0, x_0) Y_\mu(\tau_0, t) f(x_0) \leq 0.$$

Suppose that conditions 1-4 in part (a) are satisfied. Suppose also that there exists a relatively compact neighborhood Q of x_0 in E such that

$$Y_\mu(\tau_0, t) f(x_0) - \psi(\tau_0, x_0) = \max_{(\tau,x) \in [\tau_0, \tau_0+\delta] \times \bar{Q}} (Y_\mu(\tau, t) f(x) - \psi(\tau, x)).$$

Then

$$D_1^+ \psi(\tau, x) + [A_+(\tau)\psi(\tau)](x) - V(\tau_0, x_0) Y_\mu(\tau_0, t) f(x_0) \geq 0.$$

4.9 Notes and Comments

(a) The Kato class of potential functions was introduced and studied in [Aizenman and Simon (1982); Simon (1982)]. The definition of the Kato class in these papers is based on a condition used in [Kato (1973)]. Similar classes were studied in [Stummel (1956)] and [Schechter (1971)]. More information on the Kato classes of functions and measures can be found in [Johnson and Lapidus (2000); Demuth and van Casteren (2000); Gulisashvili (2002c)].

(b) Schrödinger semigroups are discussed in [Aizenman and Simon (1982); Simon (1979); Simon (1982); Carmona (1974); Chung and Zhao (1995); Blanchard and Ma (1990a); Blanchard and Ma (1990b); Davies (1997); Johnson and Lapidus (2000); Demuth and van Casteren (2000); Zhang (2001); Gulisashvili and Kon (1996); Gulisashvili (2000)]. See also [Carmona, Masters, and Simon (1990)].

(c) The Feynman-Kac formula in (4.2) goes back to Kac (see [Kac (1949); Kac (1951); Kac (1959)], see also [Kac (1979)]), who was inspired by Feynman's ideas. We refer the reader to [Johnson and Lapidus (2000); Kleinert (2004)] for more information on the Feynman integral and related topics.

(d) Kato classes and Feynman-Kac semigroups associated with general Markov processes are discussed in [Chung and Zhao (1995)].

(e) The results in Sections 4.3 and 4.4 concerning the inheritance of properties of free semigroups or propagators by their Feynman-Kac perturbations are taken from [Gulisashvili (2004b); Gulisashvili (2004c)]. For the case of Feynman-Kac propagators associated with the heat semigroup see [Gulisashvili (2005)]. See also [Ouhabaz, Stollmann, Sturm, and Voigt (1996)] for earlier results concerning time-independent perturbations of semigroups on the space L^1. Measure perturbations of semigroups of operators were studied in [Getoor (1999)].

(f) The reader may consult [Demuth and van Casteren (2000)] for more information on the integral kernels of Feynman-Kac semigroups.

(g) Viscosity solutions of partial differential equations were introduced in [Crandall and Lions (1983)] (see also [Crandall, Ishii, and Lions (1992)]). The results in Sections 4.7 and 4.8 concerning the generation of viscosity solutions by Feynman-Kac propagators are taken from [Gulisashvili and Van Casteren (2005)].

Chapter 5

Some Theorems of Analysis and Probability Theory

5.1 Monotone Class Theorems

In this section we formulate monotone class theorems for sets and functions. These theorems are due to Dynkin.

Definition 5.1 Let Ω be a set and let $\mathcal{S}$ be a collection of subsets of Ω. Then $\mathcal{S}$ is called a d-system if it has the following properties:

(a) $\Omega \in \mathcal{S}$.
(b) If A and B belong to $\mathcal{S}$ and if $A \supseteq B$, then $A \setminus B$ belongs to $\mathcal{S}$.
(c) If A_n, $n \in \mathbb{N}$, is an increasing sequence of elements of $\mathcal{S}$, then the union $\bigcup_{n=1}^{\infty} A_n$ belongs to $\mathcal{S}$.

Definition 5.2 Let Ω be a set and let $\mathcal{S}$ be a collection of subsets of Ω. Then $\mathcal{S}$ is called a π-system if it is closed under finite intersections.

It is clear that the intersection of any family of d-systems is a d-system. Let $\mathcal{S}$ be a collection of subsets of Ω. Then the smallest d-system containing $\mathcal{S}$ is called the d-system generated by $\mathcal{S}$. The next assertion is the monotone class theorem for sets.

Theorem 5.1 Let $\mathcal{M}$ be a π-system of subsets of Ω. Then the d-system generated by $\mathcal{M}$ coincides with the σ-algebra generated by $\mathcal{M}$.

Next we formulate the monotone class theorem for functions.

Theorem 5.2 Let Ω be a set and let $\mathcal{M}$ be a π-system of subsets of Ω. Let $\mathcal{H}$ be a vector space of real valued functions on Ω satisfying the following condition:

(i) The constant function 1 belongs to $\mathcal{H}$.
(ii) For any $A \in \mathcal{M}$, $\chi_A \in \mathcal{H}$.

(iii) If f_n, $n \in \mathbb{N}$, is an increasing sequence of non-negative functions in $\mathcal{H}$ such that $f = \sup_{n \in \mathbb{N}} f_n$ is finite (bounded), then $f \in \mathcal{H}$.

Then $\mathcal{H}$ contains all real valued functions (all real valued bounded functions) on Ω, which are $\sigma(\mathcal{M})$-measurable.

Theorems 5.1 and 5.2 are often used in the following setting. Let Ω be a set, and let $(E_i, \mathcal{E}_i)_{i \in I}$ be a family of measurable spaces, indexed by a set I. Suppose that for every $i \in I$, a π-system $\mathcal{S}_i$ of subsets of E_i generating $\mathcal{E}_i$ is given. Suppose also that for every $i \in I$, f_i is a mapping from Ω into E_i. Then the following two assertions hold.

Theorem 5.3 *Let $\mathcal{M}$ be the collection of all sets of the form $\bigcap_{i \in J} f_i^{-1}(A_i)$, where $A_i \in \mathcal{S}_i$, $i \in J$, and J is a finite subset of I. Then $\mathcal{M}$ is a π-system, and moreover,*

$$\sigma(\mathcal{M}) = \sigma(f_i : i \in I).$$

Theorem 5.4 *Let $\mathcal{H}$ be a vector space of real valued functions on Ω such that the following conditions hold:*

(i) *The constant function 1 belongs to $\mathcal{H}$.*
(ii) *If h_n, $n \in \mathbb{N}$, is an increasing sequence of non-negative functions in $\mathcal{H}$ such that $h = \sup_n h_n$ is finite (bounded), then h belongs to $\mathcal{H}$.*
(iii) *$\mathcal{H}$ contains all products of the form $\prod_{i \in J} \chi_{A_i} \circ f_i$, where $A_i \in \mathcal{S}_i$, $i \in J$, and J is a finite subset of I.*

Then $\mathcal{H}$ contains all real valued functions (all bounded real valued functions) which are measurable with respect to the σ-algebra $\sigma(f_i : i \in I)$.

In our presentation of the Markov property, $I = [0, T]$, $E_t = E$ for all $t \in [0, T]$ where E is the state space of the process, and the maps f_t, $t \in [0, T]$, are the state variables X_t.

We refer the reader to [Blumenthal and Getoor (1968); Sharpe (1988)] for more information on the monotone class theorems.

5.2 Kolmogorov's Extension Theorem

Let E be a locally compact second countable Hausdorff topological space equipped with the Borel σ-algebra $\mathcal{E}$. Let $I = [0, T]$. For every finite subset J of the set I, let $\mathbb{P}_J$ be a probability measure on the measurable space $(E^J, \mathcal{B}_{E^J})$. A family $\{\mathbb{P}_J : J \subset I, J \text{ finite}\}$ of such probability measures is called a projective or consistent family on $E^I = \prod_{t \in I} E$, provided that for any pair $J \subseteq K$ of finite subsets of I, and for any set $A \in \mathcal{B}_{E^J}$, the equality

$$\mathbb{P}_J(A) = \mathbb{P}_K \left\{ (\omega_j)_{j \in K} \in E^K : (\omega_j)_{j \in J} \in A \right\} \tag{5.1}$$

holds. For $J \subset K$ as above, define the projection

$$p_J^K : E^K \to E^J \tag{5.2}$$

by the following:

$$p_J^K \left((\omega_j)_{j \in K} \right) = (\omega_j)_{j \in J}.$$

Then equality (5.1) can be rewritten as follows:

$$\mathbb{P}_J(A) = \mathbb{P}_K \left\{ \left(p_J^K \right)^{-1} (A) \right\}.$$

Note that it is not necessary to assume that the sets in (5.2) are finite. The set E^I will be equipped with the σ-algebra $\mathcal{F}$ generated by the mappings $\{p_J^I : J \subset I, J \text{ finite}\}$. Now we are ready to formulate Kolmogorov's extension theorem.

Theorem 5.5 *Let $\{\mathbb{P}_J : J \subset I, J \text{ finite}\}$ be a a projective family on E^I. Then there exists a unique probability measure $\mathbb{P}_I$ on $(E^I, \mathcal{F})$ such that*

$$\mathbb{P}_I \left(\left(p_J^I \right)^{-1} (A) \right) = \mathbb{P}_J(A)$$

for every finite subset J of the set I and every $A \in \mathcal{B}_{E^J}$.

The proof of Kolmogorov's extension theorem can be found in [Bhattacharya and Waymire (1990)], pages 92-93.

5.3 Uniform Integrability

Let $(S, \mathcal{A}, \nu)$ be a measure space.

Definition 5.3 A family of functions $\{f_j : j \in J\}$ in $L^1(S, \mathcal{A}, \nu)$ is called uniformly integrable if for every $\varepsilon > 0$ there exists $\delta > 0$ such that

$$\sup_{j \in J} \int_A |f_j| \, d\nu \leq \varepsilon$$

whenever $A \in \mathcal{A}$ and $\nu(A) \leq \delta$.

It is clear that if the family $\{f_j : j \in J\}$ is uniformly integrable, and if $\{g_j : j \in J\}$ is such that for every $j \in J$, $|g_j| \le |f_j|$ ν-almost everywhere, then the family $\{g_j : j \in J\}$ is uniformly integrable. It is also clear that Cauchy sequences in $L^1(S, \mathcal{A}, \nu)$ are uniformly integrable. Next we give an example of a family of functions which is not uniformly integrable. Let $f \ge 0$ be a function from the space $L^1(\mathbb{R}^d, \mathcal{B}_{\mathbb{R}^d}, m)$ where m is the Lebesgue measure on $\mathbb{R}^d$. Suppose $\int f(x) dm(x) > 0$ and $\lim_{n \to \infty} n^d f(nx) = 0$ for all $x \ne 0$, and put $f_n(x) = n^d f(nx)$, $n \in \mathbb{N}$. Then the sequence f_n is not uniformly integrable (see [Meyer (1966)] for more information on the uniform integrability of families of functions).

5.4 Radon-Nikodym Theorem

Let $(S, \mathcal{A})$ be a measurable space, and let ν_1 and ν_2 be two measures on $\mathcal{A}$. By definition, the measure ν_2 is absolutely continuous with respect to the measure ν_1 if for every $A \in \mathcal{A}$ with $\nu_1(A) = 0$, the condition $\nu_2(A) = 0$ holds. A measure ν on $\mathcal{A}$ is called σ-finite if there exists an increasing sequence $S_n \in \mathcal{A}$, $n \in \mathbb{N}$, such that $S = \bigcup_{n \in \mathbb{N}} S_n$ and $\nu(S_n) < \infty$ for all $n \in \mathbb{N}$. Now we are ready to formulate the Radon-Nikodym theorem.

Theorem 5.6 *Let $(S, \mathcal{A})$ be a measurable space, and let ν_1 and ν_2 be two measures on $\mathcal{A}$. Suppose that the measure ν_1 is σ-finite, and the measure ν_2 is absolutely continuous with respect to the measure ν_1. Then there exists a nonnegative $\mathcal{A}$-measurable function f such that $\nu_2(A) = \int_A f d\nu_1$ for all $A \in \mathcal{A}$.*

The function f is unique in the following sense: If f_1 and f_2 are two functions satisfying the conditions in Theorem 5.6, then $f_1 = f_2$ almost everywhere with respect to the measure ν_1. The function f in Theorem 5.6 is called the Radon-Nikodym derivative of the measure ν_2 with respect to the measure ν_1. If $\nu_2(S) < \infty$, then the Radon-Nikodym derivative of a σ-finite measure ν_1 belongs to the space $L^1(S, \mathcal{A}, \nu_1)$.

An important corollary of the Radon-Nikodym theorem concerns the existence of conditional expectations.

Corollary 5.1 *Let $(\Omega, \mathcal{F}, \mathbb{P})$ be a probability space, and let $\mathcal{F}_0$ be a sub-σ-algebra of $\mathcal{F}$. Suppose that $F : \Omega \to [0, \infty]$ is a random variable from the space $L^1(\Omega, \mathcal{F}, \mathbb{P})$. Then there exists a random variable $G \in L^1(\Omega, \mathcal{F}_0, \mathbb{P})$ such that $\mathbb{E}[F\chi_A] = \mathbb{E}[G\chi_A]$ for all $A \in \mathcal{F}_0$.*

The random variable G is called the conditional expectation of the random variable F given the σ-algebra $\mathcal{F}_0$. It is defined $\mathbb{P}$-almost surely, and is denoted by $G = \mathbb{E}[F \mid \mathcal{F}_0]$. We refer the reader to [Folland (1999)] for more information on the Radon-Nikodym theorem.

5.5 Vitali-Hahn-Saks Theorem

Let $(S, \mathcal{A})$ be a measurable space, and let ν be a signed real-valued measure on $\mathcal{A}$. By $|\nu|$ will be denoted the variation of ν. By Corollary 5 on p.127 in [Dunford and Schwartz (1988)], the measure ν is of bounded variation. The Vitali-Hahn-Saks theorem concerns setwise convergent sequences of signed measures of bounded variation which are absolutely continuous with respect to a non-negative measure λ.

Theorem 5.7 *Let $(S, \mathcal{A})$ be a measurable space, and let λ be a non-negative finite measure on $\mathcal{A}$. Suppose that ν_n is a sequence of signed real-valued measures on $\mathcal{A}$ such that for every $n \geq 1$, the measure ν_n is absolutely continuous with respect to the measure λ and the limit $\lim_{n \to \infty} \nu_n(A)$ exists for every $A \in \mathcal{A}$. Then $\lim_{|\nu(A)| \to 0} \nu_n(A) = 0$ uniformly for $n \geq 1$.*

The following corollary to Theorem 5.7 is useful. It is due to Nikodým.

Corollary 5.2 *Let ξ_n be a sequence of signed real-valued measures on $(S, \mathcal{A})$ such that the limit $\xi(A) = \lim_{n \to \infty} \xi_n(A)$ exists for all $A \in \mathcal{A}$. Then the sequence $\{\xi_n\}$ is uniformly countably additive and the set function ξ is countably additive on $\mathcal{A}$.*

We refer the reader to [Dunford and Schwartz (1988)] for more information on the Vitali-Hahn-Saks theorem.

5.6 Doob's Inequalities

The next assertion contains Doob's inequalities for right-continuous martingales (see, e.g., [Revuz and Yor (1991)]).

Theorem 5.8 *Let X_t, $0 \leq t \leq T$, be a right-continuous martingale on $(\Omega, \mathcal{F}, \mathcal{G}_t, \mathbb{P})$. Then for all $p \geq 1$ and $\lambda > 0$,*

$$\lambda^p \mathbb{P}\left[\sup_{t \in [0,T]} |X_t| \geq \lambda\right] \leq \mathbb{E}\left[|X_T|^p\right].$$

Moreover, for all $p > 1$,

$$\left\| \sup_{t \in [0,T]} |X_t| \right\|_p \leq \frac{p}{p-1} \|X_T\|_p.$$

Bibliography

Acquistapace, P. (1993). Abstract Linear Nonautonomous Parabolic Equations: A Survey, in: Differential Equations in Banach Spaces: Proc. of the Bologna Conference (eds. G. Dore, A. Favini, E. Obrecht, and A. Venni), *Lect. Notes in Pure and Appl. Math.* **148**, Marcel Dekker, Inc., New York, pp. 1–19.

Aebi R. (1996). Schrödinger Diffusion Processes, *Probability and its Applications*, Birkhäuser, Basel.

Aizenman, N. and Simon, B. (1982). Brownian motion and Harnack's inequality for Schrödinger operators, *Comm. Pure Appl. Math.* **35**, pp. 209–273.

Albeverio, S., Blanchard, Ph., and Ma, Z. M. (1991). Feynman-Kac semigroups in terms of signed smooth measures, in: *Proc. Random Partial Differential Equations*, Internat. Series of Numerical Math., **102**, Birkhäuser, Boston, pp. 1–31.

Albeverio, S. and Ma, Z. M. (1991). Additive functionals, nowhere Radon and Kato class smooth measures associated with Dirichlet forms, *Osaka J. Math.* **29**, pp. 247–265.

Applebaum, D. (2004). Lévy Processes and Stochastic Calculus, *Cambridge Studies in Advanced Mathematics* **93**, Cambridge University Press.

Aronson, D. G. (1967). Bounds on the fundamental solution of a parabolic equation, *Bull. Amer. Math. Soc.* **73**, pp. 890–896.

Aronson, D. G. (1968). Non-negative solutions of linear parabolic equations, *Ann. Scuola Norm. Sup. Pisa (3)* **22**, pp. 607–694.

Barndorff-Nielsen, O. E., Mikosch, T., Resnick, S. I. (Editors) (2001), Lévy Processes, Birkhäuser.

Bernstein, S. (1932). Sur les liaisons entre les grandeurs aléatoires, *Vehr. des intern. Mathematikerkongr.*, Band 1, Zürich.

Bertoin, J. (1996). Lévy Processes, Cambridge Univ. Press, Melbourne, NY.

Beznea, L. and Boboc, N. (2000). Excessive kernels and Revuz measures, *Probab. Theory Relat. Fields* **117**, pp. 267–288.

Beznea, L. and Boboc, N. (2004). Fine densities for excessive measures and the Revuz correspondence, *Potential Analysis* **20**, pp. 61–83.

Bhattacharya, R. N. and Waymire E. C. (1990). Stochastic Processes with Applications, *Wiley Series in Probability and Mathematical Statistics*, A Wiley-

Intescience Publication, John Wiley & Sons, New York.

Blanchard, Ph. and Ma, Z. M. (1990a). Semigroup of Schrödinger operators with potentials given by Radon measures, in: Stochastic Processes–Physics and Geometry (eds. S. Albeverio, et al.), World Scientific Publishing Co., Inc., Teaneck, NJ, pp. 160–195.

Blanchard, Ph. and Ma, Z. M. (1990b). New results on the Schrödinger semigroups with potentials given by signed smooth measures, in: Stochastic Analysis and Related Topics, II (Silivri, 1988) Lect. Notes in Math. **1444**, Springer-Verlag, Berlin, pp. 213–243.

Blumenthal, R. M. and Getoor, R. K. (1968). Markov processes and potential theory, Pure and Applied Mathematics **29**, Academic Press, New York.

Bourbaki, N. (1965). Éléments de Mathématiques, Livre III, topologie générale, 2nd edition, Chapter 9, Hermann, Paris.

Carlen, E. A. (1984). Conservative diffusions, Comm. Math. Physics **94**, pp. 293–315.

Carmona, R. (1974). Regularity properties of Schrödinger and Dirichlet semigroups, J. Funct. Anal. **17**, pp. 227–237.

Carmona, R., Masters, W. C., and Simon, B. (1990). Relativistic Schrödinger operators: asymptotic behavior of the eigenfunctions, J. Funct. Anal. **91**, pp. 117–142.

Cerrai, S. (2001). Second Order PDE's in Finite and Infinite Dimension. A Probabilistic Approach, Springer-Verlag, Berlin.

Chicone, C. and Latushkin, Y. (1999). Evolution Semigroups in Dynamical Systems and Differential Equations, American Mathematical Society, Providence, RI.

Chung, K. L. (1982). Lectures from Markov Processes to Brownian Motion, Grundlehren der Mathematischen Wissenschaften **249**, Springer-Verlag, New York.

Chung, K. L. and Doob, J. L. (1965). Fields, optionality and measurability, Amer. J. Math. **87**, pp. 397–424.

Chung, K. L. and Walsh, J. B. (1969). To reverse a Markov process, Acta Math. **123**, pp. 225–251.

Chung, K. L. and Walsh, J. B. (2005). Markov Processes, Brownian Motion, and Time Symmetry, Grundlehren der Mathematischen Wissenschaften **249**, Second edition, Springer-Verlag, New York.

Chung, K. L. and Williams R. J. (1990). Introduction to Stochastic Integration, Probability and its Applications, Birkhäuser, Boston, MA.

Chung, K. L. and Zambrini, J.-C. (2003). Introduction to Random Time and Quantum Randomness, Monographs of the Portuguese Mathematical Society **1**, New edition, World Sci. Publishing, River Edge, NJ.

Chung, K. L. and Zhao, Z. X. (1995). From Brownian Motion to Schrödinger's Equation, Grundlehren der Mathematischen Wissenschaften **312**, Springer-Verlag, Berlin.

Crandall, M. G. and Lions, P. L. (1983). Viscosity solutions of Hamilton-Jacobi equations. Trans. Amer. Math. Soc. **277**, pp. 1–42.

Crandall, M. G., Ishii, H., and Lions, P.-L. (1992). User's guide to viscosity solutions of second order partial differential equations, *Bull. Amer. Math. Soc. (N.S.)* **27**, pp. 1–67.

Cruzeiro, A. B. and Zambrini, J.-C. (1994). Euclidean Quantum Mechanics. An outline, A. L. Cardoso et al. (eds.), Stochastic Analysis and Applications in Physics, Kluwer Academic Publishers, pp. 59–97.

Cruzeiro, A. B., Wu, Liming, and Zambrini, J. C. (2000). Bernstein processes associated with a Markov process, in: Stochastic Analysis and Mathematical Physics, ANESTOC'98, Proceedings of the Third International Workshop, Ed. R. Rebolledo, Trends in Mathematics Series, Birkhäuser, Boston, 2000.

Davies, E. B. (1997). L^p-spectral theory of higher-order elliptic differential operators, *Bull. London Math. Soc.* **29**, pp. 513–546.

Dellacherie, C. and Meyer, P.-A. (1978). Probabilities and Potential, North-Holland, Amsterdam.

Demuth, M. and van Casteren, J. A. (2000). Stochastic Spectral Theory for Self-adjoint Feller Operators: A functional integration approach, *Probability and its Applications*, Birkhäuser, Basel.

Doob, J. L. (2001). Classical Potential Theory and Its Probabilistic Counterpart, *Classics in Mathematics*, Reprint of the 1984 edition, Springer-Verlag, Berlin.

Dressel, F. G. (1940). The fundamental solution of the parabolic equation, *Duke Math. J.* **7**, pp. 186–203.

Dressel, F. G. (1946). The fundamental solution of the parabolic equation, II *Duke Math. J.* **13**, pp. 61–77.

Dunford, N. and Schwartz, J. T. (1988). Linear Operators, Part I: General Theory, A Wiley-Interscience Publication, John Wiley & Sons, New York.

Durrett, R. (1984). Brownian Motion and Martingales in Analysis, *Wadsworth Mathematics Series*, Wadsworth International Group, Belmont, CA.

Durrett, R. (1991). Probability: Theory and Examples, *The Wadsworth & Brooks/Cole Statistics/Probability Series* Wadsworth & Brooks/Cole Advanced Books & Software, Pacific Grove, CA.

Durrett, R. (1996). Stochastic Calculus, A practical introduction, *Probability and Stochastic Series*, CRC Press, Boca Raton.

Dynkin, E. B. (1960). Theory of Markov Processes, Pergamon Press, Oxford-London-New York-Paris.

Dynkin, E. B. (1965). Markov processes. Vols. I, II, *Grundlehren der Mathematischen Wissenschaften* **121**, **122**, Academic Press Inc., Publishers, New York.

Dynkin, E. B. (1973). Regular Markov processes (Russian), *Uspekhi Mat. Nauk* **28**, pp. 35–64. English translation in: Markov Processes and Related Problems of Analysis, *London Math. Soc. Lecture Notes Ser.* **54**, Cambridege Univ. Press, Cambridge, 1982, pp. 187–218.

Dynkin, E. B. (1982). Markov Processes and Related Problems of Analysis, *London Mathematical Society Lecture Note Series* **54**, Cambridge Univ. Press, Cambridge.

Dynkin, E. B. (1994). An Introduction to Branching Measure-Valued Processes, *CRM Monograph Series* **6**, American Math. Society, Providence, Rhode

Island.

Dynkin, E. B. (2000). Selected papers of E. B. Dynkin with commentary, Edited by A. A. Yushkevich, G. M. Seitz and A. L. Onishchik, Amer. Math. Soc., Providence, RI.

Dynkin, E. B. (2002). Diffusions, Superdiffusions and Partial Differential Equations, *American Mathematical Society Colloquium Publications* **50**, Amer. Math. Soc., Providence, RI.

Dynkin, E. B. and Yushkevich, A. A. (1969). Markov Processes: Theorems and Problems, Plenum Press, New York.

Eberle, A. (1999). Uniqueness and Non-uniqueness of Semigroups Generated by Singular Diffusion Operators, Springer, *Lecture Notes in Math.* **1718**, Springer-Verlag, Berlin.

Eidel'man, S. D. (1969). Parabolic Systems, North-Holland Publishing Company, Amsterdam-London, Wolters-Noordhoff Publishing, Groningen.

Eidel'man, S. D. and Zhitarashu, N. V. (1998). Parabolic Boundary Value Problems, Birkhäuser, Basel.

Engel, K.-J. and Nagel, R. (2000). One-Parameter Semigroups for Linear Evolution Equations, *Graduate Texts in Mathematics*, **194**, With contributions by S. Brendle, M. Campiti, T. Hahn, G. Metafune, G. Nickel, D. Pallara, C. Perazzoli, A. Rhandi, S. Romanelli and R. Schnaubelt, Springer-Verlag, New York.

Ethier, S. N. and Kurtz, T. G. (1986). Markov Processes: Characterization and convergence, *Wiley Series in Probability and Mathematical Statistics: Probability and Mathematical Statistics*, John Wiley & Sons Inc., New York.

Fabes, E. B. (1993). Gaussian upper bounds on fundamental solutions of parabolic equations; the method of Nash, in: Dirichlet Forms (Varenna, 1992), *Lecture Notes in Math.* **1563**, Springer, Berlin, pp. 1–20.

Fabes, E. B. and Stroock, D. W. (1986). A new proof of Moser's parabolic Harnack inequality using the old ideas of Nash, *Arch. Rat. Mach. Anal.* **96**, pp. 327–338.

Folland, G. B. (1999). Real Analysis: Modern techniques and their applications, A Wiley-Interscience Publication, John Wiley & Sons Inc., New York.

Freidlin, M. (1985). Functional Integration and Partial Differential Equations, *Ann. of Math. Studies* **109**, Princeton University Press, Princeton, N.J.

Friedman, A. (1964). Partial Differential Equations of Parabolic Type, Prentice-Hall Inc., Englewood Cliffs, N.J.

Friedman, A. (1975). Stochastic Differential Equations and Applications, Vol. 1, *Probability and Mathematical Statistics* **28**, Academic Press [Harcourt Brace Jovanovich Publishers], New York.

Friedman, A. (1976). Stochastic Differential Equations and Applications, Vol. 2, *Probability and Mathematical Statistics* **28**, Academic Press [Harcourt Brace Jovanovich Publishers], New York.

Fukushima, M., Ōshima, Y., and Takeda, M. (1994). Dirichlet Forms and Symmetric Markov processes, *de Gruyter Studies in Mathematics* **19**, Walter de Gruyter & Co., Berlin.

Getoor, R. K. (1999). Measure perturbations of Markovian semigroups, *Potential*

Anal. **11**, pp. 101–133.
Gihman, I. I. and Skorohod, A. V. (1974). The Theory of Stochastic Processes I, Springer-Verlag, Berlin.
Gihman, I. I. and Skorohod, A. V. (1975). The Theory of Stochastic Processes II, Springer-Verlag, New York.
Gihman, I. I. and Skorohod, A. V. (1979). The Theory of Stochastic Processes III, Springer-Verlag, Berlin.
Glover, J., Rao, M., and Song R. (1993). Generalized Schrödinger semigroups, Seminar on Stochastic Processes (Seattle, WA, 1992), *Progr. Probab.* **33**, Birkhäuser, Boston, pp. 143–172.
Glover, J., Rao, M., Sikić, H., and Song, R. (1994). Quadratic forms corresponding to the generalized Schródinger semigroups, *J. Funct. Anal.* **125**, pp. 358–378.
Goldstein, J. A. (1985). Semigroups of Linear Operators and Applications, Oxford Mathematical Monographs, Oxford University Press, Oxford.
Gulisashvili, A. (2000). Sharp estimates in smoothing theorems for Schrödinger semigroups, *J. Funct. Anal.* **170**, pp. 161–187.
Gulisashvili, A. (2002a). On the heat equation with a time-dependent singular potential, *J. Funct. Anal.* **194**, pp. 17–52.
Gulisashvili, A. (2002b). Classes of time-dependent measures and the behavior of Feynman-Kac propagators, *C. R. Math. Acad. Sci. Paris* **334**, pp. 445–449.
Gulisashvili, A. (2002c). On the Kato classes of distributions and the BMO-classes, in: Differential Equations and Control Theory, *Lect. Notes in Pure and Applied Mathematics* **225** (S. Aizicovici and N. Pavel, eds.), Marcel Dekker, New York, pp. 159–176.
Gulisashvili, A. (2004a). Free propagators and Feynman-Kac propagators, *Seminar of Mathematical Analysis*, Proceedings, Universities of Malaga and Seville (Spain), September 2003-June 2004, Daniel Girela Álvarez, Genaro Lopez Acedo, Rafael Villa Caro (editors), Secretariado de Publicaciones, Universidad de Sevilla, pp. 47–64.
Gulisashvili, A. (2004b). Markov processes and Feynman-Kac propagators, *Centre de Recerca Matemàtica Preprint Series* bf 573, April 2004, 48 pages.
Gulisashvili, A. (2004c). Markov processes, classes of time-dependent measures, and Feynman-Kac propagators, submitted for publication.
Gulisashvili, A. (2005). Nonautonomous Kato classes of measures and Feynman-Kac propagators, *Trans. Amer. Math. Soc.* **357**, pp. 4607–4632.
Gulisashvili, A. and Kon, M. (1996). Exact smoothing properties of Schrödinger semigroups, *Amer. J. Math.* **118**, pp. 1215–1248.
Gulisashvili, A. and Van Casteren, J.A. (2005). Feynman-Kac propagators and viscosity solutions, *J. Evol. Equ.* **5**, pp. 105–121.
Howland, J. S. (1974). Stationary scattering theory for time-dependent Hamiltonians *Math. Ann.* **207**, pp. 315–335.
Ikeda, N. and Watanabe, S. (1989). Stochastic Differential Equations and Diffusion Processes, Second edition, *North-Holland Mathematical Library* **24**, North-Holland, Amsterdam.
Il'in, A. M., Kalashnikov, A. S., and Oleinik, O. A. (1962). Linear Equations of

the Second Order of Parabolic Type, *Russian Math. Surveyes* **17**, pp. 1–143.
Itô, K. (1951). On Stochastic Differential Equations, *Mem. Amer. Math. Soc.* **4**, Amer. Math. Soc., Providence, R.I.
Itô, K. (1987). Selected papers, Edited and with an introduction by S. R. S. Varadhan and Daniel W. Stroock, Springer, New York.
Itô, K. and McKean, Jr., H. P. (1965). Diffusion Processes and Their Sample Paths, *Grundlehren der Mathematischen Wissenschaften* Band **125**, Academic Press Inc., Publishers, New York.
Jacob, N. (2001). Pseudo Differential Operators and Markov Processes. Vol. I, Fourier Analysis and Semigroups, Imperial College Press, London.
Jacob, N. (2002). Pseudo Differential Operators and Markov processes. Vol. II, Generators and Their Potential Theory, Imperial College Press, London.
Jacob, N. (2005). Pseudo Differential Operators and Markov Processes. Vol. III, Markov Processes and Applications, Imperial College Press, London.
Jacod, J., and Shiryaev, A. N. (1987). Limit Theorems for Stochastic Processes, Springer-Verlag, 1987.
Jamison, B. (1970). Reciprocal processes: The stationary Gaussian case, *Ann. Math. Statist.* **41**, pp. 1624–1630.
Jamison, B. (1974). Reciprocal processes, *Z. Wahrscheinlichkeitstheorie und Verw. Gebiete* **30**, pp. 65–86.
Jamison, B. (1975). The Markov processes of Schrödinger, *Z. Wahrscheinlichkeitstheorie und Verw. Gebiete* **32**, pp. 323–331.
Johnson, G. W. and Lapidus, M. L. (2000). The Feynman Integral and Feynman's Operational Calculus, *Oxford Mathematical Monographs*, Oxford Science Publications, Clarendon Press, Oxford.
Kac, M. (1949). On distributions of certain Wiener functionals, *Trans. Amer. Math. Soc.* **65**, pp. 1–13.
Kac, M. (1951). On some connections between probability theory and differential and integral equations. In: *Proc. Second Berkeley Symposium on Mathematical Statistics and Probability* (ed. J. Neyman), University of California Press, Berkeley, pp. 189–215.
Kac, M. (1959). Probability and Related Topics in Physical Sciences, With special lectures by G. E. Uhlenbeck, A. R. Hibbs, and B. van der Pol. *Lectures in Applied Mathematics, Proceedings of the Summer Seminar, Boulder, Colorado* **1**, Interscience Publishers, London-New York.
Kac, M. (1979). Mark Kac: Probability, Number Theory, and Statistical Physics, *Mathematicians of Our Time* **14**, Selected papers, Edited by K. Baclawski and M. D. Donsker, MIT Press, Cambridge, Mass.-London.
Kahane, J.-P. (1997). A century of interplay between Taylor series, Fourier series and Brownian motion, *Bull. London Math. Soc.* **29**, pp. 257–279.
Kahane, J.-P. (1998). Le mouvement brownien, *Séminaires et Congrès* **3**, pp. 123–155.
Karatzas, I. and Shreve, S. (1991). Brownian Motion and Stochastic Calculus, *Graduate Texts in Mathematics* **113**, Second edition, Springer-Verlag, New York.
Kato, T. (1973). Schrödinger operators with singular potentials, *Israel J. Math.*

13, pp. 135–148.

Kleinert, H. (2004). Path Integrals in Quantum Mechanics, Statistics, Polymer Physics, and Financial Markets, Third edition, World Scientific Publishing Co. Inc., River Edge, N.J.

Kolmogorov, A. N. (1931). Über die analytischen Methoden in der Wahrscheinlichkeitsrechnung, *Math. Ann.* **104**, pp. 415–488. English translation in: Selected Works of A. N. Kolmogorov, Vol. II, Kluwer Acadamic Publishers, pp. 62–108.

Kolmogorov, A. N. (1933). Zur Theorie der stetigen züfälligen Prozesse, *Math. Ann.* **108**, pp. 149–160. English translation in: Selected Works of A. N. Kolmogorov, Vol. II, Kiuwer Acadamic Publishers, pp. 156–168.

Kuznetsov, S. E. (1982). Nonhomogeneous Markov processes (Russian), *Sovremennye Problemy Matematiki* **20**, VINITI, Moscow, pp. 37–178. English translation in: *J. Soviet Math.* **25** (1984), pp. 1380–1498.

Ladyženskaja, O. A., Solonnikov, V. A., and Ural'ceva, N. N. (1968). Linear and Quasilinear Equations of Parabolic Type, *Translations of Mathematical Monographs* **23**, Amer. Math. Soc., Providence, R.I.

Lamberton, D. and Lapeyre, B. (1996). Introduction to Stochastic Calculus Applied to Finance, Chapman & Hall/CRC, London.

Liggett, T. M. (2005). Interacting Particle Systems, Springer, New York.

Liskevich, V. and Semenov Y. (2000). Estimates for fundamental solutions of second-order parabolic equations, *J. London Math. Soc. (2)* **62**, pp. 521–543.

Liskevich, V., Vogt, H. and Voigt, J. (2005). Gaussian bounds for propagators perturbed by potentials, submitted for publication.

Métivier, M. and Pellaumail, J. (1980). Stochastic Integration. *Probability and Mathematical Statistics*, Academic Press [Harcourt Brace Jovanovich Publishers], New York.

Métivier, M. and Viot, M. (1987). On weak solutions of stochastic partial differential equations, *Lecture Notes in Math.* **1322**, Springer-Verlag, pp. 139–150.

Meyer, P.-A. (1966). Probability and Potentials, Blaisdell, Waltham.

Nagasawa, M. (1993). Schrödinger Equations and Diffusion Theory, *Monographs in Mathematics* **86**, Birkhäuser, Basel.

Nagasawa, M. (2000). Stochastic Processes in Quantum Physics, *Monographs in Mathematics* **94**, Birkhäuser, Basel.

Nagel, R. (1995). Semigroup methods for non-autonomous Cauchy problems, in: Evolution Equations, eds. G. Ferreyra et al. *Lect. Notes Pure Appl. Math.* **168**, Marcel Dekker, New York, pp. 301–316.

Nagel, R. and Nikel, G. (2002). Wellposedness for Nonautonomous Abstract Cauchy Problems, in: Progress in Nonlinear Differential Equations and Their Applications, **50**, Birkhäuser Verlag, Basel, pp. 279–293.

Nash J. (1958). Continuity of solutions of parabolic and elliptic equations, *Amer. J. Math.* **80**, pp. 931–954.

Nelson, E. (1967). Dynamical Theories of Brownian Motion, Math. Notes, Princeton University Press, Princeton, N.J.

Nelson, E. (1985). Quantum Fluctuations, *Princeton Series in Physics*, Princeton

University Press, Princeton, N.J.

Nickel, G. (1997). Evolution semigroups for nonautonomous Cauchy problems, *Abstract and Applied Analysis* **2**, pp. 73–95.

Øksendal, B. (1998). Stochastic Differential Equations, *Universitext*, Fifth edition, Springer-Verlag, Berlin.

Ouhabaz, E. M., Stollmann, P., Sturm, K.-Th., and Voigt, J. (1996). The Feller property for absorption semigroups, *J. Funct. Anal.* **138**, pp. 351–378.

Pazy, A. (1983). Semigroups of Linear Operators and Applications to Partial Differential Equations, Springer-Verlag, New York.

Porper, F. O. and Eidel'man, S. D. (1984). Two-sided estimates of the fundamental solutions of second-order parabolic equations, and some applications, *Uspekhi Mat. Nauk* **39**, pp. 107–156 (in Russian). English translation in *Russian Math. Surveyes* **39**, (1984), pp. 119–178.

Privault, N. and Zambrini, J.-C. (2004). Markovian bridges and reversible diffusion processes with jumps, *Ann. I. H. Poincaré - PR* **40**, pp. 599–633.

Privault, N. and Zambrini, J.-C. (2005). Euclidean Quantum Mechanics in the momentum representation, *J. Math. Phys.* **46**, 25 p.

Protter, P. (2005). Stochastic Integration and Differential Equations, Second Edition, *Stochastic Modelling and Applied Probability* **21** Springer-Verlag, Berlin.

Räbiger, F., Rhandi, A., Schnaubelt, R., and Voigt, J. (2000). Non-autonomous Miyadera perturbations, *Differential Integral Equations* **13**, pp. 341–368.

Revuz, D. and Yor, M. (1991). Continuous Martingales and Brownian Motion, *Grundlehren der Mathematischen Wissenschaften* **293**, Springer-Verlag, Berlin.

Roelly, S. and Thieullen, M. (2002). A characterization of reciprocal processes via an integration by parts formula on the path space, *Probab. Theory Related Fields* **123**, pp. 97–120.

Roelly, S. and Thieullen, M. (2005). Duality formula for the bridges of a Brownian diffusion: applications to gradient drifts, *Stoch. Proc. Appl.* **115**, pp. 1677–1700.

Rogers, L. C. G. and Williams, D. (2000a). Diffusions, Markov Processes, and Martingales. Vol. 1. Foundations. *Cambridge Mathematical Library*, Cambridge University Press, Cambridge.

Rogers, L. C. G. and Williams, D. (2000). Diffusions, Markov Processes, and Martingales. Vol. 2. Itô Calculus. *Cambridge Mathematical Library*, Cambridge University Press, Cambridge.

Royden, H. L. (1988). Real Analysis, Third edition, Macmillan Publishing Company, New York.

Sato, K. (2000). Levy Processes and Infinitiely Divisible Distributions, Cambridge University Press, Cambridge.

Schechter, M. (1971). Spectra of Partial Differential Operators, North-Holland, Amsterdam.

Schnaubelt, R. (2000/2001). Well-posedness and asymptotic behavior of non-autonomous linear evolution equations, *Tübinger Berichte zur Funktionalanalysis* **10**, pp. 195–218.

Schnaubelt, R. (2000). Semigroups for nonautonomous Cauchy problems, in: K. Engel and R. Nagel, One-Parametes Semigroups for Linear Evolution Equations, Springer-Verlag, pp. 477–496.
Schnaubelt, R. and Voigt, J. (1999), The non-autonomous Kato class, *Arch. Math.* **72**, pp. 454–460.
Schoutens, W. (2003). Lévy Processes in Finance. Pricing Financial Derivatives, John Wiley & Sons.
Schrödinger (1931). Über die Umkehrung der Naturgesetze. *Sitzungsber. Preuss. Akad. Wiss. Berlin Phys. Math. Kl.* **8/9**, pp. 144–153.
Semenov, Yu. A. (1999). On perturbation theory for linear elliptic and parabolic operators; the method of Nash, Applied Analysis, J. R. Doroh, G. Riuz Goldstein, J. A. Goldstein, M. M. Tom, Editors, AMS, Providence, RI, *Contemporary Mathematics* **221**, pp. 217–284.
Sharpe, M. (1988). General Theory of Markov processes, *Pure and Applied Mathematics* **133**, Academic Press Inc., Boston, MA.
Simon, B. (1979). Functional Integration and Quantum Physics, *Pure and Applied Mathematics* **86**, Academic Press Inc. [Harcourt Brace Jovanovich Publishers], New York.
Simon, B. (1982), Schrödinger semigroups, *Bull. Amer. Math. Soc. (N.S.)* **7**, pp. 445–526.
Sobolevskii, P. E. (1961). Equations of parabolic type in a Banach space, *Trudy Moscov. Math. Obsc.* **10**, pp. 297–350 (Russian); English transl.: *Amer. Math. Soc. Transl.* **49** (1965), pp. 1–62.
Stannat, W. (2004). Time-dependent diffusion operators on L^1, *J. Evol. Equ.*, **4**, pp. 463–495.
Stein, E. and Weiss, G. (1971). Introduction to Fourier Analysis on Euclidean Spaces, *Princeton Mathematical Series* **32**, Princeton University Press, Princeton, N.J.
Stroock, D. W. (1987). Lectures on Stochastic Analysis: Diffusion Theory, *London Mathematical Society Student Texts* **6**, Cambridge University Press, Cambridge.
Stroock, D. W. (1993). Probability Theory, an Analytic View, Cambridge University Press, Cambridge.
Stroock D. W. (2003). Markov processes from K. Itô's perspective, *Ann. of Math. Stud.* **155**, Princeton Univ. Press, Princeton, NJ.
Stroock D. W. (2005). An Introduction to Markov Processes, *Graduate Texts in Mathematics* **230**, Springer-Verlag, Berlin, 2005.
Stroock, D. W. and Varadhan, S. R. S. (1979). Multidimensional Diffusion Processes, *Grundlehren der Mathematischen Wissenschaften* **233**, Springer-Verlag, Berlin.
Stummel, F. (1956). Singulare elliptische differentialoperatoren in Hilbertschen Räumen, *Math. Ann.* **132**, pp. 150–176.
Sturm, K.-Th. (1994). Harnack's inequality for parabolic operators with singular low order terms, *Math. Z.* **216**, pp. 593–611.
Tanabe, H. (1960a). Remarks on the equations of evolution in a Banach space, *Osaka Math. J.* **12**, pp. 145–166.

Tanabe, H. (1960b). On the equations of evolution in a Banach space, *Osaka Math. J.* **12**, pp. 363–376.

Tanabe, H. (1997). Functional Analytic Methods for Partial Differential Equations, Marcel Dekker, Inc., New York.

Thieullen, M. (1993). Second order stochastic differential equations and non gaussian reciprocal diffusions, *Probab. Theory Related Fields* **97**, pp. 231–257.

Thieullen, M. (1998), Reciprocal diffusions and symmetries, *Stochastics Stochastics Rep.* **65**, pp. 41–77.

Thieullen, M. (2002). Reciprocal diffusions and symmetries of parabolic PDE: the nonflat case, *Potential Anal.* **16**, pp. 1–28.

Thieullen, M. and Zambrini J. C. (1997). Probability and quantum symmetries 1. The theorem of Noether in Schrodinger's Euclidean quantum mechanics, *Ann. Inst. H. Poincaré* **67**, 3, pp. 297–338.

Truman, A. and Davies, I. M. (1988). Stochastic Mechanics and Stochastic Processes, *Lecture Notes in Mathematics* **1325**, Truman, A. and Davies, I. M. (editors), Springer-Verlag, Berlin.

van Casteren, J. A. (2000). Some problems in Stochastic Analysis and Semigroup Theory, Progress in Nonlinear Differential Equations and Their Applications, Vol. 42, Birkhäuser-Verlag, Basel, pp. 43–60.

van Casteren, J. A. (2002). Markov pocesses and Feller semigroups, *Conferenze del Seminario di Matematica dell'Università di Bari* **286**, Aracne, Roma.

Voigt, J. (1986). Absorption semigroups, their generators, and Schrödinger semigroups, *J. Funct. Anal.* **67**, pp. 167–205.

Voigt, J. (1995), Absorption semigroups, Feller property, and Kato class, in: Partial Differential Operators and Mathematical Physics (Holzhau, 1994). *Oper. Theory Adv. Appl.* **78**, pp. 389–396, Birkhäuser, Basel.

Wilansky, A. (1983). Topology for Analysis, Robert E. Krieger Publishing Co. Inc., Melbourne, FL.

Yeh, J. (1995). Martingales and Stochastic Analysis, *Series on Multivariate Analysis* **1**, World Scientific, Singapore.

Zhang, Qi (1996). On a parabolic equation with a singular lower order term, *Trans. Amer. Math. Soc.* **348**, pp. 2811–2844.

Zhang, Qi (1997). On a parabolic equation with a singular lower order term II. The Gaussian bounds, *Indiana Univ. Math. J.* **46**, pp. 989–1020.

Zhang, Qi (2003). A sharp comparison result concerning Schrödinger heat kernels, *Bull. London Math. Soc.* **35**, pp. 461–472.

Zhang, T. S. (2001). Generalized Feynman-Kac semigroups, associated quadratic forms and asymptotic properties, *Potential Analysis* **14**, pp. 387–408.

Index

π-system, 7, 325
σ-algebra, 2
 $\mathcal{G}_T^{S,V}$, 138
 generated by a stopping time, 134
d-system, 325

additive functional A_V, 256
 exponential estimate, 269
additive functional A_μ, 257
 existence, 261
 exponential estimate, 269
 uniqueness, 266
admissible family of stopping times, 156
 examples, 170
approximation in the potential sense, 260, 286
augmentation, 8

backward Feynman–Kac propagator
 L^s-boundedness, 288
 $(L^s\text{-}L^q)$-smoothing property, 292
 L^∞-boundedness, 285
 strong BUC-property, 293
 strong Feller property, 293
backward Feynman-Kac propagator, 284
 BUC-property, 295
 continuity properties, 296
 Duhamel's formula, 308
 Feller-Dynkin property, 295
 integral kernels, 299

backward flow conditions, 101
backward Kolmogorov representation, 79
backward propagator, 101
 Feller, 124, 252
 Feller–Dynkin, 253
 left generators, 107
 right generators, 108
backward transition function, 11
Brownian bridge, 93
Brownian motion, 90
 orthogonal invariance, 91
 scale invariance, 91
 standard, 90
 translation invariance, 91
Burkholder–Davis–Gundy
 inequalities, 237

Cauchy bridge, 97
Cauchy process, 96
 scale invariance, 96
Chapman-Kolmogorov equation, 10, 62
 for densities, 13
Choquet capacitability theorem, 180
Choquet capacity, 179
completion of a σ-algebra
 with respect to a family of measures, 30
 with respect to a measure, 28
conditional expectation, 3, 328

derived density, 68
differential operator
 divergence form, 222
 non-divergence form, 217
diffusion process, 233, 251
 covariance, 233
 drift, 233
 generator, 233
 martingale characterization, 233
 strong Markov property, 255
distribution, 2
Doob's inequality, 264, 273

entrance-exit law, 76
entry time, 178

Feynman-Kac formula, 279, 280
filtration, 7
final condition, 218
final value problem, 217
finite-dimensional distributions, 6, 15
floor and ceiling functions, 1
flow conditions, 101
Fokker–Planck equation, 119
forward Kolmogorov representation, 79
free backward propagator, 105
 weak right generators, 115
 weak left generators, 112
free propagator, 106
function space, 104
 BC, 104
 BUC, 104
 C_0, 104
 $L^p_{\text{loc},u}$, 280
 L^r, 105
 $L^r_{\mathcal{E}}$, 104
functional
 additive, 193, 255
 Kac, 194
 multiplicative, 194
fundamental solution, 218
 weak, 224, 225

Gaussian density, 68
Gaussian estimates, 219

Gronwall lemma, 241

hitting time, 178
Howland semigroup
 associated with a backward propagator, 121
 associated with a free backward propagator, 123
 associated with a free propagator, 123
 associated with a propagator, 123
 associated with backward Feynman–Kac propagators, 305

independent σ-algebras, 3
inheritance problem, 306
initial condition, 218
initial value problem, 218
initial-final distribution, 65
Itô integral, 235
Itô's formula, 250

Kato class, 281
 associated with a Howland semigroup, 305
 associated with a time-homogeneous Markov process, 283
 equivalent characterizations, 281
Kato classes of functions
 non-autonomous, 196, 201
 weighted non-autonomous, 204
Kato classes of time-dependent measures
 non-autonomous, 196, 201
 weighted non-autonomous, 204
Khas'minski's Lemma, 269
Kolmogorov's backward equation, 114
Kolmogorov's criterion, 247
Kolmogorov's extension theorem, 14, 327
Kolmogorov's forward equation, 118

locally compact space, 4

Index

marginal distributions, 15
Markov process, 8, 15
  progressively measurable, 255
  reciprocal, 55
  separable, 33
Markov property, 8, 9
  reciprocal, 9
martingale, 27
Martingale Convergence Theorem, 27
martingale problem, 253
measure space
  complete, 29
monotone class theorem
  for functions, 325
  for sets, 325

Nikodým's theorem, 329

partial ordering $\preceq$, 156
path properties of Markov processes
  continuity, 51
  one-sided continuity, 44
  progressive measurability, 42
  separability, 37
path properties of reciprocal processes
  continuity, 83
  one-sided continuity, 82
path space, 6
pinned measure, 88, 299
  associated with Brownian motion, 92
potential
  of a function, 196, 201
  of a time-dependent measure, 196, 201
probability space, 2
projective system of measures, 6
propagator
  Feller–Dynkin, 172
  Feynman-Kac, 284
  left generators, 110
  locally uniformly bounded, 102
  right generators, 109
  separately strongly continuous, 102
  strongly continuous, 102
  uniformly bounded, 102

pseudo-hitting time, 178

Radon–Nikodym theorem, 143, 302
Radon-Nikodym theorem, 328
random variable, 2
random variables
  independent, 3
reference measure, 13, 104

sample path, 6, 25
sample space, 2
Schrödinger operator, 279
Schrödinger representation, 79
Schrödinger semigroup, 279
space-time process, 18, 23
  sample space, 19
standard process, 185
standard realization of a Markov process, 14
stochastic differential equation, 235, 237
  unique solvability, 238
stochastic process, 8
  adapted, 7, 27
  continuous, 26
  left-continuous, 26
  measurable, 33
  modifications, 7
  progressively measurable, 34
  quasi left-continuous, 175
  right-continuous, 26
  right-continuous with left-hand limits, 26
  standard, 177
  stochastically continuous, 26
  strongly stochastically continuous, 26
  time reversed, 9
stochastic processes, 5, 15
  indistinguishable, 7
  stochastically equivalent, 6, 15
stopping time, 134
  terminal, 138
strong BUC-condition, 124
strong Feller property, 124
strong Markov process, 149

strong Markov property
  with respect to families of stopping
    times and measures, 140,
    150, 157
  with respect to hitting times, 189
strongly subadditive function, 179
submartingale, 27
supermartingale, 27, 55

time reversal, 11
time shift operators, 17
time-dependent measure, 195
transition function, 10, 11
  normal, 10
  reciprocal, 62
  time-homogeneous, 17

transition probability density, 13
transition probability function, 10
  backward, 11
two-parameter filtration, 7

uniform integrability, 327
uniform parabolicity condition, 217
Urysohn's Lemma, 4

viscosity solutions, 311, 320, 321
Vitali–Hahn–Saks theorem, 118
Vitali-Hahn-Saks theorem, 329

weak solutions, 223, 225
Wiener space, 91